A CULTURAL HISTORY OF PLANTS

VOLUME 5

A Cultural History of Plants
General Editors: Annette Giesecke and David J. Mabberley

Volume 1
A Cultural History of Plants in Antiquity
Edited by Annette Giesecke

Volume 2
A Cultural History of Plants in the Post-Classical Era
Edited by Alain Touwaide

Volume 3
A Cultural History of Plants in the Early Modern Era
Edited by Andrew Dalby and Annette Giesecke

Volume 4
A Cultural History of Plants in the Seventeenth and Eighteenth Centuries
Edited by Jennifer Milam

Volume 5
A Cultural History of Plants in the Nineteenth Century
Edited by David J. Mabberley

Volume 6
A Cultural History of Plants in the Modern Era
Edited by Stephen Forbes

A CULTURAL HISTORY OF PLANTS

IN THE NINETEENTH CENTURY

VOLUME 5

Edited by David J. Mabberley

BLOOMSBURY ACADEMIC
LONDON • NEW YORK • OXFORD • NEW DELHI • SYDNEY

BLOOMSBURY ACADEMIC
Bloomsbury Publishing Plc
50 Bedford Square, London, WC1B 3DP, UK
1385 Broadway, New York, NY 10018, USA
29 Earlsfort Terrace, Dublin 2, Ireland

BLOOMSBURY, BLOOMSBURY ACADEMIC and the Diana logo are trademarks
of Bloomsbury Publishing Plc

First published in Great Britain 2022

Copyright © Bloomsbury Publishing Plc, 2022

David J. Mabberley and contributors have asserted their rights under the Copyright,
Designs and Patents Act, 1988, to be identified as the Authors of this work.

Series design by Raven Design
Cover image: A design by John Henry Dearle produced for Morris and Co., the decorative art
company of William Morris. (Photo by © Historical Picture Archive/CORBIS/Corbis via Getty Images)

All rights reserved. No part of this publication may be reproduced or transmitted
in any form or by any means, electronic or mechanical, including photocopying,
recording, or any information storage or retrieval system, without
prior permission in writing from the publishers.

Bloomsbury Publishing Plc does not have any control over, or responsibility for, any third-party
websites referred to or in this book. All internet addresses given in this book were correct at the
time of going to press. The author and publisher regret any inconvenience caused if addresses have
changed or sites have ceased to exist, but can accept no responsibility for any such changes.

A catalogue record for this book is available from the British Library.

Library of Congress Control Number: 2021932844

ISBN:	HB:	978-1-4742-7351-0
	Set:	978-1-4742-7359-6

Series: The Cultural History Series

Typeset by Integra Software Services Pvt. Ltd.
Printed and bound in Great Britain

To find out more about our authors and books visit www.bloomsbury.com
and sign up for our newsletters.

CONTENTS

LIST OF ILLUSTRATIONS	vi
SERIES PREFACE *Annette Giesecke and David J. Mabberley*	x
Introduction *David J. Mabberley*	1
1　Plants as Staple Foods 　　*Claudia Ciotir*	27
2　Plants as Luxury Foods 　　*Patrick Hunt*	43
3　Trade and Exploration 　　*Mark Nesbitt*	67
4　Plant Technology and Science 　　*Anne Osbourn*	85
5　Plants and Medicine 　　*Monique S. J. Simmonds*	105
6　Plants in Culture 　　*Roy Vickery*	127
7　Plants as Natural Ornaments 　　*Clemens Alexander Wimmer*	147
8　The Representation of Plants 　　*H. Walter Lack*	171
NOTES	196
BIBLIOGRAPHY	198
NOTES ON CONTRIBUTORS	230
INDEX	232

ILLUSTRATIONS

0.1 Sugar beet works and refinery at Waterloo, Belgium, 1830–60. Lithograph drawn by E. Toovey. Photo by SSPL/Getty Images   2

0.2 Botanischer Garten, Berlin, 1914. Photo by Zander & Labisch/ullstein bild/Getty Images   6

0.3 View of the Crystal Palace in Hyde Park, which housed the Great Exhibition in 1851. Photo by Guildhall Library and Art Gallery/Heritage Images/Getty Images   7

0.4 Alexander von Humboldt's 1807 *Tableau physique*—Zentralbibliothek Zürich/Wikimedia Commons   11

0.5 Charles and Emily Darwin. Chalk drawing by Ellen Sharples (1769–1849), 1816. Wikimedia Commons   15

0.6 Darwin's greenhouse built in 1863. *The Century Illustrated Monthly Magazine*, January 1883. Courtesy of Hathi Trust   16

0.7 Heavily bearded John Muir (to President Roosevelt's left) in front of the Grizzly Giant (*Sequoiadendron giganteum*), now some three thousand years old, in Mariposa Grove, at Yosemite Valley, California in 1903. Photo by PhotoQuest/Getty Images   25

1.1 Maize grown in North America during the period 1800–1920s: a) Northern Dent—Purple with Yellow Spot (Manitoba, Canada); b) Yellow Northern Flint (Quebec, Canada); c) Reid Yellow Dent (Indiana); d) Krug Dent (Iowa). Courtesy of the Anderson-Cutler Maize Collection, Missouri Botanical Garden   29

1.2 Wheat cultivars grown in Europe during the period 1800–1920: a) 'Beloturka'; b) 'Belle de Talavera'; c) 'Chidham'; d) 'Shirreff Squarehead.' From Vilmorin 1880   31

1.3 Combined seeder and disc harrow in the 1920s. Courtesy of University of Saskatchewan, University Archives and Special Collections, MG 108, H.A. Lewis Fonds, II B 12   33

1.4 Potato harvest in Scotland in early 1920s. Original contributor John Clark Maddison and image courtesy of www.ayrshirehistory.com   40

2.1 Wine industry in California, picking grapes, *c.* 1900. Turrill and Miller, photo postcard. Chronicle/Alamy Stock Photo   46

2.2 The pleasures of chocolate. French postcard for Chocolat Suchard, dated *c.* 1900. Photo by Universal History Archive/Universal Images Group via Getty Images   51

ILLUSTRATIONS vii

2.3 The Comfort Sugar Maple Tree in Autumn, approximately five hundred years old. The oldest maple tree in Canada. Pelham, Ontario. Courtesy of Design Pics Inc/Alamy Stock Photo — 55

2.4 A standing portrait of Edmond Albius, a former slave, who invented an efficient method of vanilla pollination, standing next to vanilla crops, 1863. From the New York Public Library. Photo via Smith Collection/Gado/Getty Images — 59

2.5 The first place where Coca-Cola could be bought was at the soda fountain in Jacobs' Pharmacy, 1886, in Atlanta, Georgia. Photo by John van Hasselt/Sygma via Getty Images — 63

3.1 Propagating Garden, Washington, DC. Prepared in 1858 to receive Robert Fortune's tea plants from China. Courtesy of the Smithsonian Institution (*Report of the Commissioner of Patents for the year 1858. Agriculture.* Washington, DC) — 69

3.2 Santos Museum of Economic Botany, Adelaide Botanic Garden. Courtesy of Mark Nesbitt — 72

3.3 Flowers of *Cattleya labiata gaskelliana*. Illustration from *Reichenbachia: Orchids illustrated and described*, by the nurseryman Frederick Sander (1888–94). Courtesy of New York Public Library — 75

3.4 The Amazonian rubber tree (*Hevea brasiliensis*), its flower and fruit segments bordered by six scenes illustrating its use by humans. Colored lithograph, *c.* 1840. Courtesy of Hamza Khan/Alamy Photo — 78

3.5 Nine scenes showing tea cultivation and preparation on an Indian plantation. Engraving by T. Brown, *c.* 1850, after J. L. Williams. Courtesy of the Wellcome Collection, London — 81

4.1 Schleiden's drawings of cells and embryonic structures from certain plant species. A single free cytoblast is shown (3), with a cytoblast with a cell forming on it, at low and higher resolution (4,5). From Schwann, *Microscopical researches into the accordance in the structure and growth of plants* (1847)—Plate I. Image from the John Innes Archives, courtesy of the John Innes Foundation — 87

4.2 Crown gall of alfalfa [*Medicago sativa*]. From O. T. Wilson in the *Botanical Gazette* (1920) — 89

4.3 Anatomy of *Nepenthes distillatoria* showing vascular tissue and stomata. From *Ladies' botany, or a familiar Introduction to the study of the natural system of botany* by John Lindley (1840, vol. 2, 2nd edn., London: Ridgway)—a section of Plate XLVIII. Image from the John Innes Historical Collections, courtesy of the John Innes Foundation — 92

4.4 Starch picture of Jan Ingen-Housz on a geranium (*Pelargonium*) leaf, made using a simplified version of Molisch's method (Molisch 1914). From Gest (1991) — 94

4.5 Drawing of mitosis by Walther Flemming, from Flemming (1882). Image from the John Innes Historical Collections, courtesy of the John Innes Foundation — 98

4.6	Rowland Biffen (1874–1949), one of the great pioneers of modern plant breeding, examines a giant ear of wheat furnished by the University of Cambridge agricultural students to convey playfully aspirations for the future of wheat breeding. Image from the John Innes Archives, courtesy of the John Innes Foundation	99
4.7	Sugar beet experiments, nineteenth century. Illustration of sugar beet grown under different conditions and with different fertilizers. Artwork originally from Turgan (1860). Courtesy of the Science Photo Library	101
4.8	A laboratory at the Merck chemicals factory in Darmstadt around 1900. Image reproduced with permission of Merck: Merk-Archiv, Y01/ea-2122	102
5.1	*Cinchona ledgeriana* (*C. calisaya*), Ceylon. Royal Botanic Gardens, Kew. Permission from the Wellcome Collection. Credit: Wellcome Collection. Attribution 4.0 International CC BY 4.0 (https://wellcomecollection.org/works/p3jffpgf)	110
5.2	*Coffea arabica*—fruiting stem. Watercolor, *c.* 1823. Credit: Wellcome Collection. CC BY (https://wellcomecollection.org/works/td95hg54)	111
5.3	*Erythroxylum coca*—gathering the coca plant in Bolivia. Credit: Wellcome Collection. CC BY (https://wellcomecollection.org/works/t59rv6pq?query=Erythroxylum%20coca)	113
5.4	*Papaver somniferum*. Credit: Wellcome Collection (https://wellcomecollection.org/works/pnu6j3n2)	116
5.5	*Strychnos nux-vomica*. Credit: Wellcome Collection (https://wellcomecollection.org/works/s3rawysv/images?id=dsddfju6)	118
6.1	"Firing at the apple-tree," in Devonshire, etching, *Illustrated London News*, January 11, 1851: 28. Photo by Roy Vickery	128
6.2	"Love makes sweet use of the language of flowers." Postcard printed in France, sent by Harold Carver, "on active service," to his wife in Green End, Hertfordshire, England, September 1919. Photo by Roy Vickery	136
6.3	"Language of Flowers, China Rose for Affection etc." Postcard, produced by H.M. & Co., London, printed in Germany, mailed in Glasgow, Scotland, August 1904. Photo by Roy Vickery	137
6.4	"Language of Flowers, Heliotrope for Devotion." "Trichromatic postcard by J. Welch & Sons, Portsmouth, printed at our works in Belgium," sent to Mr. G. Balcombe, Icklesham, Sussex, England, August 1906. Photo by Roy Vickery	139
6.5	"Love's Symbols, Forget-me-not." Postcard, printer unknown, mailed in New Hampshire, USA, September 1911. Photo by Roy Vickery	140
6.6	"For Luck! A Sprig o' Real White Heather." Postcard, produced by The Cynicus Publishing Co. Ltd, Tayport, Fife, Scotland, mailed in Elgin, Scotland, 1911. Photo by Roy Vickery	141

6.7	"Hearty Christmas Greetings." Postcard, printer unknown, sent to Master Franklyn North, Havant, Hampshire, England, December 23, 1906. Photo by Roy Vickery	146
7.1	Grouping of fir and birch. From Izabella Czartoryska, *Myśli różne o sposobie zakładania ogrodów*, 1805	149
7.2	Arnold von Regel, Design for Hungerberg, 1881. From Регель, *Арнольдъ: Изящное садоводство и художественные сады: историко-дидактический очеркъ*, 1896	151
7.3	Beatrice Parsons, Group of flowering shrubs at South Lodge, Horsham. From J. G. Millais, *Rhododendrons in which is set forth an account of all species of the genus Rhododendron and the various hybrids*, 1917	152
7.4	Archimedian Mower by Williams & Cie. From *Annuaire du Commerce*, 1878	158
7.5	Numbers of woody plant species introduced to Britain according to Loudon (1838)	163
7.6	Maples by Fritz Graf von Schwerin. From *Mitteilungen der deutschen dendrologischen Gesellschaft*, vol. 5, 1896	167
8.1	Gustav Klimt, *Der Kuss*, 1908. Österreichische Galerie Belvedere, Vienna. Photo by DeAgostini/Getty Images	172
8.2	Alphonse Mucha, Sarah Bernhardt, *La Plume*, 1897. Alphonse Mucha Museum, Prague. Photo by Fine Art Images/Getty Images	174
8.3	Vincent van Gogh, *Zonnebloemen*, 1889. Van Gogh Museum, Amsterdam. Photo by DeAgostini/Getty Images	176
8.4	John Everett Millais, *Ophelia*, 1852. Tate Britain, London. Photo by DeAgostini/Getty Images	177
8.5	Paul Gauguin, *Vaïraumati téi oa*, 1892. Pushkin Museum, Moscow. Photo by DeAgostini/Getty Images	179
8.6	Egon Schiele, *Selbstbildnis mit Lampionsfrüchten*, 1912. Leopold Museum, Vienna. Photo by DEA/E. LESSING/De Agostini via Getty Images	180
8.7	Joseph Maria Olbrich, *Secession Building in Vienna*, 1898. Photo by Alan John Ainsworth/Heritage Images/Getty Images	181
8.8	William Morris, Blackthorn wallpaper, c. 1890. Victoria and Albert Museum, London. Photo © Historical Picture Archive/CORBIS/Corbis via Getty Images	187
8.9	Ferdinand Bauer, *Cyclamen persicum*, c. 1792. Courtesy of Sherardian Library, University of Oxford	190
8.10	Henry Fox Talbot, *Astrantia minor*, 1838. Photo by The Royal Photographic Society Collection/Victoria and Albert Museum, London/Getty Images	194

SERIES PREFACE

The connectedness of humans to plants is the most fundamental of human relationships. Plants are, and historically have been, sources of food, shelter, bedding, tools, medicine, and, most importantly, the very air we breathe. Plants have inspired awe, a sense of well-being, religious fervor, and acquisitiveness alike. They have been collected, propagated, and mutated, as well as endangered or driven into extinction by human impacts such as global warming, deforestation, fire suppression, and over-grazing. *A Cultural History of Plants* traces the global dependence of human life and civilization on plants from antiquity to the twenty-first century and comprises contributions by experts and scholars in a wide range of fields, including anthropology, archaeology, art history, botany, classics, garden history, history, literature, and environmental studies more broadly. The series consists of six illustrated volumes, each devoted to an examination of plants as grounded in, and shaping, the cultural experiences of a particular historical period. Each of the six volumes, in turn, is structured in the same way, beginning with an introductory chapter that offers a sweeping view of the cultural history of plants in the period in question, followed by chapters on plants as staple foods, plants as luxury foods, trade and exploration, plant technology and science, plants and medicine, plants in (popular) culture, plants as natural ornaments, and the representation of plants. This cohesive structure offers readers the opportunity both to explore a meaningful cross-section of humans' uses of plants in a given period and to trace a particular use—as in medicine, for example—through time from volume to volume. The six volumes comprising *A Cultural History of Plants* are as follows:

Volume 1: *A Cultural History of Plants in Antiquity* (c. 10,000 BCE–500 CE)
Volume 2: *A Cultural History of Plants in the Post-Classical Era* (500–1400)
Volume 3: *A Cultural History of Plants in the Early Modern Era* (1400–1650)
Volume 4: *A Cultural History of Plants in the Seventeenth and Eighteenth Centuries* (1650–1800)
Volume 5: *A Cultural History of Plants in the Nineteenth Century* (1800–1920)
Volume 6: *A Cultural History of Plants in the Modern Era* (1920–present).

By way of guidance to our readers, it should be noted that the plant names used in these volumes accord with those in the fourth edition of *Mabberley's Plant-book* (Cambridge University Press, 2017). When they are discussed, individual plants are identified using their common names and, at their first mention in each chapter, with their scientific names: e.g. bay laurel (*Laurus nobilis*). As is recommended for general works such as this, the authorities to whom the scientific names are attributed (e.g. *Laurus nobilis* L., where L. identifies Linnaeus as the identifying authority) have been omitted.

Annette Giesecke and David J. Mabberley,
General Editors

Introduction

DAVID J. MABBERLEY

BACKGROUND

The population of the world in 1800 is estimated to have been just under one billion people; by 1920 it was approaching double that. This was in large part due to reduced infant mortality, but also improvements in sanitation and healthcare, specifically vaccination. The trends toward industrialization and urbanization in the eighteenth century accelerated greatly, while communication through improved roads and postal systems, newspapers, the colossal rise of railways and shipping including the opening of the Suez (1869) and Panama (1914) Canals, with the beginnings of radio in 1895 (Marconi) and air travel (the English Channel flown in 1909), set in train the route to globalization. Much of this was due to the inventiveness of those of European descent—from the Voltaic battery (1800), Trevithick's locomotive (1802), and Popham's telegraph (1816) through the first calculating machine (1822) to photography, motor cars, aeroplanes, and tanks.

Indeed, the increasing sophistication of armaments and warfare, from the British use of shrapnel (1803), the production of revolvers (1836) and practical rifles with the first bolt-action ones used by the Prussians from 1848 onward, to chemical weapons and aircraft machine guns after 1900, meant that the period bounded by the Napoleonic Wars, at the beginning, and the aftermath of the cataclysm that was the Great War (the First World War), at the end, saw a revolution in the way conflicts were waged. And plant products figured in this too, as vegetable glycerin(e) used in the production of nitroglycerin was a byproduct in copra production from coconut (*Cocos nucifera*). As coconut was the most important plant source of glycerin at the time, its cultivation was of strategic importance until the explosive was superseded from 1945 onward. By treating glycerine with nitric acid, nitroglycerin was first synthesized in Italy in 1847, and commercialized by Alfred Nobel (1833–96; whose fortune was used for the Nobel Prizes) in Sweden in 1867; Nobel combined it with diatomaceous earth (derived from algae) to make dynamite, and later synthesized gelignite and cordite. A large amount of nitroglycerin was used in the Great War, but it was also used medicinally, even in treating Nobel's own heart condition.

POLITICAL HISTORY

Until 1914 the French Revolution (1789), according to historian David Thomson, was likely considered the most important event in modern Europe, comparable with the Reformation of the sixteenth century and the religious wars of the seventeenth, like them destroying "the old order" in politics, economics, social life, and thought (Thomson 1966: 49). From 1793 there was almost continuous war until 1815, a global conflict with the British Royal Navy seeing action from the Arctic to the Pacific, while Napoleon occupied Berlin, Vienna, Warsaw, Rome, Madrid, and Lisbon. From 1800–3, he, as First Consul, reorganized France and later northern Italy, Dutch and Swiss republics and, with the Czar, he put together a new constitution for Germany. Although crowning himself Emperor in 1804, he swept away feudalism in all the countries under his control; he planted his relations as kings across Europe and effectively launched a French colonial empire. In "peacetime" he acquired Louisiana (North America) from the Spanish (Blanning 2007: 654, 659) and sold it to the United States in 1803. Despite his abortive attempt to invade Britain in 1805, he had great military triumphs on the Continent, but was set back by Nelson at Trafalgar (1805).

There followed an economic boycott of Britain, but two-thirds of British exports went to the rest of the world. This "Continental System" banned imports of British goods to Europe occupied by Napoleon and his allies, though it was widely evaded, even by Napoleon's Empress Joséphine, who openly bought enormous quantities of expensive plants from London for her garden at Malmaison, near Paris (Mabberley 2019: 162–3). More significantly, Russia still relied on Britain, sending there 91 percent of her flax (*Linum usitatissimum*) exports and 73 percent of hemp (*Cannabis sativa*); half the ships docking at St. Petersburg were British (Blanning 2007: 662–3). It has been argued that the shortage of hemp for French shipping was a critical factor at Trafalgar, but, in general, Britain's naval supremacy throttled Napoleon, particularly with regard to sugar, such that Napoleon promoted European-grown sugar beet (*Beta vulgaris* Sugar Beet Group) in the

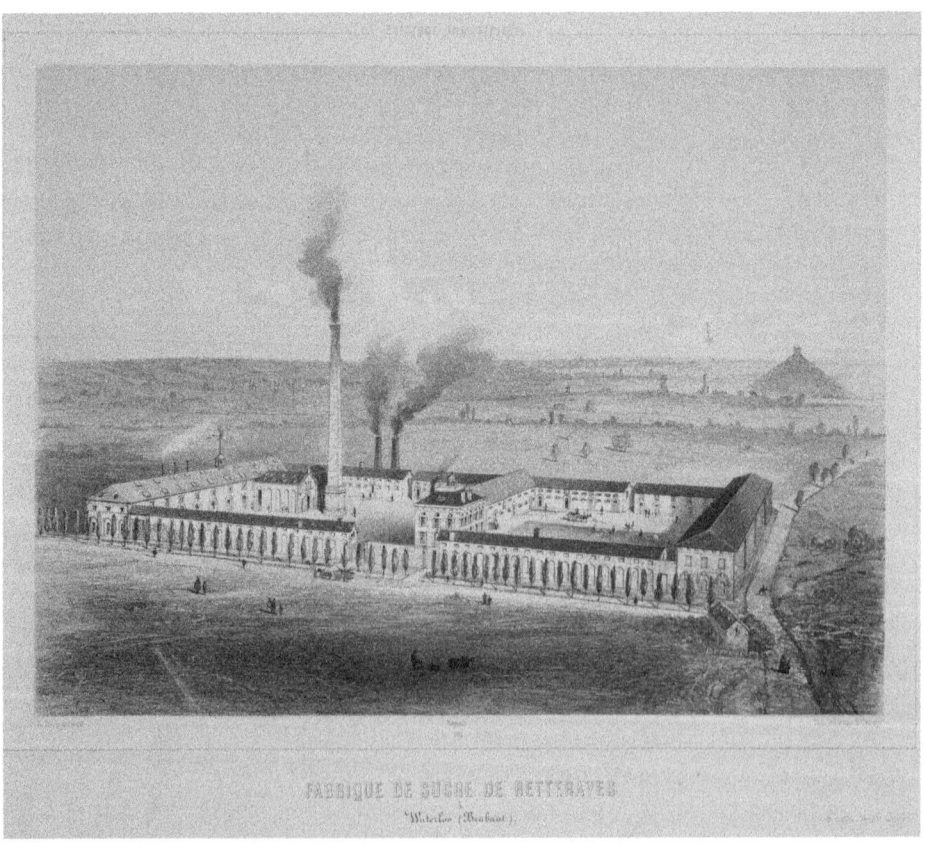

FIGURE 0.1 Sugar beet works and refinery at Waterloo, Belgium, 1830–60. Lithograph drawn by E. Toovey. Photo by SSPL/Getty Images.

stead of tropical cane sugar (*Saccharum officinarum*) grown in the Caribbean (Jiggens 2012: 125). In 1809 the US Congress boycotted trade with both Britain and France and, although the "Continental System" was abandoned in 1810, the Tsar's refusal to join Napoleon's blockade was an excuse for Napoleon to invade Russia—with disastrous consequences for the French (Schweikart and Allen 2004: 2). There followed the allied occupation of France and the abdication and banishment of Napoleon, though he was to return, landing in the south of the country in March 1815, but leading to his defeat at Waterloo and his second abdication in June.

By the end of the Napoleonic Wars, France still dominated western and southern Continental Europe and as far as Russia to the east, but the British domination of the "world overseas" was confirmed (Blanning 2007: 653). In this Europe of 1815 there was a marked contrast between the forces of change, notably increasing population, industrialization, and urbanization, besides nationalism and political ideas stemming from the Revolution and Napoleon, on the one hand, and, on the other, the conservative powers of the church, monarchy, and landowning aristocracy, joined by a widespread wish for peace after twenty-five years of war in Europe. In the ensuing century, liberalism, democracy, and socialism led to a fundamental change between state and citizens, the American ideals of 1776—and the later French Revolution, with effective parliamentary government, besides socialist experiments like the establishment of New Harmony in Indiana, USA (Thomson 1966: 83).

Spain lost her American empire during the Napoleonic Wars and afterwards—it was all gone by 1825, the year Brazil became free from Portugal—and the French lost many of their overseas posts. Although Britain was now without her American colonies, she had close ties with Canada and India, while there were new colonies in Australia, such that by 1815 she had bases or lands in all continents including Africa (at the Cape of Good Hope, won from the Dutch in 1814, though the Netherlands became increasingly enriched by exploitation of the Dutch East Indies, modern Indonesia). China's "Century of Humiliation" began in 1839 with the first Opium War which, as with all subsequent wars, she lost, ceding or leasing territory to invading powers, besides opening ports to foreign trade.

In 1812–15 there was war between the United States and a Britain supporting the First North Americans against United States' expansion, leading to the burning of Washington in 1814, followed by the defeat of all First North Americans east of the Mississippi and the relentless occupation of their territories to Florida (1818–20 from the Seminoles, and Spanish) and then all the way to the Pacific coast by 1848, with the annexation of the Oregon territory and California (gold had been discovered there in 1847), New Mexico, Utah, and Nevada. In 1867 Alaska was bought from Russia; in 1898 Hawaii was "annexed" and in 1903 the Canal Zone of Panama was purchased. Although there were internal national revolutions in Europe, there was no war of great importance until 1854, so this was one of the longest periods of peace in modern Europe. But between 1854 and 1878 there were six major conflicts including the Crimean War (1854–6) and the Franco-Prussian War of 1870.

ECONOMIC HISTORY

In 1800 the population of Europe was some 180 million, in 1914 some 460 million (Thomson 1966: 113), probably the result of falling death rates due to medical advances while diseases of domestic animals and cultivated crops were better controlled. Overall, famine, plague, and disease were diminishing, slavery was being abolished, and illiteracy was falling. Sanitation standards and factory regulation greatly increased and between 1870 and 1900 real wages rose by a half in western Europe (Thomson 1966: 428).

From 1815 to 1914, more than forty million people migrated to the United States, Canada, Australia, and colonial possessions in Africa and elsewhere, which territories were largely filled with the "overflow" from Europe, becoming the "NeoEuropes" (Crosby 1986: esp. ch. 12). The smallest continent had generated a third of the human race; European civilization and cultivation practices were exported worldwide. The agricultural revolution was very marked in the United States, with even higher population growth rates and seemingly unlimited land to bring into cultivation. Important overall was the use of winter root crops like beet (*Beta vulgaris*) and turnips (*Brassica rapa* Rapifera Group) with green crops like alfalfa (*Medicago sativa*) and clovers (*Trifolium* spp.), while the three-field rotation (one third fallow at any one time) was replaced by a four-course rotation, using all the land every year and providing feed to maintain livestock throughout the winter. With increasingly efficient transport, the "food baskets" of the United States, Canada, and, later, Australia became increasingly important in feeding Europeans. But all this expansion resulted in the destruction of aboriginal land management as in the burning and plowing of the prairie of the American Midwest with catastrophic consequences in the formation of the "Dustbowl" in the twentieth century. Even more devastating was that in Australia with the superposing from 1788 onward of European farm practices on another fire-managed landscape, elaborated by a 60,000 year-old civilization, which may have had sophisticated cultivation techniques utilizing native grasses and other plants: the disregarding of this management has led to the dust storms and cataclysmic fires of the twenty-first century (Pascoe 2018: passim).

In the years 1815–16 there were bad harvests across Europe, while the post-war depression led to the failure of thousands of banks and businesses, high taxation, and unemployment. In Britain the Corn Laws were introduced to maintain high prices for local producers through heavy tariffs on imported grains, thereby propping up the land-owning classes there: the laws were not repealed until 1846 after two years of the Great Famine in Ireland, the collapse of the potato (*Solanum tuberosum*) crop there leading to a need for supplies of cheap food from overseas. The repeal is now seen as a major impetus to the growth of free trade in general. Then, during the American Civil War (1861–5) food shortages in the Confederacy had a significant effect on the outcome of the war and there was economic stress in England as the supply of raw cotton from the South dried up (Schweikart and Allen 2004: passim).

Because British foreign export trade had tripled between 1789 and 1815, the expanding British Empire was to become a huge market for goods like cotton yarns and fabrics. This was an important impetus to the spread of mechanization of the inventions of the eighteenth century, leading to the notorious mills of northern England, and Britain becoming "the workshop of the world," importing raw materials and foodstuffs for a growing population. In 1807–8 the slave trade ended, but the economy of the United States was still reliant on slavery; by 1810 the all-important cotton production, largely by slaves, reaching 178,000 bales (Schweikart and Allen 2004: 107). In 1830, three-quarters of Britain's raw cotton came from the United States, and, by mid-century, half a million people worked in the British cotton industry alone, with textiles as a whole employing double that there (Thomson 1966: 179ff.). France, by contrast, could feed her people, so the imperative for overseas trade was a lesser one, while Germany, as a set of largely independent states, was held back too. By mid-century 60 percent of world shipping was

British and, until the 1860s at least, Britain had international economic supremacy. In many senses the nineteenth century was "the British century," Britain's influence failing only in the twentieth century—with the reluctance to accept her decline still affecting Britain's local politics today.

As set out by Mark Nesbitt in this volume, the pivotal importance of botanic gardens in the prosperity of the Empire cannot by overestimated, nor can the consequential movements of peoples to grow crops introduced from other parts of the world via them: Indians being moved to Malaysia to tap rubber and Pacific Ocean islanders to Australia to grow sugar cane are just two examples following in the mold of earlier centuries. The perils of genetically narrow populations of plants grown in monocultures led not only to the Great Famine in Ireland (potatoes) but also the collapse of coffee (*Coffea arabica*) through coffee-berry disease in Malaysia (and subsequent replacement with South American rubber, *Hevea brasiliensis*), once their natural pests and diseases from their places of origin caught up with them.

Between 1842 and 1858 a treaty system with China was put in place, Japan was "opened" in 1854, and the rule of the East India Company in India was ended by the British government in 1859, leading to Queen Victoria (of German stock) becoming Empress of India in 1876 (Thomson 1966: 239). Britain paid for grain and other imports with manufactures, shipping, and insuring—effectively as an industrial state. Of the cotton goods exported, a quarter went to Asia, especially India; by 1860 Asia and Africa together taking half. Cotton imports to France from the United States made Le Havre prosperous while the Netherlands became the entrepôt for tropical products like coffee, tea (*Camellia sinensis*), cane sugar, and spices. In 1871 increasing free trade with Britain's traditional rival France included reducing duty on French wines and brandy (Thomson 1966: 255ff.).

In 1875 less than a tenth of Africa was European colonies; by 1895 only a tenth was not. The assumed superiority of the Christianized West underlay the purloining of the natural resources of "unexplored" territories as the right of the "discoverer," often with a veneer of respectability afforded by the thirst for furtherance of knowledge of the Creation—but exploitation was really the motive. Concomitant with this, the museums, art galleries, zoos, and botanic gardens, public and private, in the home countries of the victorious invaders of Asia, Australia, and Africa filled with colonial trophy acquisitions, not a few of them looted. This colonialism was merely a crescendo in a long-established trend, in that by 1815 there had already been four hundred years of continuous European imperialism; Vladimir Lenin (1870–1924), following the British economist John Hobson (1858–1940), attributed this expansion to new economic forces, to "excessive capital in search of investment," hence seeking lucrative overseas enterprises (Thomson 1966: 489). Indeed, the generation after 1870 has been labeled "the age of imperialism" with a need for fresh markets for industrial output more likely to have been more significant than seeking raw materials or offloading population, though Africa and Asia had the raw materials—including cotton, rubber (transplanted from Brazil), and vegetable oils. As European nations could satisfy their own markets as they industrialized, surpluses had to go elsewhere, especially after 1880; between 1909 and 1913, for example, 36 percent of annual investment of British capital went to British overseas territories. But there was more to it, as David Thomson, in elegant alliteration has noted, "It was not just that trade followed the flag, but that the flag accompanied the botanist and buccaneer, the Bible and the bureaucrat, along with the banker and the businessman" (1966: 498).

FIGURE 0.2 Botanischer Garten, Berlin, 1914. Photo by Zander & Labisch/ullstein bild/Getty Images.

By the end of the century most of the Far East and Pacific, too, had gone to the British, French, Germans, Dutch, and Americans, so that war between any of these powers would inevitably lead to global conflict. In 1871 there was a "balance of power" between major European states, but by 1890 mutual fear of aggression and an arms war led to a series of crises culminating in the Great War of 1914–18 in which Russia lost two million men,

Germany nearly as many, and the British Empire at least a million. In all, ten million people died, far more than the less than 4.5 million in all wars of the previous hundred years (Thomson 1966: 349ff., 574).

By 1914 Germany was importing a fifth of her food, but tariffs were already being raised by the exporting countries (Thomson 1966: 376ff., 558). At the outbreak of war, France and Britain, joined by the United States in 1917, halted the shipping of all goods, including food, to Germany and her allies. The blockade of Germany led to many plant-based improvisations, notably bread made with turnips and the addition of potatoes to wheat flour (Thomson 1966: 576). Nettle fibre (*Urtica dioica*) was used for German military uniforms while a substitute for cotton was cellulose, synthesized in laboratories; other *ersatz* materials led to the development of plastics, rayons, and other synthetics (Mabberley 2017: 955). But the end of the Great War coincided with the worst recorded influenza pandemic (Spanish flu, 1918–20) during which another fifty million to a hundred million people (1 percent of the world's population) died, some twenty-five million in the first six months.

SCIENCE AND THE ARTS

The achievements and optimism of empire were perhaps most ostentatiously manifest in the First International Exhibition (1851), the Great Exhibition, held in London, to be followed by many others in Continental capitals. The Great Exhibition with its enormous Crystal Palace, designed by the gardener, Joseph Paxton (1803–65), was a demonstration not only of industrial might and colonial power but also of rampant consumerism appropriate to the burgeoning middle class.[1] It was strongly supported by Queen Victoria's Prince Consort (and cousin), Albert of Saxe-Coburg and Gotha (1819–61), who did much to promote science and technology, his legacy being "Albertopolis," the surviving complex of museums in South Kensington, London. The period culminating in the Great War also saw unprecedented advances in science and mathematics from Mendeleev's Periodic Table of the elements (1869) to Rutherford's discovery of the atomic nucleus (1911) and Rontgen's of X-rays (1895), to Marie Curie's theory of radioactivity and Albert Einstein's theory of relativity (1916). The increasing

FIGURE 0.3 View of the Crystal Palace in Hyde Park, which housed the Great Exhibition in 1851. Photo by Guildhall Library and Art Gallery/Heritage Images/Getty Images.

importance of science and its practical applications contributed to the broadening of university curricula and the professionalization of science, not least the establishment of scientific public servants, in Britain an early botanical one being Robert Brown (1773–1858), first keeper of botany ("underlibrarian") at the British Museum from 1827, and the establishment of Kew Gardens as an at least partly science-based organization in 1840 (Mabberley 1985a: 262-8).

The increasing rate of change in medicine, science, technology, and war was mirrored by enormous changes in the arts. Concomitant with advances made then—and now largely taken for granted—the literature of the nineteenth century still entrances us. From Goethe and Schiller, and Blake to Chateaubriand, Mme de Staël, Coleridge, Wordsworth, Southey, and Audubon through Hegel, Byron, Washington Irving, and Jane Austen to Shelley and Keats, then Pushkin, Balzac, Stendhal, Hugo, Gogol, Lermontov, Carlyle and Macaulay, Mill, and Sand to Hawthorne, Longfellow, the brothers Grimm, and Faraday, Hans Anderson, Dumas *père et fils*, and Kierkegaard; to Dickens, Thackeray, Poe, and the Brontes; Ruskin and Browning to Thoreau, Melville, Whitman, Arnold, and Turgenev, then Baudelaire, Mrs Gaskell, Feydeau, Zola, George Eliot and Trollope, Jules Verne, Lewis Carroll, Dostoyevsky, Swinburne, Tolstoy, and Neruda; Rimbaud, Mark Twain, Henry James, Christina Rossetti, Flaubert, Daudet, Marx, Gilbert and Sullivan, and Ibsen; Robert Louis Stevenson, Kipling, D'Annunzio, and Nietzsche, Conan Doyle, and H. G. Wells, Oscar Wilde, Hardy, Bram Stoker, Strindberg, Beatrix Potter, and J.M. Barrie—and Freud; and, after 1900, Thomas Mann, Gide, Gorky, Jack London and Kenneth Grahame to E. M. Forster, Conrad, D. H. Lawrence, and T. S. Eliot.

And the same can be said for the abiding appreciation of music and the arts of the period: Cherubini to Beethoven and Constable, to Weber, Schubert and Turner, Delacroix, Bellini, and Paganini, to Rossini, Mendelssohn and Chopin; to Corot and Landseer and Schumann; then Millais, Rosetti, and Verdi to Wagner and Gounod; and on to Berlioz and Brahms, Whistler, then Dvorak and Faure, Monet, Pissaro, Gauguin, and van Gogh, followed by Elgar, Glazunov, and Nielsen, Degas, Munch, and Renoir. The imperial adventure in the tropical world led to the exotica in the work of Douanier Rousseau and Gauguin, that of the Middle East to Leighton and Alma-Tadema. Impressionism retained realism and representational art but at the beginning of the twentieth century came Fauvism with the strong colors of Matisse, for example, and Cubism through Picasso and Braque leading to abstract art. In a few cases, as in what is now the Czech Republic, did this find expression in architecture, but it also had roots in literature, notably in the work of Gertrude Stein who repeated phrases, using them like building blocks, her writing greatly influencing Picasso.

Although Art Nouveau in its various regional facies relied heavily on naturalism, especially plant form, Futurismo (1909) in Italy reveled in the new technology and youth rather than tradition and age; airplanes and cars epitomized speed. Then came the anticapitalist Dadaism, a reaction of nonsense and absurdity to the Great War and first called "anti-art" in 1913, though it had some precursors in Jarry's *Ubu Roi* (1896) and was manifest in music such as Satie's ballet, *Parade* (1916–17).

Despite the long-term importance of all this artistic creativity, Silvanus Thompson (1851–1916), the great physicist, could nonetheless declare, "The nineteenth century has, intellectually, been the golden age, not of art or of poetry, not of drama or of adventure, but of science. It has been an epoch distinguished by a galaxy of men [sic] who made it great, and who, whether the world recognises it or not, were great men" (1910: 1213). In the generation before 1914, Darwinism (see below) was the most "far reaching and controversial" of all theories put forward in science. By 1871 Darwin could write, "He who is not content to look, like a savage at the phenomena of nature as disconnected, cannot any longer believe that man is the work of a separate act of creation …" (Thomson 1966: 433).

With this repositioning of humans in Western thought and radical innovation across all spheres, during international tumult of an unprecedented kind, the period has presented huge challenges to the authors in this volume. It is inevitable that they have had to be even less comprehensive in their coverage than have those in earlier volumes in this series.

HUMBOLDT, LAMARCK, AND DARWIN—AND PLANTS

The nineteenth century in Europe and the United States saw a boom in popular scientific literature such as fortnightlies and quarterlies, besides several competing monthlies with botanical novelties, as well as magazines covering gardening and farming. There were also popular natural histories besides weekly reviews like *Athenaeum* in London, newspapers commenting on science, and publications of learned societies besides specialist publications on botany, zoology, and geology. There was money to be made to satisfy the largely middle-class thirst for information. Moreover, as Darwin's biographer, Janet Browne (1995: 206) has pointed out, natural history collections, live as well as preserved, were marketable through agents and auction houses, notable being J. C. Stevens (Allingham 1924) in London, while important collectors sent out their own agents to purchase or collect. This was at a time when ostentatious consumerism with a focus on objects made these things more attractive—exotic garden plants and dried herbarium specimens, besides rocks, pinned insects, stuffed animals, aquariums, collections of shells. Natural history collections had become status symbols.

Science in general was a surprisingly public spectacle and natural history was especially popular—"in a miscellaneous spectrum of stage shows, art exhibitions, pageants, theatres, circuses, panoramic displays, fireworks, magic lanterns, freak shows, funfairs and the crammed cases of civic museums" (Browne 1995: 206). In London, for example, many exhibitions, including spectacular plant specimens or wax models of them, were mounted in such places as the Egyptian Hall in Piccadilly or in the pleasure grounds of Ranelagh and Vauxhall. On the other hand, in 1800 the Royal Society of London (equivalent to an academy of sciences elsewhere) was still only halfway through the presidency of Sir Joseph Banks (1743–1820), its members needing only to be interested in science (sometimes patrons), rather than practitioners, and not necessarily even natural science (Hall 1984: 1, 4). Other learned bodies were at heart for their members' enjoyment, so there arose a need to promote discussions between true scientists and to promote science generally. Robert Brown was a key figure in the attempt to reform the Royal Society in 1830; the British Association for the Advancement of Science was founded in 1831 and held annual meetings in different towns and cities, that at Oxford in 1832 being attended by Brown, "the greatest botanist of this or any other country" as was opined at the time, among others receiving a DCL (Mabberley 1985a: 279–80, 287). Nonetheless the Oxford establishment lamented how the university had "truckled sadly to the spirit of the times" in receiving the "hodge podge of philosophers" among them (Howarth 1931: 7–9, 46; Mabberley 1985a: 287–8).

The Continent was ahead of Britain or the United States in bringing together public interests and science, with the establishment of the Deutscher Naturforscher Versammlung, founded by Lorenz Oken (1779–1851), professor of natural history at Jena from 1807, and of physiology at Munich from 1827. In 1817 Oken started a monthly journal of science, *Isis*, in which annual meetings were proposed. The first such were held at Leipzig in 1822 and at that in Heidelberg in 1829, the president, the physiologist Friedrich Tiedemann (1781–1861), was able to declare in his address to the "cultivators of natural science and medicine,"

> Whereas in former times men regarded the inquisition of nature as a pleasant but useless employment, and as a harmless pastime for idle heads, they have of late years become daily more convinced of its influence upon the civilisation and welfare of nations, and the leaders of the public are everywhere bestirring themselves for the erection of establishments to promote its extension and advancement.[2]

Indeed, the meeting in Berlin in 1828 had been organized by the Prussian government, dispelling any political undertones to the venture by having Alexander von Humboldt (1769–1859) as president. His name is relatively little-known today, paradoxically because he was a polymath but, by the end of his life, everyone was becoming a specialist, such that the "interdisciplinary approach" was to be lost. On top of this, so much of Humboldt's contribution has become so mainstream rather than being associated with any particular breakthrough, because his broad worldview has become absorbed into the main corpus of modern thought. Perhaps, also, as his biographer Andrea Wulf has it, his being German may have contributed to his being sidelined, as a victim of anti-German feeling in the Great War; a time when, in America, German books were burned in a bonfire in Cleveland, Ohio, while Humboldt Street in Cincinnati became Taft Street (Wulf 2015: 235). And, today, the reluctance of many Anglophone, particularly American, scholars to study or cite anything scientific not written in English has reinforced such neglect.

Humboldt and Charles Darwin were undoubtedly the most significant nineteenth-century figures in biological thought; both were heavily influenced by their encounters with plants in different parts of the world. Humboldt was inspired first by the botanical garden in Bern, Switzerland, and greatly moved as early as 1790 by meeting Georg(e) Forster (1754–94), who had traveled on James Cook's Second Voyage to the Pacific. Not only did that encounter lead to Humboldt's own pioneering travels and researches in South America, but also provided the stimulus for his ground-breaking essay on plant geography *Essai sur la géographie des plantes; accompagné d'un tableau physique des regions équinoxiales* (1807) dedicated to Goethe. He had noted the similarity between tropical high-altitude and temperate vegetation types and corresponded with and became a mutual admirer of Robert Brown, who was the first to discuss the phytogeography of Australia and then extended his approach to Africa (Mabberley 1985a: 200, 398). Humboldt himself argued that Africa and America were once connected and later separated by subterranean forces, foreshadowing the theory of Continental Drift, introduced by German geophysicist, Alfred Wegener (1880–1930) but not broadly accepted until the 1960s. Humboldt, Brown, and, later, Swiss banker-botanist Augustin Pyramus de Candolle (1778–1841) and his son, Alphonse (1806–93), concentrated on the systematic ordering of facts about plants, but the teasing out of biogeographical patterns was not merely descriptive, for throughout the process that ordering was accompanied with "speculative narration" and postulation of hypotheses based on deductive principles (Juàrez-Barrera et al. 2018).

In 1834 Humboldt wanted to bring together all natural phenomena in a single "book on Nature," so at sixty-five he embarked on *Cosmos*, encompassing geology as

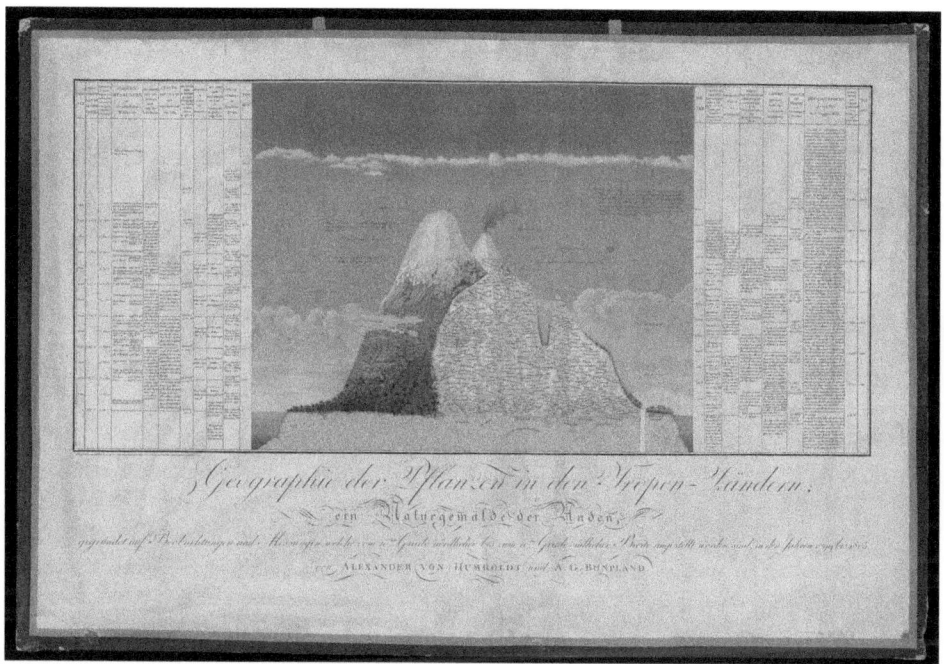

FIGURE 0.4 Alexander von Humboldt's 1807 *Tableau physique*—Zentralbibliothek Zürich/Wikimedia Commons.

well as biology, a sketch of the physical description of the universe, effectively a treatise built on lectures he had been giving in Berlin. His holistic approach was to show the interconnectedness of the natural world, in modern terms, as a global ecosystem. Through a wide correspondence comprising thousands of letters, he collected a mass of information from colleagues, including Brown. In essence, he portrayed a world in flux, such that he has been labeled a "Pre-Darwinian Darwinian." The first volume was published in 1845, the twenty thousand copies of this German edition selling in two months, and was then translated into English and many other European languages; the second volume appeared in 1847, the third 1850–1, and the fourth in 1856 (Mabberley 1985a: 200, 398). He finished the text of the fifth volume, on the earth and distribution of plants, but died in 1859 before publication.

Much of Humboldt's work tied the ecological devastation he witnessed in South America to slavery and other aspects of colonialism. *Cosmos* had a great effect on both Thomas Jefferson in the United States and, particularly, Simon Bolivar—and hence in the liberation of South American countries, such that Humboldt's name is much better known there than it is in the United States or even Europe (Wulf 2015: 152–5). Humboldt's work influenced the poetry of Byron, Wordsworth, and Coleridge and the writings of Americans like Edgar Allan Poe (1809–49) and Walt Whitman (1819–92), also likely the painting of Frederic Edwin Church (1826–1900) (Wulf 2015: 280). Perhaps most importantly, Henry Thoreau's *Walden* was a personal response to Humboldt. With his hat used as a "botany box," Thoreau (1817–62) explored, noting the dates of first flowering or leafing of plants (phenological data useful to climatologists today)[3] and read Humboldt, such that *Walden* was a local *Cosmos* (Wulf 2015: 257). But Humboldt, like Brown, scarcely addressed evolution directly. By contrast, Charles Darwin's grandfather, Erasmus Darwin (1731–1802), doctor, inventor, poet, and botanist, wrote books in which he expressed his concerns about a wide range of topics including slavery and the biological control of pests. In his posthumous poem "with philosophical notes," *The Temple of Nature: Or, the Origin of Society* (1803), he argued for the origin of organic life in the sea. He also clearly grasped what is now termed sexual selection, "The final cause of this contest amongst the males seems to be, that the strongest and most active animal should propagate the species, which should thence become improved" (King-Hele 1977: 243).

Erasmus Darwin's *Phytologia* (1800) was a kind of synthesis of practical botany and covered advances in plant nutrition and agricultural chemistry. He argued that plants "breathed" through stomata (as was confirmed thirty years later) and, correctly, that carbon dioxide was the "food" of plants, effectively anticipating Justus von Liebig's work on photosynthesis in the 1840s. He understood the importance of nitrates in plant nutrition, and also phosphorus (King-Hele 1977: 278). He also set out Thomas Knight's experimental results in plant breeding with the intimations of what is now known as Mendelian inheritance (Darlington 1961: 12–13). His work no doubt prompted that of the cleric, Thomas Malthus (1766–1834), on population, while his *Loves of the Plants,* a popularization in a poem of Linnaeus' sexual system of plant classification, translated into French in 1800, is said to have moved Jean-Baptiste Pierre Antoine de Monet, chevalier de Lamarck (1744–1829), soldier and botanist, to consider the transformation of species.

In 1809, Lamarck's *Philosophie Zoologique* followed Erasmus Darwin in some respects, and although adding nothing really new, presented a cogent theory of evolutionary change in plants and animals, namely that variation has direction imposed on it before it arises (soft heredity), as opposed to random variation followed by selection (hard heredity)—in other words, "design," purposeful with benevolent or divine guidance versus random without moral or meaning (Darlington 1961: 15). This was flattering to humans and to God and was consonant with old ideas fitting the notions of working hard and a purpose in life. Lamarck fully developed his ideas by 1815, but his speculations received little recognition. Although setting out what seemed "obvious," however, it led to questioning, not in Napoleonic France where a deal had been done with the Pope, but in England, by William Lawrence (1783–1867), William Wells, and James Prichard (a Quaker), medical practitioners, who rejected soft heredity, turning to Erasmus Darwin's interest in competition (Darlington 1961: 19). Lawrence's *Natural History of Man* (1819) was opposed by the Church and the universities it controlled, such that he suppressed his own book—but there were nine illicit editions.

CHARLES DARWIN AND NATURAL SELECTION

The Scottish geologist, Charles Lyell (1797–1875), studying fossils, had concluded that life appeared at several places and times, not just once, and argued that the earth was much older than had been concluded by biblical scholars—and that extinction was possible. Moreover, as the climate of the earth changed, so would the distributions of plants and animals. Charles Darwin was influenced by Lyell and was inspired by Humboldt's work, which he read as a Cambridge undergraduate (Browne 1995: 131). At Cambridge he attended botanical lectures by John Henslow (1796–1861) rather than those required for theology, his chosen subject; he became Henslow's favorite pupil and protégé, in effect his teaching assistant. Darwin traveled to Wales with the geologist Adam Sedgwick (1785–1873), like Henslow, in holy orders. Inspired by Humboldt, Darwin wanted to go to the Canary Islands, but Henslow put him forward (after considering himself and others) for Captain Robert Fitzroy's *Beagle* world voyage.

Beagle was in the Galapagos, Ecuador, in September 1835, where Darwin wrote, "The botany of this group is fully as interesting as the zoology" (Allan 1977: 98). A key evolutionary hint there came from the vice-governor, Nicholas Lawson (Nicolai Olaus Lossius, 1790–1851), who pointed out to Darwin that the different islands in the group had different giant tortoise species (whose shells Lawson was using as flowerpots), but the importance of this did not dawn on Darwin until much later (Browne 1995: 302, 304). On his return from the voyage, Darwin prepared his account of his travels and started his famous notebooks, one of them being "B," his thoughts on "transformism." As the birds of the Galapagos were perplexing, he pressed Henslow to work on the plants he had collected—to no avail.

Darwin's writing style reflected Humboldt's, so perhaps unsurprisingly Humboldt praised Darwin's published *Narrative* (1839), whose main fault was the lack of botany, but it made his name (Wulf 2015: 226). The influence continued in that Darwin's most famous passage, the "entangled bank" in *Origin of species* (1859), seems to have been a reworking of Humboldt prose (Wulf 2015: 234):

> It is interesting to contemplate a tangled bank, clothed with many plants of many kinds, with birds singing on the bushes, with various insects flitting about, and with worms crawling through the damp earth, and to reflect that these elaborately constructed forms, so different from each other, and dependent upon each other in so complex a manner, have all been produced by laws acting around us. These laws, taken in the largest sense, being Growth with reproduction; Inheritance which is almost implied by reproduction; Variability from the indirect and direct action of the conditions of life, and from use and disuse; a Ratio of Increase so high as to lead to a Struggle for Life, and as a consequence to Natural Selection, entailing Divergence of Character and the Extinction of less improved forms. Thus, from the war of nature, from famine and death, the most exalted object which we are capable of conceiving, namely, the production of the higher animals, directly follows. There is grandeur in this view of life, with its several powers, having been originally breathed into a few forms or into one; and that, whilst this planet has gone circling on according to the fixed law of gravity, from so simple a beginning endless forms most beautiful and most wonderful have been, and are being evolved.

Like Humboldt, Darwin maintained a vast correspondence, delighting in "facts" from horticulturists and animal breeders, the changes they wrought through selection being analogous to his hypothesized "natural" process. Darwin benefitted from the surge of interest in plants—clubs and societies, glass houses and heating systems, expeditions sent out by nurserymen, magazines—with an improved international postal system allowing access to information and the expertise of nurserymen, breeders, and practical gardeners. In his notebooks Darwin began postulating experiments to test generation of variation from isolated plants and at the end of his first notebook (completed February 1838) he was concluding that transmutation of species had occurred where populations were isolated, giving rise to "varieties" and thence species (Allan 1977: 122–4).

On September 28, 1838, Darwin read Malthus' *Essay on the Principle of Population* (1798), in his autobiography attributing this as a critical insight. He set about a whole research program on breeding and hybridizing plants and animals under domestication in the light of Malthus (Browne 1995: 409), started a new notebook on "Questions and Experiments," corresponded with an old friend of Henslow, Dean William Herbert (1778–1847), a vicar in Yorkshire, an authority on bulbs and hybridizing, and writing up his experiments for *Gardeners' Chronicle*, founded in 1841. Darwin's notes are full of the importance of crossing plants, performing transplant experiments, and studying heterostyly. In November 1841 Robert Brown recommended Darwin read Christian Konrad Sprengel's *Das entdeckte Geheimnis der Natur in Bau und in der Befruchtung der Blumen* (Nature's discovered secret in the propagation and fertilization of plants, 1793) in which it was shown that in many instances pollen needed to be taken from one flower to another, though not reaching the conclusion that outcrossing was advantageous in terms of generating variation.

Indeed, contrary to popular belief, Darwin largely worked on plants, their geography, morphology, and physiology, and kept up an enormous correspondence with botanists. His family had a strong tradition in botany, not only in his grandfather Erasmus (see above) but also in Erasmus' brother, Robert Waring Darwin (1724–1816), who published his *Principia botanica* in 1787. Charles Darwin's father, also Robert Waring Darwin (1766–1848), a physician, was a great horticulturist; indeed, Charles Darwin's first portrait, aged six, depicts him and sister Emily Catherine (1810–66) with a pot of South African flowering bulbs, *Lachenalia aloides*, grown by his father. Charles Darwin worked in his own garden and surrounding countryside, writing seven major works on plants, spending more time studying them than doing anything else. Although these books have been somewhat eclipsed by his other work, it was with plants that he tested his theory of Natural Selection (Allan 1977: 138). Critical was his *Variation of animals and plants under domestication* (1868) but perhaps he was somewhat side-tracked by *Insectivorous plants* (1875), *The various contrivances by which orchids are fertilised by insects* (1862), *The movements and habits of climbing plants* (1865), *The effects of cross- and self-fertilization* (1876), *The different forms of flowers on plants of the same species* (1877), and *The power of movement in plants* (1880)—all with highly original observations and experiments.

By 1842 Darwin was using the term "natural selection": "Who, seeing how plants vary in a garden [and] what blind foolish man has one in a few years, will deny [what] an all-seeing being in thousands of years could effect … either by his own direct foresight or by an intermediate means" (Browne 1995: 437). It was to a botanist, Joseph Hooker (1817–1911), who examined the Galapagos plants Henslow neglected to study, that Darwin first put his transmutation ideas, writing on January 11, 1844,

> I determined to collect blindly every sort of fact, which cd bear any way on what are species—I have read heaps of agricultural and horticultural books, and have never ceased collecting facts—at last gleams of light have come, and I am almost convinced (quite contrary to opinion I started with) that species are not (it is like confessing a murder) immutable. Heaven forfend me from Lamarck nonsense of a "tendency to progression", "adaptations from the slow willing of animals" &c.—but the conclusions I am led to are not widely different from his—though the means of change are wholly so—I think I have found out (here's presumption) the simple way by which species become exquisitely adapted to various ends.

Later in 1844 an amanuensis wrote out Darwin's manuscript "Sketch" in 231ff. in two halves, "On the variation of Organic Beings under Domestication and in the Natural State" and "On the Evidence favourable and opposed to the view that species are naturally formed races descended from common stocks" (Allan 1977: 141).

FIGURE 0.5 Charles and Emily Darwin. Chalk drawing by Ellen Sharples (1769–1849), 1816. Wikimedia Commons.

The prevailing geological theory had already embraced the idea of change in and by nature, but was anathema to natural theology: geology embraced paleontology, which had been studied systematically from about 1800. By mid-century there was, among scientists, a generally agreed succession of life on earth, and it was accepted that distributions of modern organisms were the results of historical processes (Coleman 1977: 63). In 1844, the year of Darwin's "Sketch," there appeared Robert Chambers' *Vestiges of Creation*, an anonymous potboiler essentially of Lamarckian ideology, which brought the matter to the reading public. Robert Chambers (1802–71) and then the respected Herbert Spencer

THE GREENHOUSE IN WHICH MR. DARWIN'S EXPERIMENTS AND OBSERVATIONS WERE MADE.

FIGURE 0.6 Darwin's greenhouse built in 1863. *The Century Illustrated Monthly Magazine*, January 1883. Courtesy of Hathi Trust.

(1820–93), the first a popularizer, the second more a writer for the intelligentsia, again leaning on Lamarckism, were both free of "the establishment" being self-taught and self-employed, their work generating enormous interest in evolution.

Vestiges stunned Darwin with its published assertion that species are not immutable but went further than he did, in including humans. It quoted the current literature on plant embryology and cells, paleobotany and although largely concerned with animals the author (at the time speculated to be Lyell or even Prince Albert) cited paleontologist Richard Owen (1804–92), "Araucariae and cycadeous plants likewise flourish on the Australian continent, where marsupial quadrupeds abound, and thus appear to complete a picture of an ancient condition of the earth's surface, which has been superseded in our hemisphere by other strata and a higher type of mammalian organisation" ([Chambers] 1850: 221). The importance of progression, Chambers' "progressive development,"

> under the providence of God, the results, *first*, of an inherent impulse in the forms of life to advance, in definite times, by generation, through grades of organization terminating in the highest dicotyledons and vertebrata ... *second*, of another inherent impulse connected with the vital forces, tending in the course of generations, to modify organic structures in accordance with external circumstances, as food, the nature of the habitat, and the meteoric agencies, these being the "adaptations" of the natural theologian.
>
> (13–31)

In discussing the cruciferous vegetables as descendants of the wild *Brassica oleracea* in a general critique on the "intransibility of species," Chambers cited cases where wheat was claimed to be transformed to rye, oats into rye, barley, or even wheat (144–5). This stretched the credulity of most (but not all) scientists, including Robert Brown, who "smiled sarcastically, and remarked that such a transmutation might be very convenient [sic]" (Mabberley 1985a: 329), but Chambers accepted Lamarckian evolution for variations within, for example, birds, though not the major "great grades of organization" ([Chambers] 1850: 154). Chambers wrote,

> The nucleated vesicle [i.e. a cell] is contemplated as the fundamental form of all organization, the meeting-point between the inorganic and organic—the end of the mineral and beginning of the vegetable and animal kingdoms, which thence start in different directions, but in a general parallelism and analogy. The nucleated vesicle is itself a type of mature and independent being, as well as the starting point of the foetal progress of every higher individual in creation, both animal and vegetable.
>
> (154)

In a discussion of the Carboniferous flora he also pointed out, "certainly storing up mineral masses which were in long subsequent ages to prove of the greatest service to the human race, even to the extent of favouring the progress of its civilization." His book was a bestseller, running to twenty more editions, and was translated into German in 1851; it was hotly debated and pamphlets opposing it were printed, bringing out the critics of evolution of every kind—to Darwin's chagrin (Browne 1995: 462).

In 1852, while Darwin still stewed, the French botanist, Charles Naudin (1815–99) published in *Revue horticole* what was in effect Darwin's theory, but it was ignored by Darwin, who, by contrast, in the 1830s had copied into manuscript Edward Blyth's published articles. Early criticism of Darwin's work was to note that he had not acknowledged such predecessors, but, in fact, he had begun such a list (ten names) for his unpublished "big book" on the subject—and this was eventually printed in the first German edition of *Origin of species* of 1860 with eighteen names, increased to thirty in the third English edition (1861) and even further in later ones (Berra 2015). Notable among them was William Wells (1757–1817), whose paper on evolution was read to the Royal Society as early as 1813, but it referred only to human beings; by the 1866 edition, Darwin was to admit that Wells "distinctly recognizes the principle of Natural Selection." In fact, as Cyril Darlington argued, Darwin did not even give credit to his grandfather, perhaps for personal reasons, so that with Naudin's work ignored, Lawrence's censured by the church, and Chambers' by the academic world, "evolution" was very soon synonymous with "Darwinism" and so, perhaps, it has largely remained—in all but French eyes (Darlington 1961: 63).

But it was another "outsider" who precipitated Darwin's publishing *Origin of species* (1859). The explorer-naturalist, Alfred Russell Wallace (1823–1913) was from a relatively humble background and was therefore perhaps more likely a candidate than Darwin to have revolutionary ideas to reform society. Wallace, working in Borneo, wrote "On the Law that has regulated the Introduction of New Species" was published without much fanfare in *Annals and Magazine of Natural History* in 1855. Encouraged by Lyell and Hooker, Wallace—who, in February 1858, was ill with malaria but recalled Malthus' work, seeing, as had Darwin, its importance—wrote to Darwin sending him a manuscript paper, which in essence was Darwin's theory. What Darwin rather disingenuously called a "bolt from the blue," stirred him to action, leading to the reading of Wallace's paper (without the author's consent) and Darwin's "Extracts from an unpublished Work on Species" on July 1, 1858, at a special meeting of the Linnean Society of London, called to elect a new member of the Council to replace the recently deceased Robert Brown (Mabberley 1985a: 389; Moody 1971). George Bentham (1800–84), a banker-botanist, was elected but, after a eulogy of Brown, Bentham's scheduled paper (paradoxically on the fixity of plant species) was replaced by the Darwin and Wallace contributions, followed by five more papers. The very long meeting, no doubt tiring and boring the audience, led to the only published notice Darwin could recall, the "verdict was that all that was new in them [his and Wallace's papers] was false, and what was true was old." The President later famously wrote (Browne 2002: 40–2),

> The year which has passed ... has not, indeed, been marked by any of those striking discoveries which at once revolutionize, so to speak, the department of science on which they bear; it is only at remote intervals that we can reasonably expect any sudden or brilliant innovation which shall produce a marked and permanent impress on mankind.

However, it led to Darwin's rapidly publishing an "abstract" of his intended great work and calling it *On the origin of species by means of Natural Selection, or the preservation of favoured races in the struggle for life*. It was published in an edition of 1250 on November 24, 1859, with five hundred copies taken by the subscription-based Mudie's Library; in early January 1860 a second edition of three thousand was issued. There were to be four more editions in Darwin's lifetime, the fifth (1869) being the first to use Spencer's term, "the survival of the fittest."

Prominent in Darwin's argumentation were many plant examples, such as increasing size of fruit through "roguing" in horticulture, an analogy with non-human-induced change. The *Gardeners' Chronicle* of December 31, 1859 was "much impressed with its importance and have moreover found it to be so dependent on the phenomena of horticultural operations, for its facts and results, and so full of experiments that may be repeated and discussed by intelligent gardeners and of ideas that may sooner fructify in their minds than in those of any other naturalists ...." Indeed, a month later Darwinian principles were being applied to improvement of sugar, wheat, and cotton in the colonies.

Unlike *Vestiges*, there was nothing in *Origin* on humans or the origin of life. Nonetheless, in Darwin's lifetime his book was discussed in more than a hundred journals and newspapers in Britain alone (Browne 2002: 103–4), Darwin himself collecting 347 reviews and 1,571 general articles. The most quoted discussion since was the Oxford meeting of June 30, 1860, a debate between Thomas Henry Huxley (1825–95) and Bishop Samuel Wilberforce (1805–73), primed by Richard Owen, with Joseph Hooker, though it was little reported at the time (Moore 1979: 61). Both "sides" felt they had won, the Darwinians concluding that Hooker had saved the day, but the important thing was that it had happened at all, with public figures in a prominent public forum, something that Wallace could not have pulled off (Browne 2002: 124). It was in hindsight seen as a "battle" as part of the "military milieu" of the period, with science versus theology, the Darwinian "revolution," led by "Darwin's bulldog" (a label not then used), Huxley, being preceded by the Crimea War and the Indian Mutiny and to be followed by the Boer War, while in America this "revolution" was postponed by the Civil War (Moore 1979: 51).

But the scientific "combatants" were often in holy orders, as exemplified by the lesser-known but earlier contretemps at the Cambridge Philosophical Society on May 7, 1860, when Adam Sedgwick, canon of Norwich, yet professor of geology, spoke violently against Darwin, but Henslow, a Suffolk rector and professor of botany, defended him. Indeed, the military metaphor is inappropriate; moreover, religious observance was predominant in Fellows of the Royal Society surveyed in 1874, their not perceiving any great conflict between faith and their science (Moore 1979: 82–3). In 1864 a "Declaration" of harmony between the two was circulated. At the same time the "Xclub" including Huxley and Hooker was devoted to "science pure and free, untrammelled by religious dogmas," whereas "The Victoria Institute or Philosophical Society of Great Britain," founded after it, wanted to uphold the principles of the Declaration. It has even been argued that Christian men untrained in science were more open-minded on Darwinism than those Christians trained as scientists (Moore 1979: 89ff., 94).

Darwin's critics, opposing his scientific explanation in using a hypothesis that had not been "proved," included rector and ornithologist, Francis Orpen Morris (1810–93), who wrote anti-Darwinian tracts that were distributed on railway bookstalls in Britain from 1869 to 1890, while clergyman-scientist Enoch Burr (1818–1907) in the United States considered Darwin's work "not merely the most noted, plausible, influential, and violent enemy of Theism in our day but what is its only possible enemy for all ages to come" (Moore 1979: 196, 198). Others were more accommodating, arguing a divine agency as directing evolution, though denying Natural Selection, the botanist-clergyman George Henslow (1835–1925), son of Darwin's mentor, John Henslow, arguing for "directivity," a force inserted into the earliest protoplasm by the Creator. He wrote in his *Present-day rationalism critically examined* (1904), "If the Argument from Design be not restored, that of Adaptation under Directivity takes its place; and [William] Paley's argument [*Natural theology or evidences of the existence and attributes of the Deity*, 1802], readapted by Evolution, becomes as sound as before; and, indeed, far strengthened as being strictly in accordance with the facts [sic]" (Moore 1979: 221). Asa Gray (1810–88), a botanist at Harvard, was Darwin's most prominent advocate in America but also a devout Christian able to reconcile these, writing "Natural selection is not the wind that propels the vessel, but the rudder which, by friction, now on this side and now that, shapes the course" (Moore 1979: 269, 276).

The importance of *Variation* was in countering the critics of *Origin* who had said that Darwin had not explained the underlying cause of variability, so the book was essentially a sequel in the place of the comprehensive volume he had aspired to write, but never did. (Henry Charles) Fleeming Jenkin (1833–85), a Scottish engineer, had pointed out in his 1867 review of *Origin* that any variation, crucial to Darwin's theory, could hardly spread in an outbreeding population as it would be swamped by "blending inheritance." The problem of blending inheritance led to a "Neo-Lamarckism" movement especially promoted by Herbert Spencer with a great following in the United States, where neo-Lamarckians were strongest, with neo-Darwinians in Britain, but in a later edition of *Origin*, Darwin wrote, "I am convinced that Natural Selection has been the most important, but not the exclusive, means of modification," and in the sixth edition (1876) admitted a role for inheritance of acquired characteristics (Moore 1979: 152, 175). Indeed, in 1857 just before publication of *Origin*, Darwin could write to Asa Gray of both external forces and hereditary ones, so confusing the two, despite denying soft heredity in the introductory chapters of *Origin*.

On the other hand, Wallace had actually repudiated Lamarck, though Darwin did not, and, moreover, promoted pangenesis, where factors of inheritance (gemmules, i.e. from the soma), were passed on (Darlington 1961: 40ff.). Perversely, Darwin's experiments with, for example, peloric flowers gave the familiar Mendelian ratios derived from sex-cells, but he was not a mathematician (Allan 1977: 230). Long before, Thomas Knight had read a paper involving pea-breeding as early as 1823, but did not record the numbers that would have led to the ratios, though by 1799 he had discussed what would now be called dominance and was the first to apply the science of hybridization to plant improvement (Roberts 1929: 88–90). In 1820, John Goss (1787–1833), a Devon schoolmaster, crossed peas, demonstrating segregation from heterozygotes and, in France, in the 1820s, Augustin Sageret (1763–1851) crossed musk melon (*Cucumis melo* Reticulatus Group) with cantaloupe and found that characters did not blend (Roberts 1929: 121).

The seminal paper of abbot Řehoř (Gregor, né Johann), Mendel (1822–84) of Brno, published locally in 1866, was "discovered" only in 1900 and quoted almost simultaneously by three European botanists—firstly Hugo De Vries in the Netherlands, then Carl Correns in Germany, and Erich von Tschermak in Austria (Roberts 1929: 286ff.). Mendel's work was famously botanical, the breeding of lines of peas like John Goss before him, but he also experimented with bee-breeding. Mendel, who had studied, among other things, mathematics at the University of Vienna 1851–3, wrote (Roberts l.c.),

> Those who survey the work done in this department will arrive at the conviction that, among all the numerous experiments made, not one has been carried out to such an extent and in such a way as to make it possible to determine the number of different forms under which the offspring of hybrids appear, or to arrange these forms with certainty according to their separate generations, or definitely to ascertain their statistical relations.

He used the terms "dominant" and "recessive" and realized the importance of "characters." The melding of his work with Darwin's in the twentieth century led to the scientific orthodoxy of today.

Assessment

Darwin brought science to confront existing beliefs and philosophies and provided a process for evolutionary change, first clear in his mind by about 1844. Malthus brought to both Darwin and Wallace, independently, the importance of competition between organisms, combined with variation, "The preservation of favourable variations and the rejection of injurious variations, I call Natural Selection," Darwin wrote, "the element of descent is the hidden bond of connexion which naturalists have sought under the term Natural System," of classification. The existence of "natural" classifications of organisms was grist to Darwin's mill, in that they could, in his thinking, only be explained by evolution. As has been pointed out, there is copious evidence that taxonomy was very important to Darwin, who wrote a taxonomic treatise on barnacles; his deference to the plant taxonomist, Joseph Hooker, reinforces this (Winsor 2009). The fact that this is not widely understood is an example of the "image problem" taxonomy continues to have, namely that for a long time the idea was that naturalists were "mindlessly following the sterile system of Plato and Aristotle." Indeed, by Darwin's time, many European naturalists had arrived at classifications based on relatedness, reflecting embryological similarities or indeed historical derivation from a common ancestor. With acceptance of his theory, such "natural classifications" become "evolutionary" ones, reflecting species' genealogy while "archetypes" became "ancestors." By the 1880s, embracing new lines of study, it was realized that cell and tissue structure, as well as gross morphology, could provide useful characters for classification (Stevens 1994: 249).

With perhaps the exception of the technological innovations of electricity and telegraphy, few advances of the period have had such an impact as Darwinism (Browne 2002: 378). Marx certainly read Darwin's work when preparing *Das Kapital* and inscribed a copy "his sincere admirer Karl Marx" (Browne 2002: 378, 403). In 1862 Marx wrote to Engels, "It is remarkable how Darwin rediscovers among beasts and plants the society of England, with its division of labour, competition, opening up of new markets, inventions and the Malthusian struggle for existence" (Browne 2002: 187–8). Darwin's theory fitted the commercial entrepreneurial ethos of untrammeled competition in industrializing Europe as manifest in the successful careers of the men in his own family, with supporting evidence from familiar country pursuits like gardening and animal breeding (Browne 1995: 390).

Darwin's evidence also came from fossils excavated in building canals and railways characteristic of this hectic period, but biogeography was one of the main planks of his evolutionary theory; plant geography, Darwin wrote, was a "key-stone of the laws of creation."[4] At the time, botany, like astronomy, was strong because of its evident importance to the British imperial program, with Kew's Hooker father and son the two most powerful botanists of the century. Joseph used Kew as a vehicle to support Darwin's idea, so Darwinism was soon studied in Kolkata, Cape Town, and Sydney (Browne 2002: 131). The geographical breadth and government contact engendered by Kew consolidated this scientific shift in the Empire and via Asa Gray to America. The simultaneous publication of the Darwin and Wallace papers (1858) and the rushed *Origin of species* (1859), drawing together all branches of biology and providing an explanation, meant that it was a staple of textbooks by the 1880s. Outstanding Darwinian botanists influential though their teaching and publications included, at Oxford, Arthur Harry Church (1865–1937), whose original, all-embracing theory of the evolution of plants was crystallized in his *Thalassiophyta and the subaerial transmigration* of 1919, beginning with the famous words, "The beginnings of botany are in the sea" (Mabberley 2000: 30,

32–6, 57–82, 94–6, 121–4). But only in the 1930s was Darwinism ineradicably fixed in biological thought, with the acceptance of the slow changes advocated by Darwin, as opposed to major jumps as had been proposed by de Vries based on his work on evening primroses (*Oenothera* spp.), which have a very exceptional type of inheritance where whole genomes are passed unchanged to gametes, the resulting outcrossed zygotes with alarmingly "new" types of morphology.

According to Mary Winsor (2009) it was Darwin's demonstrating the lack of fixity of species that led to his being called the Newton of Biology, but she argues for the Copernicus of Biology too, if it is indeed true that Darwin was the first to propose branching evolutionary patterns. Copernicus proposed a Sun-centered system for which Newton provided the mechanism: prejudices supporting an immobile earth were very like those favoring unchanging species. As with Copernicus' theories, many things were explicable by Darwin's hypothesis of common descent. On Darwin's being buried in Westminster Abbey, the *Pall Mall Gazette* noted of "the greatest Englishman since Newton," who had given "the same stir, the same direction to all that is most characteristic in the intellectual energy of the nineteenth century, as did Locke and Newton in the eighteenth" and no one, according to *The Times*, "wielded a power over men and their intelligences more complete than that which for the last twenty-three years has emanated from a simple country house in Kent."

On the other hand, it has been cogently argued that modern evolutionists have fallen into the trap of the Whiggish view of history, setting up Darwin as a hero and pulling out of the period events and opinions that support modern thought, this "progessionism" towards the pinnacle of present understanding being similar according to that which pervaded nineteenth-century thought.[5] Evolutionary ideas settled naturally into Victorian "progressionism" with Man as the ultimate product, such that religious views could in at least some cases reconciled with "scientific" ones. But, despite having catalyzed the debate, Darwinism was not the center of the evolutionary movement as much as many present-day Darwinists would argue. It was Darwin's position in society and the scientific milieu—the journal *Nature* was in part founded as a vehicle for his views—that led to a general currency of evolutionary ideas, something Wallace could never have achieved on his own. As early as the late 1860s, Cambridge University Tripos papers now had questions on the subject, but the "non-Revolution" is best illustrated by the fact that neo-Lamarckian ideas, following the now-controversial biologists Louis Agassiz (1807–73) and Edward Cope (1840–97) among others, became established in America—and ideas of orthogenesis and "overdevelopment" were commonplace. The recent stress on the importance of the "Mendelian Revolution" has also been questioned, the new heroes being those concerned with population genetics after our period—in short, the true "revolution" did not take place until the 1930s.

ECOLOGY

Much of the advance in biological science in the nineteenth century (and since) derived from tropical experience, where naturalists were confronted with a fresh and bewildering new milieu (Mabberley 1992: 2–3). Indeed, it was the tropics that are argued to have triggered both Charles Darwin and Alfred Russell Wallace to formulate their converging views on the mode of evolution by Natural Selection, where Darwin stressed the biotic relationships, Wallace the physical environment, in essentially ecologically based theories of evolution.

In the eighteenth century, it had been widely believed that, though the tropics had a number of peculiar organisms, they were not particularly rich in species, at least in species of plant. Indeed, Linnaeus himself seems to have believed that the tropical flora was rather homogeneous and limited, perhaps reflecting the collections of pantropical weeds around settlements and the widespread littoral species readily accessible to travelers and brought back as specimens to Europe. Access to the canopy of rain forest was rarely obtained and familiar tropical phenomena such as cauliflory, the bearing of flowers on the trunks of trees, were so little understood that a mahogany relation in Java with cauliflorous flowers was pigeon-holed as a parasite by Linnaeus' pupil Pehr Osbeck, who in consequence named in *Melia parasitica*, the *Epicharis parasitica* (Meliaceae) of today. More extensive traveling and the penetration of the continents led to a discarding of these ideas and, by the turn of the nineteenth century, attitudes were changing rapidly, partly due, at least, to advances in attempts to put some geographical order into the hordes of plants and animals being brought back to Europe.

Linnaeus had some six thousand or so plant species in his *Species plantarum* of 1753, but this was outstripped may times over by 1800; by 1920, literally hundreds of thousands of plant species, the great bulk of them from the tropics, had been named, though many still lurked, unnamed, in herbarium cabinets (Mabberley 1991). For ecological and other studies, plant inventories, local and global, were needed. More and more updated editions of Linnaeus' work appeared and increasingly new compendia based on the natural system, principal being Augustin-Pyramus de Candolle's unfinished *Prodromus systematis naturalis regni vegetabilis* (1824–73) and the intensively detailed *Das Pflanzenreich* (1900–) of Adolf Engler (1844–1930), also unfinished, besides hordes of local floras, their compilation, in the case of the tropical ones, fitting the imperial project.

Early ideas and much original thinking on plant distribution were brought together following two early-nineteenth-century expeditions: Alexander von Humboldt climbing the Ecuadorean volcano, Mount Chimborazo (6,300 m and, at the time, thought to be the world's highest peak) and noting the correlation between climate and vegetation type as he ascended, and Robert Brown, on Matthew Flinders' voyage to Australia, preparing an early phytogeographical account of that continent and, later, tropical Africa (see above). The term *Okologie*, ecology, was coined in 1866 by the zoologist Ernst Haeckel (1834–1919), who also coined the German word from which "phylogeny" derives (and, incidentally, was the first, in 1914, to use the term "First World War"). Although most classical ecological studies were carried out in the more accessible north temperate zone, it is perhaps surprising to learn that it can be convincingly argued that the scientific study of what we now call ecology began in the tropics of South America.

It is true that, as in much of science, some of these early ecological ideas were not altogether new, for the seventeenth-century microscopist, Antonie van Leeuwenhoek (1632–1723), developed the idea of food chains, and, from the observations of shellfish in Dutch canals, came to some idea of competition between organisms, while in the following century, Georges Louis Leclerc, Comte de Buffon (1707–88) in France was groping towards the concept of what is now termed "succession." Nonetheless, these figures did not, in these areas of inquiry, affect the mainstream of scientific thought as it was passed on to the nineteenth century. Only then was the intellectual world receptive.

Although "oecology" largely derived from plant geography, for example Oscar Drude's *Handbuch der Pflanzengeographie* (1890), the key figure in its advancement is held to have been Eugen Warming (1841–1924), a Dane who spent three years writing what would now be considered an ecological survey, an account of the forest around the Brazilian village of Lagoa Santa. Warming returned to Europe to give lectures in ecology in the University of Copenhagen and wrote the first textbook (1895) specifically devoted to ecology, but, because in 1898, Andreas F.W. Schimper (1857–1901) published his *Pflanzengeographie auf physiologischer Grundlage*, a book to become a standard text in English (1909) as well as German, leaning heavily on Warming's work though without even a footnote to acknowledge it, Warming's ground-breaking contribution has been largely overlooked.

In the early twentieth century plant ecology became more fully established in the English-speaking world, with Arthur Tansley (1871–1955), strongly influenced by Warming (and also undoubtedly Arthur Harry Church), later introducing the word ecosystem (coined by Arthur Clapham [1904–90]) into biology. Tansley spearheaded the foundation of the first ever ecological committee, leading to the British Ecological Society, and the *Journal of Ecology* (both 1913 to present). He forged links with American ecologists and the Ecological Society of America (1915 to present), notably Henry Cowles (1869–1939) and Frederic E. Clements (1874–1945). Clements, a neo-Lamarckian, published his influential *Plant succession* in 1916, establishing the idea of vegetation development leading to self-sustaining climax vegetation, analogous to the development of an organism.

Conservation

When Humboldt was in South America (Venezuela), he began to think in terms of climate-change through human intervention, the forest-clearing by colonists leading to erosion and drying up of streams (Wulf 2015: 57ff.). He noted, "The wooded region acts in a threefold manner in diminishing the temperature; by cooling shade, by evaporation, and by radiation." Following his journeys to Central Asia he emphasized in a lecture in St. Petersburg in 1829 the effects of deforestation on climate (Wulf 2015: 215). On the other hand, his suggestion that the Amazon should be deforested on a large scale to improve agriculture was part of the European view of how tropical possessions might well be developed (Mabberley 1992: 221ff.). With the establishment of colonies as opposed to mere trading-posts came the need for production and exports over and above the earlier economic relationships between the developed world and the tropics. By the time of the establishment of the Brazilian Republic (1889), the indigenous people had few rights and the rubber boom brought entrepreneurs, railroads and telegraph lines. The pattern was repeated throughout the tropics and the resulting dire conservation issues are well known. Striking examples of timber trees are the "genetic erosion" of mahogany (*Swietenia mahagoni*) in tropical America and the almost complete removal of red cedar (*Toona ciliata*) in Australia, largely in the nineteenth century.

It can be argued that, as with the scientific study of ecology, conservation also has tropical roots in that the amateur and later professional naturalists employed in the colonial services, particularly in India from the end of the eighteenth century onwards, advocated conservation measures (Mabberley 1992: 253ff.). Combined with the initially conservative silvicultural policies deriving from German practice, such apparent brakes on development were perhaps acceptable to the distant European governments because they had the effect of preventing ecological changes that might have led to crop failures and consequently jeopardized the suzerainty of the colonial powers.

In the northern hemisphere, parts of the forest of Fountainebleau, southeast of Paris, France, were set aside for conservation as early as 1861. Although the first national park in the United States—and likely the world, Yellowstone, was declared in 1872, this was largely to conserve its geothermal and other geological features. It was followed in 1879 by the (Royal) National Park (the first use of the term "national park"), south of Sydney, Australia. Practical plant conservation measures began in England with the establishment of the Parkinson Society for Lovers of Hardy Flowers in 1884, one of its aims being "to prevent extermination of wild flowers" (Gager 1940), followed by the Society for the Promotion of Nature Reserves (1912). In 1902 the Wild Flower Preservation Society of America was formed in New York.

The most famous advocate for plant conservation in North America in the nineteenth century was John Muir (1838–1914), a botanist, whose interests began in the home garden in Dunbar, Scotland, and when a farmer in Wisconsin, then as a university student (Gisel 2008: xxii–xxviii). He was engrossed by Humboldt's writings before traveling to California where, encouraged by Asa Gray, he collected specimens, but also wrote accounts of the ecology, preparing articles for newspapers and magazines (Gisel 2008: 5; Wulf 2015: 315ff.). He visited Canada and Cuba and made three journeys to Alaska, "Nature's own reservation," believing that the beauty of the wilderness was more important than exploiting its resources (Gisel 2008: 151). First visited by Europeans in 1857 with the giant sequoias a major tourist attraction, especially from 1879, land at Yosemite in California was set aside for preservation and public use by Abraham Lincoln in 1864. Muir wrote of the giant sequoias, "Greatest of trees, greatest of living things,"

FIGURE 0.7 Heavily bearded John Muir (to President Roosevelt's left) in front of the Grizzly Giant (*Sequoiadendron giganteum*), now some three thousand years old, in Mariposa Grove, at Yosemite Valley, California in 1903. Photo by PhotoQuest/Getty Images.

and Yosemite, "the grandest of all the special temples of Nature ... It must be the *sanctum sanctorum* of the Sierras!" (Gisel 2008: 95). His advocacy led to the Yosemite National Park we know today. Muir was also the founding president of the Sierra Club (1892), now one of the strongest environmental pressure-groups in the world.

In the Pacific Northwest, the Hudson Bay Company had a major logging enterprise at Fort Vancouver from 1827, a good market for the company's timber later being railway sleepers (ties), used in the expanding railroad network. By 1890 there was no "frontier" left, but lumber companies, like Weyerhauser, were attempting to conserve what forest had survived and, by 1900, reafforestation had begun. In 1902 Kimberly-Clark became the first paper-producer to initiate long-term forest management, before the federal government did anything about it.

CONCLUSION

The biological science of the nineteenth century saw off "essences" and "humors," replacing them with cells, nuclei, and molecules. The need for "design" to explain the living world evaporated. The new "biology" was an endeavor to move beyond classification and biogeography, building on human physiology from medicine to all organisms, such that for much of the century "biology" and "physiology" were almost synonyms (Coleman 1977: 2). At the beginning of the century it was France in the vanguard but by the 1840s German universities were the founts of innovation and were to remain so until the Great War. In the never-ending pursuit of "What is life?" there was the "Pantheistic vitalism" of German nature-philosophers at the beginning of century, but by the 1850s there were radical mechanists or physiological materialists, English physiologists still espousing forms of vitalism. By then there was a move to make biology an experimental science, particularly in Germany where a reductionist ethos held that living processes could be reduced to chemistry and physics (Coleman 1977: 13).

By mid-century, beginning with botanical research, the cell was recognized as the critical unit of organic function with Charles Brisseau de Mirbel (1776–1854) claiming cells were ubiquitous in plants (Coleman 1977: 17). Robert Brown of the British Museum named the nucleus and, in the 1820s, using a single-lensed microscope, accurately described division in plant sex cells, a process now known as meiosis (Mabberley 1985a: 288, 297–9). Nonetheless, through to the 1860s there were still doubts as to whether the nucleus was merely an artifact but, by about 1875, it was generally agreed that the cell was a recognizable entity, possessing a nucleus containing chromosomes (named in 1888) and cytoplasm (Coleman 1977: 30). By 1884–5 several cytologists independently confirmed that the nucleus was the key to intergenerational continuity and by about 1902, with the "discovery" of Mendel's work, cytology joined the consensus to form cytogenetics.

Despite much stress by some authors on the significance of animal physiology with its anthropomorphic connotations, we find that in the work of Erasmus Darwin, Lamarck, Humboldt, Brown, Darwin, Mendel, and beyond, the study of plants was clearly crucial in the major scientific advances of the period. But the public mind, as ever, has since embraced those creatures most allied to us as poster-boys—in conservation as well as popular science. The chapters in this book demonstrate conclusively that those animals, all ultimately reliant on plants, need to be returned in our thinking to the subservient position they rightfully had in most of the nineteenth century.

CHAPTER ONE

Plants as Staple Foods

CLAUDIA CIOTIR

INTRODUCTION

During the nineteenth century globalization of many important crop plants reached new heights, largely driven by Europeans who transformed the agriculture of their colonies. Crops that had already been exchanged between the New and Old Worlds became staples displacing local ones. In many regions of Africa maize was grown instead of sorghum and other millets, and Asian rice replaced African rice, while in Europe potato ousted buckwheat and millets; this process continued for many other crops during the twentieth century (Kingsbury 2009: 99; National Research Council (NRC) 1984: 12).

Introduction of crops to new continents required cultivars that would thrive there, and breeding programs facilitated crop expansion. Cereals including maize (corn), wheat, and rice were some of the crops of principal interest (Bradshaw 2016: 16–19). Wheat was the subject of early improvement programs first in Europe, and later in the United States and Canada (Walton 1999: 31). Corn was first improved in the United States and later in Europe (Wallace and Brown 1956: 23–7). Breeding advanced following the work of the fathers of Natural Selection and statistical genetics, Charles Darwin (1809–82) and Gregor Mendel (1822–84) (Murphy 2007: 21). In England, Darwin published his *Origin of species* (Darwin 1859) and used hybridization experiments to demonstrate artificial selection in plant crops (Darwin 1876: 149, 163). In Moravia (Czech Republic), Mendel formulated the laws of inheritance that facilitated modern plant breeding (1901: 32). A major contribution was also made by Wilhelm Johannsen (1857–1927), who published the "pure line" theory of inbreeding (1903: 9, 16–18). Generating inbred lines with inherited desirable traits led to uniform cultivars with consistent traits for crop improvement or true breeding (Roll-Hansen 2009: 458, 482). Given this possibility, breeders replaced intuitive selection on variable landraces (traditional locally adapted varieties without modern crop improvement) with breeding of higher-yielding and more uniform cultivars (Bradshaw 2016). Over time, both publicly funded agencies as well as private companies became involved in the production of new crop cultivars (Murphy 2007: 3).

Concomitant with the rise of scientific plant breeding in the Americas, financial and political support for agricultural research increased. In 1862 President Lincoln established the USDA and enacted the Morrill Act which established colleges to teach agriculture and related sciences (USDA Report 1863). In 1888, each agricultural university was granted land for an experimental station (Murphy 2007: 3). Very important in the development of US agriculture was the introduction of new species and cultivars adapted to new environments, while in 1897, the USDA established the Office of Seed and Plant Introductions to develop new crops in the US. Appropriate seeds were collected and tested for decades in colleges and experimental stations and trialed by associated farmers and other growers (Kingsbury 2009: 144). Wheat was introduced to the Midwest, rice to Louisiana, soybean to Illinois and elsewhere. By the end of the nineteenth century, plant breeding turned into a worldwide social and intellectual mission with scientists exchanging ideas and seed germplasm across all continents (Olmstead and Rhode 2006: 23).[1]

In this chapter I describe changes in cultivation and breeding of staple crops from 1800–1920, largely in the West and specifically of those crops upon which large numbers of people depend for at least 25–30 percent of their daily caloric food intake (Loftas and Ross 1995: 21). I focus on four main categories: cereals, pseudo-cereals, legumes, and root crops. I discuss where and how they were grown, those parts of the plant used, methods of cultivation, and improvements made in these crops during the period. This is not an exhaustive treatment, many important crops which were regional staples in various parts of the world being excluded due to space constraints. Similarly, crops like rice, sorghum, and soybeans were staples in the Eastern Hemisphere but only minor crops in the West at this time, thus their history in the latter is not extensively discussed.

CEREALS

Maize (Zea mays)

Maize was domesticated between 10,000 and 6,250 years ago in southern Mexico (Goodman 1995: 194; Piperno 2011: S458). With other crops domesticated in Central and South America maize enabled millennia of flourishing agriculture there (Harris 2005: 21). When Europeans arrived in the Americas, it was the most important cereal; it was grown from Chile to Canada (Badu-Apraku and Fakorede 2006: 229). Maize was soon introduced into the Old World where it became similarly important (Baker 1970: 73). In the early nineteenth century it was a common crop in Europe and, by the 1930s, it was a staple in east and southern Africa too (Miracle 1965: 43). In the 1800s, there were eight major groups of maize cultivars, characterized by the structure and shape of the grain: dent maizeflint maize (see Figure 1.1, a and b), flint-dent maize, flour maize or soft maize, pod maize, pop maize or popcorn, sweet maize, and waxy maize (Badu-Apraku and Fakorede 2006: 232; Baker 1970: 76–7). The European settlers who arrived in North America used indigenous maize cultivars and planted the seeds they saved from year to year (Goodman 1995: 197–9). Unlike Native Americans who kept landraces separate, Europeans deliberately mixed them to increase yields and select for early maturity (Kingsbury 2009: 220). In the US, farmers and breeders grew different corn cultivars in adjacent lines and detasseled alternate rows. Thus, female flowers were fertilized with pollen from a different plant, generating the first generation of F1 corn hybrids (Hunt 1915: 188). Around 1850, Midwestern farmers crossed southern dents with early maturing northern flints. Descendants from these early crosses became the Corn Belt corns which were

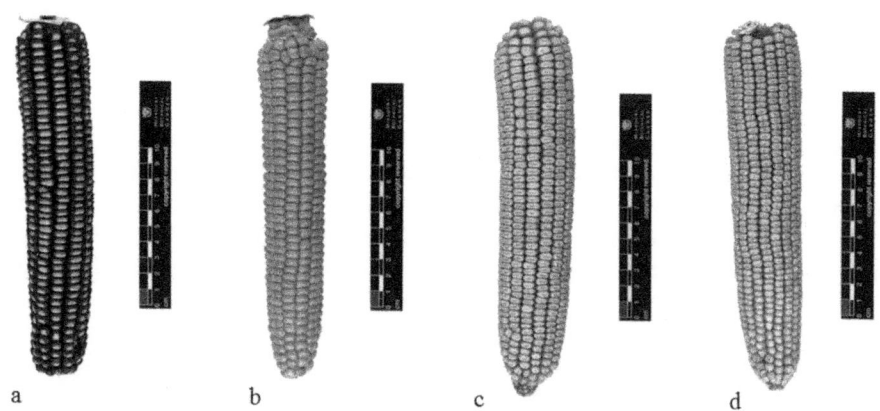

FIGURE 1.1 Maize grown in North America during the period 1800–1920s: a) Northern Dent—Purple with Yellow Spot (Manitoba, Canada); b) Yellow Northern Flint (Quebec, Canada); c) Reid Yellow Dent (Indiana); d) Krug Dent (Iowa). Courtesy of the Anderson-Cutler Maize Collection, Missouri Botanical Garden.

introduced to Europe and became established in southern latitudes (Goodman 1995: 197). However, to begin with most of this was taking place at a local level and the results were generally little-known beyond.

By 1878, few open-pollinated maize cultivars were contributing parental material to maize improvement among Midwestern farmers, but they raised corn yields to about 6,725 kg/ha (100 bushels per acre). For example, 'Reid's Yellow Dent' was selected for uniform filling of the ear, and for making husking easier (Anderson 1903: 15–16) (see Figure 1.1, c). 'Krug' maize was derived from 'Reid's Yellow Dent' and yielded 672 kg/ha more (Wallace and Brown 1956: 84) (see Figure 1.1, d), while 'Leaming' had thin ears that ripened early and grew better in unfavorable conditions (Anderson 1903: 22–4). 'Lancaster Sure Crop,' released in 1910, had an early maturation, disease resistance, and a tendency for the ear's body to break cleanly from the shank. These characteristics made it popular across the northeastern US (Wallace and Brown 1956: 88–91). Individual scientists selected these cultivars to develop inbred lines and hybrid maize (Wallace and Brown 1956: 80–98).

Early in the twentieth century George Harrison Shull (1874–1954) demonstrated the power of heterosis where single-cross hybrids between two inbred lines outperform the parents (Shull 1908: 296–301). In 1917, Donald F. Jones (1890–1963) invented the double-cross method of hybrid seed production (Crow 1998: 924). The first harvest of the double-cross hybrids yielded 7,801 kg/ha (116 bushels per acre) (Jones 1918: 248). This method in general yielded highly productive, predictable, and uniform maize plants and could combine all desirable qualities in a single maize cultivar (Baker 1970). Double-cross hybrids were released commercially as 'Burr-Leaming' in 1921 (Bradshaw 2016: 47).

In many regions of the world during the period 1800–1920 maize was sown by hand early in the season. The First Americans planted grains a yard or so apart, and colonists copied them (Hunt 1915: 230). In the US, up until the mid-1800s, maize was planted by hand, or using a hand planter, covered with a hoe or shovel plow pulled by oxen or other livestock (Hunt 1915: 218–23). By the 1870s, horses replaced oxen, as mechanical reapers, mowers, corn planters, and other equipment reached the Midwest (Fussell 1992: 145; Welch 2000). The transition from manual to mechanized labor increased the efficiency of farm production and moved agriculture to a more commercial model. Corn planters (developed after 1850) were drawn by a horse or, as early as 1875, a steam traction engine (Spielmaker 2014; Towne and Rasmussen 1960: 259). However, manual planting continued in the lowlands of Georgia and South Carolina (Towne and Rasmussen 1960: 259). Maize crops were not given extra water and weeding was done by hand or hoes (Badu-Apraku and Fakorede 2006: 233).

Maize is a high nitrogen-consumer and farmers restored soil fertility by planting it in rotation with beans, cowpea, soybean, groundnut, cotton, and tobacco crops (Badu-Apraku and Fakorede 2006: 234). Manure and lime were applied as fertilizers; calcium and magnesium lime with manure was preferred (Hunt 1915: 214–15). In Peru, for example, it was fertilized with guano collected from coastal islands (Goodman 1995: 197). Maize was harvested either fresh or dried and usually by hand. Stems were cut with sickles or knives and left to dry (Hunt 1915: 254). Manual harvest continued until the twentieth century, but corn combine harvesters were patented in 1928 (Ganzel 2007).

In North America and Europe maize was cultivated for both human consumption and animal feed (Vaughan and Geissler 1999: 6). During the period from 1800 to 1920 landraces were selected for human consumption. Cultivars used for cornmeal had large, soft, starchy grains with blue, pink, or white flour (Nesbitt 2005: 54). Sweet maize cultivars were eaten from boiled cobs or used for flour (pinole; Heiser 1981: 100), while popcorn became a popular snack (Baker 1970: 75). In Central and South America, maize grains were soaked, cooked with lime or ash then hulled. The grain was usually ground and boiled as corn meal or fermented into beer (Nesbitt 2005: 54). In Africa, flinty cultivars were consumed in northern parts of West Africa while floury ones were eaten in the south. Soft grains were commonly mixed and consumed with other cooked vegetables or meat while unripe cobs were used as a vegetable (green maize), boiled or roasted. Grain was also roasted or ground, boiled and eaten as a thick porridge (Badu-Apraku and Fakorede 2006: 230).

Wheat *(*Triticum *spp.)*

The wheats were domesticated in Southwest Asia about ten thousand years ago (Feldman et al. 1995: 186). Durum wheat, macaroni wheat (*Triticum turgidum*), originated in Turkey, Syria, Iraq, and Iran (Feldman 2001). It has long been grown as a staple crop for its high protein content mainly used to manufacture pasta products, best known in Mediterranean areas (Belay 2006a: 183; Nesbitt 2005: 52). In northern Africa and the Middle East, durum wheat has been used to manufacture couscous and bulgur, a parboiled, dried, and then crushed coarse flour (Belay 2006a: 183). In Morocco and Sicily, durum wheat was a staple food to make traditional breads. During the period 1800–1920 durum wheat was cultivated worldwide for its high yielding in dry conditions (Belay 2006a: 183; Feldman 2001: 33). In the former USSR, farmers grew durum landraces like 'Beloturka' (see Figure 1.2, a), which was also well known as a spring or autumn wheat in France (Vilmorin 1880: 140), 'Kivinka,' and 'Rusak' (Morgunov 1992: 10).

FIGURE 1.2 Wheat cultivars grown in Europe during the period 1800–1920: a) 'Beloturka'; b) 'Belle de Talavera'; c) 'Chidham'; d) 'Shirreff Squarehead.' From Vilmorin 1880.

Durum landraces were selected in Spain, Greece, and Italy (Feldman et al. 1995: 186). Southern Italy was extremely rich in durum landraces, where, in 1915, one of the most successful cultivars, 'Capelli,' was released by Nazareno Strampelli (1866–1942) (Laidò et al. 2013). 'Capelli' had a major global impact in Mediterranean durum wheat breeding and served as parent to many cultivars (Kabbaj et al. 2017). 'Rivet wheat' was an old durum cultivar developed for the cooler conditions of northern Europe (Nesbitt 2005: 52). In 1842, durum wheat cultivar 'Red Fife' from Ukraine reached Canada (Walton 1999: 31). 'Red Fife' remained a keystone wheat cultivar during the period, and represented the genetic foundation of breadwheat in Australia (e.g. 'Forward'), UK ('Yeoman'), and North America (e.g. hard spring wheats and winter wheats) (Angus 2001: 113; Olmstead and Rhode 2006: 10; Walton 1999: 31). Russian durum wheats were first brought to the US by Mark Carleton (1866–1925) in the 1900s. 'Kubanka' and 'Perodka' were resistant to rust and performed well at North Dakota Research Stations where they out-yielded all other standard cultivars of the time (Carleton 1900: 14).

Breadwheat (*Triticum aestivum*) is an annual crop, originating some seven thousand years ago in Armenia and coastal areas of the Caspian Sea in Iran (Belay 2006b: 176). As humans migrated to new areas, cultivated wheats occupied new environments—and many diverse forms arose. By the 1800s breadwheat was being cultivated in parts of Asia, North and South America, Africa, and Australia (AWB 2007: 1; Hunt 1915: 48–55; PwC 2011: 4; Vaughan and Geissler 1999: 2). In the early nineteenth century, breadwheat comprised a wide range of landraces and cultivars (Dondlinger 1908: 321). From the mid-nineteenth century to the beginning of the twentieth century, cultivars were classified in five groups: soft winter wheats, hard winter wheats, hard spring wheats, white wheats, and early wheats (Carleton 1900: 36).

Wheat and wheat bread were major staple foods of the daily diet in Russia, Ukraine, and other countries of the former USSR. Farmers developed breadwheat landraces by selecting the best spikes and sowing their grains. 'Ledianka' was an early wheat landrace in cultivation since the eighteenth century (Morgunov 1992: 5). Russian wheat landraces were well adapted to abiotic stresses, particularly to drought, and winter hardiness, and so were included in modern wheat breeding all over the world

(Merezhko: 2001: 268). For example, 'Ladoga,' introduced from Russia to the US and Canada in 1886, was important in the development of many American breadwheat cultivars such as 'Prelude,' 'Preston,' and 'Reward' (Buller 1919). Similarly, 'Orega' landrace is in the ancestry of 'Garnet,' 'Pioneer,' and 'Ruby' cultivars developed in the US and Canada (Morgunov 1992: 5). Between 1910 and 1912, experimental wheat fields and agricultural breeding stations were established in Kharkov (Ukraine), Saratov, and Samara (Volga Region), and, by 1913, Russia had developed an efficient agricultural system through land reforms and advanced agronomy practices. From 1861 to 1914, Russia exported breadwheat to many European countries (Falkus 1966: 422). During the period 1914–20, exports and wheat breeding were halted due to wars and other conflicts that left many Russians starving (Falkus 1966: 421; Morgunov 1992: 6). Breadwheat breeding in the USSR was highly influenced by Nicolai Ivanovich Vavilov (1887–1943) whose international collections provided cultivars resistant to wheat diseases and pests, tolerant of salinity, metal toxicity, drought, and low temperatures (Merezhko 2001: 281; Nabhan 2009: 62).

In western Europe popular breadwheat cultivars towards the end of the eighteenth century included 'Bellevue de Talavera' (see Figure 1.2, b) which had high gluten content and was grown in northern France and Spain until the twentieth century (Kingsbury 2009: 106; Vilmorin 1880: 48). 'Chidham' wheat was one of the most widely grown breadwheats in Britain and other European countries between 1800 and 1880 (Vilmorin 1880: 32) (see Figure 1.2, c). In England, Patrick Shirreff (1791–1876) executed wheat hybridizations and refined the technique of pure-line breeding. His cultivars such as 'Mungoswells' and 'Shirreff's Squarehead' were used in breeding programs all over Europe (Shirreff 1873: 26–9) (see Figure 1.2, d). Sir Rowland Harry Biffen (1874–1949) and his collaborators crossed 'Shirreff's Squarehead' with 'Ghirka' (a Russian rust resistant variety) and developed 'Little Joss,' a cultivar resistant to yellow rust (*Puccinia striiformis* f. sp. *tritici*). During the mid-nineteenth century, wheat from Ukraine was exported to France, where it was crossed with the established French wheats. These crosses generated cultivars such as 'Rouge de Bordeaux,' 'Japhet,' and 'Gros Bleu' with wide climatic adaptations, early ripening, medium lodging and yellow rust resistance (Bonjean et al. 2001: 140). From 1843 to 1899, wheat breeding was a primary focus of the Vilmorin Seed Company based at Verrières-le-Buisson south of Paris (Gayon and Zallen 1998: 255). Vilmorin distributed the first modern drought-tolerant French wheat 'Dattel,' a cross between French 'd'Aquitaine' and British 'Chidham' (Vilmorin 1880: 54).

Western European wheats were spring "soft wheats" which matured before frost only in the western coast areas of North America. For the Midwest and northern climates of the US and Canada, hard winter wheats were introduced from Australia, eastern Europe, Russia, central Asia, and India (Carleton 1900: 63). For example, 'Turkey Red' was introduced from Ukraine in 1874 (Dondlinger 1908: 61), and 'Forward' from Australia, because it was harvestable 15 to 20 days earlier than other wheats (Destler 1968: 205–6). In Canada, Charles Saunders (1867–1937) developed the commercial wheat 'Marquis,' the product of a cross between 'Red Fife' and 'Indian Hard Red Calcutta' (Buller 1919: 145). Launched in 1909, it was early maturing, non-shattering, and high-yielding with good bread-baking qualities (Buller 1919: 155). By 1918 'Marquis' had become the foundation of wheat production in western Canada and in the US, stretching from the Pacific to the Atlantic and setting a profitable standard for bread quality globally (Buller 1919: 170). Wheat was very important in Australia where 'Common Brown' was grown in the 1800s (O'Brien et al. 2001), but in the 1880s wheat yields were severely reduced by wheat rust. In 1901 breeder William Farrer (1845–1906) released 'Federation,' a rust-resistant cross between 'Red Fife' and Indian wheats; it represented not only a breakthrough for Australian agriculture but was also to become a germplasm source in international breeding (Olmstead and Rhode 2006: 18). Its descendant cultivars had early ripening, rust resistance, and drought- and frost-tolerance, expanding the wheat belt into dry and cold areas across Australia, Canada, and the US (O'Brien 2001: 618).

Crop rotation, manure, and mulch were used to enhance wheat production (Buller 1919: 43; Hunt 1915: 77). By 1919, US fertilizer was made of mixed organic fertilizers (vegetable and animal waste materials) and inorganic nitrogen (Curtis 1924: 440). The latter was first produced in the early 1900s in Canada and imported to the US, together with nitrogen fertilizer from Chile and potash from Germany (Curtis 1924: 442). By 1913 the synthetic nitrogen Haber-Bosch process was applied at an industrial scale in Germany and provided a new generation of fertilizers. Breeding contributed 17 percent of the increase in wheat yields between 1886 and 1890 (Olmstead and Rhode 2006: 27). Global wheat production was secured through the development of new cultivars well suited to eastern Europe, Russia, central Asia, and India. The expansion of wheat cultivation across the globe was characterized by a redistribution of the "bread baskets" from coastal moist areas towards drier areas with shorter growing seasons (e.g. Argentina, Australia, central Asia, central Canada, the central US, South Africa, and Ukraine; Kingsbury 2009: 117).

FIGURE 1.3 Combined seeder and disc harrow in the 1920s. Courtesy of University of Saskatchewan, University Archives and Special Collections, MG 108, H. A. Lewis Fonds, II B 12.

Rice *(Oryza spp.)*

Asian or paddy rice (*Oryza sativa*) is an annual grain crop domesticated from *O. rufipogon* about eight thousand years ago in at least two separate regions, southern China and northern India (Gross and Zhao 2014: 6190). In Europe, rice became a crop during the Greek and Roman empire periods (*c.* 500 BCE–500 CE) when it was grown in Spain and Italy (Zohary et al. 2012: 74). *Oryza sativa* comprises thousands of landraces and cultivars (Sweeney and McCouch 2007: 954) often classified in groups within two subspecies. *Oryza sativa* subsp. *japonica* has short thick grains and is adapted to relatively cool climates in Southeast Asia and northern China, while *O. sativa* subsp. *indica* has long thin grains and is adapted to hot, tropical climates (Zohary et al. 2012: 73). *Oryza sativa* subsp. *japonica* (sticky rice) had landraces typically grown in temperate East Asia and in upland areas of Southeast Asia (Londo et al. 2006: 9579). *Oryza sativa* subsp. *indica* (non-sticky rice) was grown primarily in lowlands throughout tropical Asia (Molina et al. 2011: 8351). Within subsp. *indica*, 'Aus,' 'Rayada,' and 'Ashina' were selected in the Himalayan foothills, Bangladesh, and India (Garris et al. 2005: 1632). The 'Aus' rices were upland types selected for drought tolerance and early maturity in Bangladesh and India (Khush 2010: 10). 'Ashina' and 'Rayada' were grown under inundation in Bangladesh and India. *Oryza sativa* subsp. *indica* rices were mostly improved by farmers, in the nineteenth century there being at least ten thousand floating rice cultivars in Bangladesh and India (Kingsbury 2009: 308). They include the aromatic rices of the Indian subcontinent, e.g. "Basmati" (Khush et al. 2010: 10).

Japan experienced an agricultural revolution that increased rice yields by as much as 1,300 kg/ha (19.33 bushels per acre) at the end of the nineteenth century (Kingsbury 2009: 65). Between 1819 and 1882 rice cultivars were collected from all over Japan, assessed, and then re-distributed to farmers, based on productivity in given areas. During this period, many of the tall and deep-water rices disappeared, the surviving ones having early maturity and grain yield improved up to two-fold (Kaneda 2010: 170). Between 1868 and 1912, experimental stations were established to assess landraces and to select cultivars for particular environments (Kingsbury 2009: 65). These institutions promoted soil management and expanded the use of fertilizers, agricultural tools, and machinery, resulting in selection of improved cultivars, increased production, and reduced labor intensity (Kaneda 2010: 172). In 1890, Japanese farmers selected two cultivars, 'Rono' or 'Shinriki' and 'Kameno-o,' that became popular throughout the whole country (Khush 1987: 196). These responded well to fertilizer, were disease resistant, shorter than older cultivars, and later became the ancestors of dwarf rice (Khush 1987: 196). In 1893, a national breeding and crop improvement program was established in Japan.

In Myanmar, the first major increases in rice production occurred under the British administration, when annual rice production rose rapidly from 44,000 tonnes (2 million bushels) in 1830 to 1 million tonnes (46 million bushels) in 1870, 5 million tonnes (230 million bushels) in 1900 and 6.5 million tonnes (299 million bushels) in 1910 (Win 1991: 20). The most grown were 'Emata' and 'Letywezin' with long slender grains and translucent kernels, 'Ngasein' with short medium grains and kernels usually translucent, 'Midon' with short, roundish, grains and kernels opaque and chalky, and 'Byat' with large, broad grains and kernels opaque and chalky (Beale 1927: 165). Asian rice was being grown in Egypt on the floodbanks of the Nile by 1847 (El Azeem et al. 2010: 376). 'Yabani 2' and 'Nabatat Asmar' were two local ones developed in the 1920s from introductions of *O. sativa* subsp. *japonica* cultivars from China, India, Italy, Japan, Spain, and the US (El Azeem et al. 2010: 380).

Rice was grown for human consumption in Indian, Chinese Italian, and Spanish cuisines (Vaughan and Geissler 1999: 8). In Italy, the first record of the traditional rice recipe "risotto" was in 1809 (Perron 2011); in Spain "paella" originated in southern Andalucia and became a popular rice dish in the early 1800s (Olver 2009). In India, 'Basmati' rice was prized for the distinctive popcorn-like aroma and food quality. Sticky rice was used in Japanese cuisine and the traditional alcoholic drink sake (Vaughan and Geissler 1999: 8). In tropical Africa, Asian rice was cooked by boiling or steaming with pulses, vegetables, meat, or fish. Rice flour was used for bread and cake mixes, as well as breakfast porridges for children (Meertens 2006: 113). Rice was also used in rice puddings and other Victorian desserts (Sumner 2004: 56). In the US, rice pudding recipes were recorded in 1796, and, by the early 1800s, griddle rice cakes were prepared for breakfast in the cuisine of the southern states (Sumner 2004: 56). Rice was also used in curry dishes of the southern US. Southern US cooks also used rice flour or boiled rice to bake yeast breads, sometimes combining rice with wheat or corn flour (Sumner 2004: 56). Dutch and British plantations introduced Indonesian rice to their daily diet. In Amsterdam and in Dutch colonies *"rijsttafel"* represented Indonesian meals which included rice and up to forty side dishes consisting of curried meat, fish, chicken, relishes, and vegetables (Leong-Salobir 2011: 78).

In early harvesting practices, rice bundles were laid in rows on the ground with the fruiting heads outward to be trampled by humans or animals (Heiser 1981: 96). Farmers would trample over harvested plants swinging a flail to beat the heads of rice, yielding grain (Clowse 1971: 129). In the early 1800s, South Carolina planters imported threshing machines from Scotland and Sicily to clean the grain (Chaplin 1993: 147–50). Rice-threshing machines became increasingly popular during the 1850s (Whitten 1982: 13). After threshing, hulls were removed and grains were polished then winnowed before milling (Tuten 2010: 20). Although most agricultural accounts of the time referred to undefined rice, it is likely that both Asian and African rice were being grown in South Carolina and other Atlantic states (Carney 2001; 144; Tuten 2010: 12).

After the American Civil War, rice production collapsed in South Carolina where slave labor was no longer available and hurricanes ravaged many plantations, but flourished in Bengal, and the Malay Archipelago, as well as Arkansas, Louisiana, and Texas (Trinkley and Fick: 23; Tuten 2010: 24). In the US, rice economy moved westward and became competitive especially in the Gulf Coast states where postwar soldiers received land parceled from former plantations. A few South Carolinian rice planters became seed producers selling both African and Asian white rice cultivars to the southwestern rice producers (Tuten 2010: 55).

African rice (*Oryza glaberrima*) is an annual crop that was domesticated in the Niger River delta of West Africa, some three thousand years ago (Linares 2002: 16360), its wild forms being known as *O. barthii* (Sharma 2010: 1). In the nineteenth century, African rice was grown in the flood plains of the Niger and Senegal rivers, Mali, Sierra Leone, and parts of Ghana and Togo (Agnoun et al. 2012: 158). African rice reached the Americas in 1647 when it was introduced to the Carolinas (Baker 1970: 93). By 1860, such rice-planting was an important part of Atlantic coast agriculture, extending from Cape Fear River in North Carolina to St. John's River in Florida (Tuten 2010: 22). African rice was also grown in South and Central America—in Brazil, Guyana, and Panama (Coclanis 2010: 414). In South Carolina, rice cultivars came from Madagascar, the 'Gold Seed Rice

Rice' being the most planted and highly esteemed across the Winyaw and Waccamaw regions (Allston 1846: 323).

Little is recorded about breeding African rice. Most cultivars of *O. glaberrima* have strongly scented, red-skinned grains; they were grown locally for their flavor properties and ability to withstand flooding, pests, and diseases (Bezançon and Diallo 2006: 107; National Research Council 1984: 21). Many cultivars were selected for compact panicles and large grains through conscious crossing and inbreeding (National Research Council 1984: 33). In the nineteenth century, upland varieties in West Africa were up to 120 cm tall, while the ones planted in irrigated conditions and floating culture reached 5 m (Bezançon 1993: 35), but during this period African rice landraces were replaced with commercial cultivars of Asian rice in most of West Africa and the US (Van Andel 2010: 2).

Sorghum *(Sorghum bicolor)*

Sorghum is one of a large group of grasses known as millets, small seeded grasses cultivated as grain crops. Millets are staple crops in the semi-arid tropics of Asia and Africa. The most important are finger millet (*Eleusine coracana*; Pradham 2013: 10–14), pearl millet (*Cenchrus americanus*, formerly *Pennisetum glaucum*), and sorghum. In the nineteenth century, sorghum was a major staple cereal of rainfed agriculture in the eastern hemisphere and the semi-arid tropics; sorghum was only a minor crop or an animal feed ingredient in the Americas. It is an annual crop that was likely domesticated five thousand years ago in the semi-arid regions of Ethiopia or Chad (Vaughan and Geissler 1999: 10). From Africa, it was spread through Southwest Asia including the Indian subcontinent to China, Europe (Mediterranean basin), and Australia (Doggett and Rao 1995: 177–9). From Africa, it was introduced to the New World in the eighteenth century and in the early twentieth century sorghum became an important crop in the Great Plains of North America (Nesbitt 2005: 57); from the US, it was introduced to South America and Australia.

In the nineteenth century, five traditional landraces of *Sorghum bicolor* were used in Africa and Asia for grain food and feed (Doggett and Rao 1995: 176–8). They were consistently improved by farmers and breeders. Sorghum landraces were named based on color and shape of their grains: Bicolor, Caudatum, Durra, Guinea, and Kafir. Bicolor landraces were developed first in Africa and Asia and included forms with small fruits almost totally enclosed by glumes (Doggett and Rao 1995: 176). Caudatum landraces had large "turtle-back" shaped grains and were widely grown for their good-quality grains in Chad, Nigeria, Sudan, and Uganda (Doggett and Rao 1995: 177). Durra cultivars were developed in Ethiopia and are adapted to semi-arid conditions. Most Durra cultivars reach maturity in both short dry and long rainy seasons and were widely grown in Egypt, Ethiopia, India, Sudan, western Asia, and along the margins of the southern Sahara (Doggett and Rao 1995: 177). Guinea landraces were developed for the humid climate and were grown in high or low rainfall regimes from West Africa, along the East African Rift, India, and coastal areas of Southeast Asia (Balole and Legwaila 2006: 169; Nesbitt 2005: 57). Kafir landraces have short grains and were grown as staple crops across Africa's eastern and southern savannas from Tanzania to South Africa. They are day-length neutral and became the most commercially traded sorghums (Balole and Legwaila 2006: 169).

In the nineteenth century, grain sorghum was harvested when the grain had hardened. The fruiting heads were cut by hand and temporarily stored in bags for further drying. Sometimes whole plants were pulled from the field and transported to households where the heads were removed later (Balole and Legwaila 2006: 172). Sweet sorghum was harvested from October to January according to when the seed was in a soft dough stage (Balole and Legwaila 2006: 172; Irvine 1974: 138). After threshing, grains were hand-hulled and ground into flour with a mortar and pestle (Irvine 1974: 140). Sorghum was a vital food grain in many parts of Africa, while the stalk and leaves were valued for building and forage (Balole and Legwaila 2006: 166). The edible types of sorghum have white endosperm and either coarse or soft grains. In northern Africa shepherds consumed Bicolor seeds mixed with milk or animal blood (Doggett and Rao 1995: 173). Coarse grains were either eaten boiled as porridge or ground to flour. Sorghum flour was also used in porridge and sometimes fermented in a sourdough bread or dumplings (Nesbitt 2005: 57). Soft grains were ground with hulls and processed to flour, then cooked on a hot plate as unleavened bread (Balole and Legwaila 2006: 166; Nesbitt 2005: 57). Some Ethiopian grain sorghums were dimpled and sweet, and they were consumed as roasted seeds or the whole panicle roasted while still green. In Sudan, cooked fermented sorghum gruel known as "nasha" was fed to infants. Sometimes gruels were mixed with other grain cereals and consumed as "uji." Green sweet sorghum stems were chewed (Doggett and Rao 1995: 173). In Africa, bitter seeds were used to flavor beer, and produce molasses (Balole and Legwaila 2006: 166).

Pseudo-cereals (Amaranthaceae [including Chenopodiaceae], Polygonaceae)

A pseudo-cereal is defined as any food crop producing a grain that exhibits nutritional content similar to cereal grains (members of the grass family, Gramineae/Poaceae) but that is produced by non-grass families (Vaughan and Geissler 1999: xvii). Buckwheat, amaranths, and quinoa were pseudo-cereal staples consumed in the nineteenth century; here I focus on buckwheat (*Fagopyrum* spp., Poylgonaceae), an important staple crop then. The Old World genus *Fagopyrum* is represented in central Asia by temperate and cold-tolerant species. *Fagopyrum esculentum* and *F. tataricum* are annual species domesticated some five thousand years ago in South China from where they were spread steadily throughout Asia as far as Japan (Campbell 1995: 409, 411; Gondola and Papp 2010: 17). Buckwheat was introduced to Europe during the Middle Ages, probably from Russia (Campbell 1997: 17). Landraces developed during this time were early-ripening and photoperiod-neutral (Fesenko et al. 2016: 99–108). Intense selection of *Fagopyrum* species during the seventeenth century made eastern Europe a secondary diversity center for buckwheat and, by the nineteenth century, common buckwheat (*F. esculentum*) was widely grown over the Northern and to some extent the Southern Hemisphere (Kreft et al. 2016: 161). Common buckwheat was a staple crop in Asia (especially China and Japan); European immigrants introduced it to the US and Canada where it was cultivated until the 1900s. Eventually, cultivation there declined due to poor response to fertilizers (Jansen 2006: 73). Tartary buckwheat (*F. tataricum*) is cultivated in India and China in cooler and harsher conditions than common buckwheat, although both species will tolerate tropical climates at higher altitudes (Joshi and Rana 1995: 85).

In Europe and Asia many traditional landraces and cultivars were used (Jansen 2006: 73; Joshi and Rana 1995: 85). In Japan, buckwheat produced up to 140,000 tonnes (6.4 million bushels) per year in the nineteenth century. Japanese farmers used it because it grew rapidly, required little labor, and tolerated infertile lands where rice was not productive (Tanaka 2016: 61–77). Himalayan farmers from India, Bhutan, and Nepal maintained 309 landraces of common buckwheat with 196 landraces of Tartary buckwheat in Nepal alone (Campbell 1997: 30). This diversity likely resulted from local preferences and the large variation in altitude and ecological conditions (Campbell 1997: 28; Joshi and Rana 1995: 120). Farmers of the alpine Himalaya grew buckwheat for grains only, while those in the lower hills developed landraces for leafy parts as well as for grains (Joshi and Rana 1995: 120). Farmers selected for larger-grained and high-yielding cultivars (Campbell 1997: 48). Tartary buckwheat was mostly cultivated at high altitudes in Asia (Farooq et al. 2016: 310). At this time few buckwheat cultivars were successfully grown in North America but included 'Common Gray,' 'Japanese,' 'Orenburg No. 6,' 'Russian No. 1,' and 'Silver Hull' (Hunt 1915: 403). Breeding improvement in common buckwheat continued in Russia and France until early in the twentieth century, at which point buckwheat was replaced in part by barley and potatoes (Campbell 1997: 48; Farooq et al. 2016: 300).

Buckwheat was used as a cereal substitute in nearly every country where cereals were cultivated (Campbell 1995: 409). In Bhutan, southern China, and Nepal, buckwheat served as a rice replacement in the staple diet (Campbell 1997: 29). In Brittany, France, buckwheat flour was used to prepare pancakes locally known as "galettes" (Nesbitt 2005: 58). In eastern Europe and Russia buckwheat grains were cooked and a porridge dish named "kasha" was served daily. In Japan buckwheat flour served to prepare noodles known as "soba" (Vaughan and Geissler 1999: 14), while, in India, it was used to make unleavened bread named "chillare." Tartary buckwheat was a common grain crop of Siberia (Nesbitt 2005: 58). In the Kulu Valley (Himachal Pradesh), India, grains were either cooked like rice or were ground and mixed with barley and wheat flour to bake "chapattis" (Joshi and Rana 1995: 87). In Europe and North America, buckwheat flour was mixed with wheat, rice, and maize flour to make breakfast cereals, biscuits, noodles, and pancakes (Campbell 1995: 409).

Legumes/pulses (Leguminosae/Fabaceae)

The members of family Leguminosae/Fabaceae are known as legumes; grain legumes (or pulses) are defined as annual, herbaceous plants whose seeds are harvested at maturity and marketed dry (Vaughan and Geissler 1999: xviii). Legume staples consumed include chickpea (*Cicer arietinum*), common bean (*Phaseolus vulgaris*), cowpea (*Vigna unguiculata*), faba or broad bean (*Vicia faba*), groundnut (*Arachis hypogaea*), lentil (*Lens culinaris*), pea (*Pisum sativum*), pigeonpea (*Cajanus cajan*), and soybean (*Glycine max*); due to lack of space here I focus on common bean, an important staple in the period.

The Common bean (*Phaseolus vulgaris*) is an annual crop, domesticated ten thousand–five thousand years ago (Debouck and Smartt 1995: 291), independently at least twice—from a Mesoamerican genepool and an Andean one (Aragão et al. 2011: 226). From Central and South America, it was taken to Europe where rapid distribution was facilitated by seed exchanges among farmers, territorial contiguity, and similarity of environments (Lioi and Piergiovanni 2013: 17–18). In Africa, it was introduced to coastal areas of Kenya, Mozambique, and Zanzibar, from where it was carried to the highlands by slave- and merchant-caravans (Wortmann 2006: 146). In the late nineteenth century, European common beans were taken back to eastern North America by the early European settlers (Hancock 2012: 155). Also in the nineteenth century, it was an established pulse crop in not only America but also Europe and many parts of tropical Africa (Wortmann 2006: 146). The Mesoamerican gene pool of the common bean had four major races: Durango, Guatemala, Jalisco, and Mesoamerican (Kelly 2010). The Durango race included pinto and Great Northern beans, which were carried by Native Americans across the central US to Alberta, Canada; the Jalisco race had small red and pink beans which were moved from the Caribbean to the east coast of North America to the Great Lakes. The Guatemala race included the climbing beans, the Mesoamerican race, the small-seeded navy and black beans (Kelly 2010: 4). In the nineteenth century, navy black beans were being grown by Native Americans around the Great Lakes and were valued by the early European settlers as a regionally adapted high-protein food source. In 1915, 'Robust Navy' was the first North American line released by Frank Spragg at Michigan University; it was selected for its hardiness and consistently uniform yield (Kelly 2010: 26) and was the first bean tinned and sent as "baked beans" to Europe, to become beloved in Britain and Ireland.

The Andean genepool (large-seeded beans) included races such as Chile, Nueva Granada, and Peru. The Chile race included vine cranberry beans, Chilean Coscorron, and Tortula types. The Nueva Granada race comprised large-seeded kidney beans, bush cranberry beans, and most snap beans, while Peru included yellow beans (Canario and Mayacoba) (Kelly 2010: 4–5). Beans from the Andean genepool were cultivated in Europe, particularly in Spain (Lioi and Piergiovanni 2013: 19). The intense selection of these cultivars made Spain a secondary center of diversification for the common bean. Following their introduction to Europe, common beans were the subject of selection for seed color, seed size, taste, plants adapted to longer day-lengths, and resistance to pests and diseases (Lioi and Piergiovanni 2013: 17–18), such that by the nineteenth century, selection had led to thousands of local landraces well adapted to various regions of Europe. Popular beans developed during the period were 'Ganxet Bean' in Catalonia, Spain, 'Prespon,' 'Florinas,' and 'Kastorias' in Greece, and 'Faggiolo del Purgatorio' in Lazio, Italy (Lioi and Piergiovanni 2013: 32–3). In Africa, landraces developed were 80 percent from the Andean pool, and 20 percent from the Mexican pool due to independent introductions from Brazil (Sauer 1993: 76).

In the nineteenth century, common beans were harvested while most pods were still green, but near physiological maturity to produce fresh seeds; pods were also harvested at maturity yielding dry seeds (Vaughan and Geissler 1999: 40). Pods of climbing types were mostly harvested by hand individually, as they matured at different times. Harvesting in tropical Africa was also mostly done by hand. Farmers dried seeds in the sun to destroy bruchids (bean weevils) and reduce moisture content (Wortmann 2006: 150). At this time, the common bean was the most important legume in Europe and North America (Debouck and Smartt 1995: 290). In many temperate and subtropical regions of the world, beans were consumed as a vegetable as young pods, green snaps and green seeds, or as dried seeds (Pearman 2005: 143; Vaughan and Geissler 1999: 40). They complemented cereal grains namely maize, rice, and millets (Harris 2005: 14; Wortmann 2006: 146). In tropical Africa, they were boiled, and often eaten with oil and seasoning. In Europe beans were mashed or made into soups and curries; they were cooked with meat pieces in traditional dishes such as French "cassoulet," and Mexican "chilli con carne" (Vaughan and Geissler 1999: 40).

Rootcrops

"Rootcrop" is a general term used for plants that are grown for below-ground vegetative tissues (roots, shoots, stems, etc.; Vaughan and Geissler 1999: 180). Rootcrops belong to different botanical families and many of them are biennial or perennial herbaceous plants grown as annuals (Vaughan and Geissler 1999: xviii). Many are the most important staples in the tropical world (Onwueme 1978: I). Root and tuber staples continuing to be consumed during the period include cassava (*Manihot esculenta*), Jerusalem artichoke

(*Helianthus tuberosus*), potato (*Solanum tuberosum*), sweet potato ("yam," *Ipomoea batatas*), taro (*Colocasia esculenta*), and true yams (*Dioscorea alata*, etc.). Here I focus on potato, internationally an important staple crop of the time.

Potato (*Solanum tuberosum*) is a perennial grown as an annual crop; it is one of the most important staples in the world, ranking fourth after wheat, corn, and rice (Sanderson 2005: 63). It originated from the Peruvian and Bolivian highlands and became widespread in prehistoric South and Central America (Sauer 1993: 147). Potato domestication likely started five thousand years ago in the region of Lake Titicaca (Simmonds 1995: 468). It was first introduced to Spain and Britain in the late sixteenth century (Vaughan and Geissler 1999: 186), though important Chilean landraces did not arrive until the 1840s. From Europe, the potato reached North America via Bermuda in 1621 (Sanderson 2005: 64). It was introduced to Australia in the late eighteenth century where it became a common crop by 1850 (Sanderson 2005: 64) and by 1900 was almost cosmopolitan.

In the early 1800s, five types of potato were known in Ireland: Black potato, Flat white kidney potato, Round red, Round white, and Yellow (Glendinning 1983: 8). These originated from *Solanum tuberosum* Andigenum Group, were sensitive to daylength, and flowered only in late autumn in Europe. Breeding was needed to overcome the adaptation to day-length (photoperiodism), virus diseases (known as potato debilitation/

FIGURE 1.4 Potato harvest in Scotland in early 1920s. Original contributor John Clark Maddison and image courtesy of www.ayrshirehistory.com.

degeneration), and potato blight. But before such could be developed in Europe, potato blight epidemics broke out in Ireland in 1840 and culminated in the Great Famine of 1845–6 (Vaughan and Geissler 1999: 186). Blight wiped out all except the cultivar 'Regent' and led to irreversible social change (Glendinning 1983: 491). During the Great Famine 8 million Irish people starved, over 1–1.5 million died, and 1 million emigrated to England, Australia, and the US (Kingsbury 2009: 118). Subsequently, disease resistance was partially achieved through repeated introduction of new cultivars of *Solanum tuberosum* Chilotanum Group from Chile (Sauer 1993: 153).

In the US, the first-developed early potato cultivars were 'Garnet Chile' and 'Rough Purple Chile' (Kingsbury 2009: 119). 'Garnet Chile' was the progenitor of many early potatoes including the most successful American cultivar of all time 'Russet Burbank' (Sauer 1993: 153). In the mid-1850s 'Rough Purple Chili' and 'Daber' varieties were introduced to England (Glendinning 1983: 491). The Scottish Plant Breeding Station program used 'Daber' to develop disease-resistant cultivars such as 'Abundance,' 'Epicure,' and 'Pepo' (Glendinning 1983: 494, 501), giving rise to some 250 cultivars that were spread throughout Europe and North America. Although they had disease resistance, these potatoes had a significantly narrowed genetic base by the 1870s. From the mid-nineteenth century onward, repeated introductions from South America played a vital role in introducing greater variability in plant characteristics. 'Up-to-Date' and 'Majestic' were developed in England at the end of the nineteenth century for early maturing, good palatability, and high-yielding characteristics. 'Majestic' was distributed worldwide and was planted commercially for over fifty years. These new cultivars allowed farmers to double their income per hectare (Kingsbury 2009: 181–2).

By the beginning of the twentieth century, the short-day requirement was bred out of the potato crop, leading to earlier maturing and heavier cropping cultivars (Vaughan and Geissler 1999: 186), but the breeding progress was slow, as potatoes need cross-pollination and new cultivars had to be selected from the variable offspring of such crosses (Kingsbury 2009: 423).

CONCLUSION

During the period of 1800–1920, staple crops were shaped by human colonization, migrations, and population expansions. Staple crops were carried to diverse regions where they took hold, often replacing more traditional local crops. Selection for increased yield, disease resistance, and local adaptation was enhanced in part by increased access to diverse breeding materials obtained through global travel and exchange, and also by a growing awareness of the power of self-fertilization and hybridization to improve crops. Further, expansion of staple crops was facilitated by mechanization and rapid progress in transportation (e.g. shipping and railways).

By the 1920s, breeding staples like maize and wheat had increasingly developed a specialized industry that transitioned from farmers to breeding institutions and the agribusiness of seed production. These institutions secured seed stocks from highly bred, hybrid, and improved selected cultivars adapted to specific soil, climate, and diseases. A clear division of labor and competition in breeding and intellectual property emerged between the state experimental stations led by the USDA, and the private seed companies, especially in hybrid maize. Early successes in plant breeding between 1800 and 1920 drove investment in agricultural research and set the groundwork for further crop development and the Green Revolution.

ACKNOWLEDGEMENTS

I would like to express my deep gratitude to Allison Miller for her generous contribution in planning, developing, and preparing earlier versions of this manuscript. Her constructive suggestions and useful critique shaped the coherence, flow, and originality of this chapter. I would like to extend my thanks to The Missouri Botanical Garden staff, librarians Mary Stiffler, Stephanie Keil, Victoria McMichael, and Dr. Linda Oestry who identified literature of the period, and ordered books and interlibrary loans on my behalf. Taryn Pelch and Jim Solomon provided access and permission to use old maize cultivars from the Anderson-Cutler Maize Collection. Mike Blomberg made the digital images for the illustrations. Finally, I would like to thank Joanna Zigouris for proofreading my manuscript.

CHAPTER TWO

Plants as Luxury Foods

PATRICK HUNT

INTRODUCTION

[London taverns are venues with] all the most delicate luxuries on earth, and where the fortuned voluptuary may indulge his appetite, not only with the natural dainties of every season, but with delicacies produced by means of preternatural ingenuity.[1]

John Roach, *Roach's London Pocket Pilot*, or *Stranger's Guide to the Metropolis* (London, 1796), 44

Luxury to some may be considered necessity to others. In the nineteenth century what was considered luxury meant it was beyond the economic reach of many. Luxury foods must fall by definition into the category of items that are not necessary staples for survival. The immense changes in the volume of trade goods in the nineteenth century reflected not just supply and demand economics but also the changing forces of technology that both met and impacted the volume of trade goods, especially because so many staples could be met by local production; others, however, like raw sugar—first luxury, then staple—could be more cheaply produced abroad and thus needed shipping for long-distance transport (O'Rourke and Williamson 1999). The role of sugar was just such a force in market economic history (Mintz 1986). Sugar, tea, coffee, and tobacco were the plant luxury products up to the nineteenth century, but these became staples in the diets of new urban working and middle classes in the nineteenth century (WTO 2013). Industrial growth also contributed much to distribution of such products because at the beginning of the nineteenth century, few sailing ships of three hundred tons were made in America; by the time clipper ships were becoming obsolete after mid-century, even the largest clippers were ten times the size of the sailing ships of 1800 and a similar pattern is true for steam engines from 1810 with 100 horsepower to 1865 with engines of 1,000 horsepower (Brady 1964: 145–203, esp. 145).

Development of a more efficient overland transportation network in Europe and North America, for example, included canals and ultimately railroads in the nineteenth century to complement the same rapid growth of international shipping, especially with multilateral trade tariff reductions after 1860, because tariffs had been the main obstacle to international trade (Kumar et al. 2013: 285–97, esp. 291–2; WTO 2013: n7, 105). The cost of luxury food products has often been tied to the distance any such products must travel from source to consumer and this was certainly true in the nineteenth century. Changes in transportation and time required to get food to markets have been major factors of change in valuation over time. The strategic importance of agriculture, for example, in nineteenth-century Germany was that it represented the largest sector of the economy, but changes in production helped to modernize Germany by transforming it from an agrarian to an industrial society, accompanied by greater markets for products abroad (Pierenkemper and Tilly 2004: 75–86). The early nineteenth century highlighted a pivotal change for world trade due to advances in ship design, navigation, and the industrial revolution that eroded the "tyranny of distance" which had complicated trade in luxury products, since shorter times reduced spoilage of perishable luxuries (WTO 2013: 46).

The new sugar beet industry in Europe from the early nineteenth century onward increased the availability of cheap sugar, especially since it had a relatively short traveling distance for home markets. In 1811 Napoleon (1769–1821) mandated production after seeing Prussian success after 1801. Although the European harvest of 1815 was excellent, it was followed by a devastating one in 1816 due to the weather across the Continent. France recovered after 1817 and even paid down its war debt—as it had profited through stripping other countries during the wars—paying off its war indemnity by 1818. Britain, which had greater war debt, also had colonial revenues while France had a male population depleted by the wars, and agricultural recovery was impacted. Financial deprivations must have severely reduced luxury plant markets in the Napoleonic Wars as well as the Franco-Prussian War of 1870–1, among others. Between wars, in France alone, sugar beet mills increased to 543 by 1837, producing thirty-five thousand tons of sugar annually; in an early example of globalization, an industrialized Germany processed much of France's beet sugar by 1880, while producing in one of the first "green revolutions" much beet sugar fertilized by imported Chilean nitrate, as in France, Great Britain, and the US (Mellilo 2012: 1028–60).

Modern luxuries may not be yesterday's and vice versa. Coffee, for example, would have still been somewhat exotic and more expensive in the early nineteenth century than in the twenty-first century when so many think of it as an essential. From the end of the eighteenth century onward in the British Empire, for example, early globalization became a force in trade, cost, and consumption:

> ... the edible produce of imperial trade pervaded British society, as these one-time luxuries moved down the scale of affordability to become semi-luxuries and, in some cases, necessities. Foods ranging from coffee to curry also became the empire's most ubiquitous symbols, and their advertisement, retail, preparation and consumption reflected and contributed to British discussions and perceptions of the empire. The importance of food to the history of early empire is incontestable. The English, and later British, penchant for sweet hot beverages helped to fuel the empire's expansion into Asia, transformed the ecosystems of large swathes of the Americas ...
>
> (Bickham 2008: 71)

Whether for consumption by the wealthy, or even by producers with a surplus at the source, the range of luxury foods includes edible **oils** such as *olive and walnut oil*; **beverages** like *tea* or *coffee*, **alcoholic** drinks such as vintage fine *wine, champagne*, or distilled aged *spirits* (whisk[e]y, brandy, cognac, bourbon, etc.); **sweets** such as *chocolates, candies, jams* and *jellies, syrups*; **condiments** such as *herbs* and *spices* like *pepper, ginger, cinnamon, nutmeg*; exotic high-protein products like **nuts**; vegetables **preserved** by *salting* or *pickling*; **fermented** indigenous traditional plant foods of the world; as well as drug or

ritual foods for religious or other purposes. This is not comprehensive but likely covers many of the luxury comestibles available in the nineteenth century.

A fascinating ancillary issue connects to new luxury food products when religious or political entrepreneurs in Quaker Societies, among Abolitionists, and in Temperance Movements refocused labor markets and changed price indices. For example, the sugar beet industry in the US, even though close to Caribbean sugar cane industry, was partly motivated to produce home beet sugar to break dependency on slave labor and curtail rum production in any way possible.

LUXURY FOOD OILS

Olive oil had a long history as olives (*Olea europaea*) provided not only food, but also oil for burning in lamps and a healthy base for medicine and cosmetics. Terracotta olive-oil amphorae seem not to have changed significantly in the Mediterranean, since at the beginning of the nineteenth century olive oil was still being shipped in terracotta until replaced by glass and metal, as preservation practices were improved in canning and related manufacturing. Because throughout the century most of the Mediterranean olive harvest was by hand, the resulting price of olive oil made it a luxury. In the nineteenth century, the olive industry contracted in Italy and France but expanded in Greece, Turkey, Tunisia, and Spain (Quiles 2006: 31). Turkey became one of the largest oil producers as many thousands of trees were planted in the area north of Pergamon and the local industry flourished, especially on the Izmir Aegean coast in Ayvalik, Turkey, which has been well studied (Terzi 2007). Spanish producers were mainly established in the south around Jaen, Cordoba, and Seville, but mechanization lagged, and Spanish olive oil was often turbid and not clarified. Milling and pressing gradually improved as conical rollers and hydraulic presses were invented. Olive oil production then either tended to quality handmade processes or to mass production (Quiles 2006: 32). The more robust producers in Provence, France, in the nineteenth century include Nicolas Alziari of Nice (1868); Marseille soaps were even then primarily olive oil based. British consumers found olive oil expensive and *Mrs. Beeton's Book of Household Management* of 1861 had it as plentiful on the Continent but only used as salad oil (Heyden 2014). Olive oil produced in San Diego, California, from trees planted there around 1784, was never intended as continental commercial production but primarily for sustaining mission life. The dearth of olive oil was thus even more pronounced in the United States, until California was persuaded by John S. Calkins to begin planting olives in the north Central Valley around 1890, using San Diego olive stock. Over fifty thousand trees were planted by 1901 (Carter 2007: 142).

French walnut (*Juglans regia*) oil, *huile de noix*, especially concentrated in the Isère Valley of France, the largest supplier, was a commodity staple throughout the nineteenth century. Rancidification was a problem if walnut oil was not protected from heat and light, but since most oil was consumed on the Continent and markets were more easily reached as the century progressed, it became less important. In California, indigenous peoples had grown native walnut trees (*J. californica*) for centuries, and it was from these old trees, up to three hundred years old, that a Californian walnut industry was seen as viable in the late nineteenth century. Some of these trees provided seed stock for the Central Valley, the first orchard planted around 1876 in Winters, Yolo County.

LUXURY PLANT-DERIVED BEVERAGES

Beverages are divided here into alcoholic and non-alcoholic. Alcoholic beverages have a long history but this summary will focus on nineteenth-century developments in the wine industry and distillation improvements that changed production and the trade. Charlemagne (742–814 CE) had stimulated viticulture in both Germany's Schloss Johannisberg with Riesling and early production in Burgundy at Aloxe-Corton with a land gift to the Abbey of Seaulieu in 775.[2] Monasteries owned vast wine properties across France, for example, until the French Revolution, when wine estates were privatized (Hunt 2013b: 47–58, 103–24).

The rise in New World wine was a somewhat brashly unexpected change. Australia began production in 1822 when Gregory Blaxland (1778–1853) began exporting it; both Hunter Valley and Barossa Valley were under intense viticulture by the 1850s (Johnson 1985: 266–9). Although Franciscan missionaries had experimented with the Mission grape in the eighteenth century, it was touch and go for a century until California's commercial production began mid-nineteenth century with Jean-Louis Vignes (1780–1862) in Los Angeles in 1831. In the Paris Centennial Exposition of 1889 twenty-seven California wines from Napa, Sonoma, and Livermore Valleys won medals, although only a dozen or so bonded California wineries were in production before 1899 (Molyneux-Berry: 1990, 282–305). South America had started viticulture in the late sixteenth century with Jesuit missionaries but Chilean and Argentinian wineries were not viable until with a European market at the end of the eighteenth century; by 1831 there were more than nineteen million vines in Chile (Stevenson 2001: 426–30ff.).

FIGURE 2.1 Wine industry in California, picking grapes, *c.* 1900. Turrill and Miller, photo postcard. Chronicle/Alamy Stock Photo.

Wine's biggest event in France in the nineteenth century was likely the 1855 Bordeaux Wine Classification of the Gironde in 1855, organized by Napoleon III (1808–73) to showcase wines mostly of the Médoc Left Bank. The tiered system ordered the best Bordeaux in five growths from first (Premier Cru) to fifth rank for the Paris Exposition Universelle, based on the reputation of the chateau and current wine trade price. The luxury wine trade had been a magnet for British taste in claret; the 1855 Classification codified what had already accepted, that Château Lafite Rothschild and Château La Tour from Pauillac, Château Margaux from Margaux, and Château Haut Brion from Graves were in this most expensive tier of First Growths. Château Petrus and others on the "Right Bank" and Sauternes such as Château d'Yquem certainly held their own as luxury wines in the nineteenth century (Coates 2004). Perhaps the most prestigious Anglophone wine merchant in the world even then was Berry Brothers (& Rudd) in St. James's, London, wine suppliers to the royal family from George III (1738–1820) and royal warrants ever since. Burgundy had its own luxury wines in Vosne-Romanée, Montrachet, Aloxe-Corton, among other Grands Crus and monopole Côte d'Or vineyard villages. Wine *negociants* in France who began producing with their own labels include Bouchard Père and Fils (1746) and Drouhin (1880), both in Beaune (Parker 1990).

As a subcategory of wine principally from Chardonnay grapes, champagne became the primary French sparkling wine. *Champagne* houses in France proliferated from the late eighteenth century to about 1850 due to increasing demand; thirty-eight of the around seventy-seven current "grandes marques" were established by 1860; only fourteen were before 1800 (Broadbent 1991: 331–53). Distillation of higher alcohol *liquor* and *cordial liqueur* also burgeoned in the nineteenth century. Monastic liqueurs had been made for centuries; for example Chartreuse in the Isère Pre-Alps, originally containing 130 plant and flower extracts, made by Carthusian monks since 1737 in a remote monastery, after 1860 until the end of the century made commercially in Fourvoirie. Although earlier Normandy monastic claims are made, Benedictine was a liqueur commercially produced in the nineteenth century by wine merchant Alexander Le Grand (1830–98) in 1863 with a secret recipe of herbs and spices said to be assembled from a much earlier monastic *grimoire* from Frécamp Abbey in Normandy. Other generic liquors like absinthe from wormwood (*Artemisia absinthium*)—its consumption greatly increasing during the wine industry's phylloxera devastation of 1863–75—became the addictive bane of literary and artistic notables toward the end of the nineteenth century, among them Baudelaire, Manet, Degas, Oscar Wilde, Van Gogh, Toulouse-Lautrec, and the poets Verlaine and Rimbaud either depicting or appearing in paintings or poems with absinthe of 144 proof (Conrad 1997). Dangerous levels of an alkaloid, thujone, made absinthe even more poisonous and likely exacerbated violent behavior. Bonal Gentiane Qina, a novel gentian and cinchona bark aperitif cordial from the Chartreuse Massif, appeared in 1865. Other liquors from the eighteenth century were popularized in the nineteenth century; the Batavia-Arrack from Java distilled from sugar cane and fermented red rice, appeared in 1862. The Italian cordial liqueur Galliano was made by Arturo Vaccari around 1896 in Livorno, Tuscany, from a sweet concoction of anise, musk yarrow (*Achillea erba-rotta* subsp. *moschata*), lavender, peppermint, and cinnamon, with a strong late infusion in vanillin.

Malt and corn whiskeys, single malt and blended, while made since at least the Middle Ages, multiplied in stills across Scotland in the early nineteenth century, especially after the Excise Act of 1823 legalized distilling, with a "whiskey boom" in the late nineteenth century, partly fueled by growing American taste. Some of the distilleries—many now famous—include the Isle of Jura Highlands (earliest 1810, now replaced), Royal Brackla (1812), Ardberg of Islay (1815), Laphroaig (1815), Mortlach (1823), Balmenach (1824), Ben Nevis (1825), Clynelish (1819, now Brora), Macallan (1824), Glenlivet (1824), Miltonduff (1824), Bladnoch (1825), Highland Park (1826), Springbank (1826), Old Pulteney (1826), Talisker (1831), Glen Scotia (1832), Lagavulin of Islay (1837, originally Ardmore, 1817), Glengoyne (1835), Edradour (1837, originally Glenforres, 1825), Glenkinchie (possibly 1837), Glen Ord (1838), Glenmorangie (1843), Royal Lochnagar (1845), Cragganmore (1870), Inchgower (1871), Glenngassaugh (1875), and many more since 1890. The continental phylloxera outbreak devastated so much French viticulture that malt whiskey experienced an added boom. American *bourbon* corn (maize) whiskey that arose in the nineteenth century associated with the South, especially Kentucky, Tennessee, and nearby regions, include Burks Distillery (now Maker's Mark; 1805), George T. Staggs (Buffalo Trace) (after 1812), both in Kentucky; George Dickel (1870), Jack Daniels (1875) both in Tennessee. *Rye* whiskey was mostly produced in Ireland.

Rum, the original distilled sugar cane drink of the Caribbean, is almost an entirely New World liquor, made famous by seventeenth-century shipping and navies, including pirates and privateers as well as the British Royal Navy. Mixed with water to make a sailor's daily ration of rum into grog, rum also had long associations with slavery in the exploitative mercantilism between the Caribbean, the US, Africa, and Britain. In the nineteenth century, rum producers included Mount Gay Rum (1703), Rhum Vieux Labbé of Haiti (1765), J. Wray and Nephew of Jamaica (1825), Bermúdez of Dominican Republic (1852), Rhum Barbancourt of Haiti (1862), Bacardi of Cuba, now Bermuda (1862), Hacienda Mercedita, now Serrallés Don Q of Puerto Rico (1865), Myer's of Jamaica (1879), Old Planteur Rum (1885), Cockspur Rum of Hanschell Innes (1886), and Brugal of Dominican Republic (1888), to name a few.

While *beer* was not necessarily a "luxury" product it was not essential for survival either. When both China and Japan resisted many Western products, Japan after the Meiji Restoration of 1868 in particular saw a rapid infatuation with Western beer, so that by the 1890s beer was a status product with the first Japanese beer hall opening in Tokyo in 1899. In the United States nineteenth-century immigrants from the northern European "Beer Belt" including Germany, Austria, Scandinavia, Bohemia, and Poland, among other countries, brought beer consumption with them, so that breweries burgeoned rapidly around the US. Around 1840 only approximately 140 American breweries operated; by 1900 the number had grown to at least 1,200. The oldest operating brewery is D. G. Yuengling and Son, established in 1829 as Eagle Brewery and tied to the Boston Beer Company, producer of Samuel Adams brands.

Tea and coffee were the dominant non-alcoholic beverages of the nineteenth century, evolving from luxury to wildly popular drinks with the expanding trade of the Dutch East India Company and the British East India Company, the earlier markets for Asian tea (*Camellia sinensis*) supplemented by Arabian coffee (*Coffea arabica*). *Tea* from the tea bush has often been named "one of the all-time great beverages of the world" but was at first an upper-class beverage due to cost and availability, but the nineteenth century increasingly saw tea made more accessible (Davidson 1999: 788). By 1836, the British had increased tea cultivation in India in order to break the Chinese monopoly. British consumption of tea involved sugar and milk, unprecedented in Asia, and black tea (fermented) surpassed green tea (unfermented) as the British Empire made tea more popular within Britain and its colonies. The British working class still considered tea a "luxury" by the 1840s, but one that had become a necessary habit, though tea was still relatively expensive until

1870, when cheap black tea from Ceylon (Sri Lanka) lowered prices (Smith 1992: 259–78). The Suez Canal (1869) also cut 25 percent from the journey out of Asia and reduced prices (Nguyen and Rose 1987–8: 43–59). Tea also took hold across Europe, but never became as popular as it was in the Anglophone world, while in many places coffee was the beverage of choice. The first (and continuing) successful London tea merchants include Berry Brothers (and Rudd) from 1698, Fortnum and Mason (1702), and Thomas Twining (1706). Others include John Horniman in Newport, Isle of Wight (1826), Joseph Tetley and Company first in Yorkshire (1837) then London (1856), and Arthur Brooke Bond and Company in Manchester (1869). Tea merchants in Paris include Ferdinand Hédiard (1832–98) who began commercially selling tea (1854), the brothers Henri and Eduoard Mariage who founded Mariage Frères Tea Company (1854), and Auguste Fauchon (1886). In St. Petersburg, Russia, Pavel Kousmitchoff (1853–1908) founded Bouquet of Flowers, what is now Kusmi Tea (1867), moving to Paris after the 1917 Revolution.

Coffee, principally from *Coffea arabica* (with *C. canephora* by the 1890s), was a relatively modern import into Western consumption, the Dutch being the major suppliers of coffee beans to Europe until Brazil surpassed them in the nineteenth century. The century did not lessen coffee's appeal, as Alfred de Musset (1810–57), George Sand (1804–76), Alexander von Humboldt (1769–1859), Honoré de Balzac (1799–1850), and Anatole France (1844–1924) were regulars, among so many others like poet Paul Verlaine (1844–96) toward the end of the century. With first plantings in 1727 but not properly established until after the revolution of 1822, Brazil became the largest coffee producer by mid-nineteenth century (Prendergast 1999: 16–20 and ff.). Some of the famous nineteenth-century coffee company names or merchants—without claiming coffee was their first or only product—that reflect the huge shipping and market changes include Michielsen Koffie (much later changed to Miko) in Turnhout, Belgium (1800), Samuel Bewley's in Dublin, Ireland (1840), Ferdinand Hédiard in Paris, France (likely by 1854), Great Atlantic and Pacific Tea Company in New Jersey, United States (1859), Julius Meinl in Vienna, Austria (1862), Matthew Algie in Glasgow, Scotland (1864), Henri Nestlé in Vevey, Switzerland (1866), Dallmayr in Munich (1870), Gustav Paulig in Helsinki, Finland (1876), Hills Brothers in San Francisco (1878), Auguste Fauchon in Paris (1886), Auguste Fayel's Café Richard in Paris, France (1892), Luigi Lavazza in Torino, Italy (1895), and Maxwell House in New York (1896), begun by grocer Joel Osley Cheek. Coffee trade saw a surge in San Francisco when Austin Herbert Hills (1851–1905) and Reuben Wilmarth Hills (1856–1934) sold coffee from a street stall in 1878, then opened a coffee and tea store called Arabian Coffee and Spice Mills in 1882, becoming Hills Brothers Coffee by 1890. Renowned established merchants who have been selling coffee for over three centuries from the same venue include Berry Brothers (& Rudd) in London (1698) and Fortnum and Mason in London (1702), among others. In summary, it is generally estimated that coffee markets and coffee's world trade expanded at least "20 fold" in the nineteenth century (Topik 2004).

SWEETS AS LUXURY PRODUCTS, MAINLY PLANT DERIVED

Chocolate (*Theobroma cacao*) is by now one of the best-known global food sweets, after being mixed with sugar and vanilla in Europe. Chocolate contains stimulants such as the alkaloids theobromine and phenylethylamine, likely also mood enhancers.[3] Chocolate was an expensive luxury when first introduced by Spanish explorers returning from Central America around 1526 but became highly commercialized in the nineteenth century, with much higher volumes through increased trade from the shipping upsurge discussed above, but also through enhanced industrial technology (Norton 2006: 660–91). Venezuela was by far the largest producer of cocoa between 1790 and 1890 by a factor of nearly a hundred relative to any other single country, increasing from 4,000 tons per annum in the 1840s to 11,000 in the 1890s (Clarence-Smith 2000: App. 3, fig. 3.1, 175).[4] Vital were innovations of the Dutch chemist Coenraad Johannes Van Houten (1801–87), such as adding alkaline salts for bitterness reduction, and transforming the chocolate industry through a cocoa-bean pressing technology that was granted a patent by King William I (1772–1843) in 1828; he had founded the Van Houten Company by 1815, and his factory in Weesp began exporting chocolate especially to Britain, France, and Germany after his processes made production more commercially viable. Chocolate was still considered a luxury even if made in higher quantity with commensurate quality by European chocolatiers, however increasingly sweetened. One of France's earliest commercial chocolatiers remains Maison À la Mère de Famille, founded in 1761, the oldest in Paris in rue du Faubourg, Montmartre. Switzerland and Belgium established high-quality production.

Swiss chocolatiers—aided by low Swiss import tariffs—began dominating the luxury chocolate market by mostly using the *cacao criollo* bean. François Cailler (1796–1852), in Vevey, turned smooth chocolate into bars by 1819 and Philippe Suchard (1797–1884) introduced abundant Swiss hydropower and used granite rollers in his Maison Suchard Chocolate production. Maison Favarger began in 1826 in Versoix near Geneva and vied with Cailler as the oldest Swiss producer. Confiseur Sprüngli began in 1836, becoming Lindt and Sprüngli in 1892 after Rudolf Lindt (1855–1909) had discovered a process called "conching" (gradually mixing in cocoa over several days and adding cocoa butter) in 1879, which softened it. Many other Swiss chocolate houses upheld high quality, with Jean Tobler (1830–1905) beginning in 1889 in Berne, but either Cailler (now part of Nestlé) or Favarger remains the oldest chocolate producer there. One of the likely oldest Belgian luxury chocolatiers still producing quality products is Charles Neuhaus, who began in 1857, followed by Felix Vanparys in 1889, both in Brussels.

The Dutch continued expanding cocoa plantations across Java and Sumatra by the early nineteenth century; one of the oldest Dutch chocolatiers still operating is Koninklijke Verkade, having begun in Zaandam in 1886. British chocolate had several early producers, largely Quakers, in Joseph Storrs Fry (1767–1835) and John Vaughan in Bristol (Fry, Vaughan and Company) who may have started around 1761 but were renamed J. S. Fry and Sons in 1822. John Cadbury (1801–89) followed in 1824 in Birmingham (Cadbury later merged in the early twentieth century with J. S. Fry). Fortnum & Mason also sold quality chocolates about the same time in Piccadilly (London), along with Charbonnel et Walker in 1875, both of them having royal patents. The German chocolate company Stollwerck began in 1839 in Cologne, while the Berlin chocolate house, Fassbender and Rausch, opened in 1890 at the southwest corner of the Gendarmenmarkt Platz.

The oldest chocolate producer in the United States was likely that of John Hannon (*c*. 1740–80), who had founded a small factory in 1764 in Dorchester, Massachusetts. Domenico Ghirardelli Chocolate began in San Francisco in 1852 after Domenico Ghirardelli (1817–94) moved there from South America; rival Étienne Guittard Chocolate began nearby about the same time when Étienne Guittard (1838–99) moved from Lyon, France, for the California Gold Rush. Much later, Hershey Chocolate Company founded

FIGURE 2.2 The pleasures of chocolate. French postcard for Chocolat Suchard, dated c. 1900. Photo by Universal History Archive/Universal Images Group via Getty Images.

by Milton Hershey (1857–1945) began producing chocolate in 1894 in Pennsylvania. Other than makers like Guittard and a few since, the quality in most commercial American chocolate was subsequently below that of established European high-quality chocolatiers and rarely used the best cocoa bean cultivars such as *cacao criollo*, instead using *cacao forestero* and *cacao trinitario*; in addition, Europe kept the minimum percentage of cocoa solids at a much higher level, as it still does.

EUROPE'S SWEET TOOTH

Many of the sweet foods of the nineteenth century were often originally exotic, sometimes even tropical, and were thus at first expensive luxury items. Many European royal estates had rare fruit trees in glasshouses or orangeries, such as can be seen at Sans Souci (Potsdam, Germany), Versailles (Paris), Kensington and Kew (London), Laeken (Belgium), Edinburgh, and Fredensborg (Copenhagen). Some of these early sweet products of the extremely wealthy were natural and others were processed, but all reflected the "sweet tooth" of the seventeenth century when sweetness was craved once cane sugar—originally from New Guinea via India—became a demanded staple with enormous tropical production. For example, per capita sugar consumption in Britain rose from 4 lbs in 1704 to 18 lbs in 1800 and 90 lbs by the 1890s, an increase of 450 percent in one century and more than 2,200 percent in two (Lawrence 2007). **Jam** and **Jelly** production involved fruit pulp with added sugar, although jams and jellies were sweeter after the seventeenth century partly due to cheaper cane sugar from slave plantations. Although there were earlier forays, preserves were mostly sixteenth- or seventeenth-century inventions resulting from boiling fruits to a reduction with added sugar and natural citrus pectin. Voltaire had corresponded in August 1771 to the Marquise de Deffand elegantly complaining how complicated it was to send a jar of apricot jam to Paris from Geneva due to opposition from Parisian confectioners, so the jam trade under tariffs was already underway for the privileged.[5] Compotes were fruits boiled with 10–15 percent sugar weight and thus had to be consumed more quickly than jams and other preserves.[6] By 1810 Paris confectioner Nicolas Appert (1749–1841) invented a process to preserve vegetables (and meats) after boiling them and enclosing them in sealed glass jars (Youngken 1922: 332–44, esp. 337). Canning of fruits such as currants, quinces, plums, and cranberries effectively began commercially in 1820 in Boston under William Underwood (1787–1864), although Peter Durand (1766–1822) began canning in Britain in 1810. *Marmalade* was made from different fruits, although citrus fruits and even peels—including orange, lemon, lime, grapefruit, mandarin, cumquat, and even bergamot—appeared most frequently as marmalade. By the nineteenth century the Spanish Seville orange (*Citrus* × *aurantium*) eventually dominated both Scottish and English marmalade.

The English word *jam* may be related to "pressing tightly" or "jamming" the preserve ostensibly into a glass container (*OED*, vol. I–J, 1933: 548). One of the most important nursery fruit growers in Britain was Thomas Rivers (1797–1877), who supplied hundreds of fruit tree cultivars all over the world from his 300-acre nursery in Sawbridgeworth, Hertfordshire. Many of Rivers' fruit trees became the parent stock of trees (apples, pears, peaches, apricots, nectarines, plums, cherries) that provided jams not only around the United Kingdom but also hybridized the orange stock that became California's most represented citrus along with the 'Czar' plums of 1874 and 'Conference' pears of 1888 besides sixty cultivars of apple (Waugh 2009). Rivers even corresponded with Charles Darwin (1809–82) about bud variation and contributed to Darwin's *Variation of Plants and Animals under Domestication* (Darwin 1868: 1:373–411). Mediterranean sources for jams included Spanish quince paste along with figs for jam.

Heavily sugared jams and cheap sugar eclipsed honey and natural fruit sugars; honey production was small-scale and still limited to collecting from beehives, and could not readily be industrialized. From 1874 onward, one notable British jam producer—famous for marmalade—was Frank Cooper (1844–1927) in Oxford; his marmalade was taken by Scott's Antarctic expeditions at the beginning of the twentieth century.

Spain introduced many unique *vegetable preserves* with jams of tomato, caramelized onion, pepper, etc. throughout the nineteenth century. Some jams, compotes, pastes, or preserves elsewhere were also unusual.[7] Jam production from fruit in the British diet by the late nineteenth century was cheapened such that the ratio of fruit pulp to sugar was reduced to 1:3 (*Guardian* 2007). One of the popular Scottish purveyors of jams and preserves, especially marmalade, was John Gray and Company of Glasgow, which provided such jams throughout the British Empire and globally by 1860. Indian *chutney* has been a "jelly" (from the word *gelatin* or its adjective *gelatinous* via French *gelée*) made from mango (see below) for centuries, a major relish in Victorian cuisine; chutney makers from the nineteenth century included recipes made for Western palates in "Major Grey's Chutney," certainly used by Crosse and Blackwell by the early 1800s, along with the old Bengal Club chutney.

Luxury fruits in the nineteenth century were mostly from tropical climates. A representative luxury fruit, the *pineapple* (*Ananus comosus*) was most likely originally brought from Caribbean Guadeloupe to Spain by early explorers. As symbols of wealth and extravagance, carved wooden pineapple finials in the wealthiest Georgian homes attested that a single pineapple in London a generation before, grown in pineapple houses in the early eighteenth century, could cost the equivalent of $8,000 today. Italian confectioner Domenico Negri opened a confectionery shop in 7–8 Berkeley Square, Mayfair, London around 1760 under the name of *The Pot and Pineapple*.[8] In 1798 William North, merchant in Lambeth, advertized in *The Morning Chronicle* of London that he had the "largest selection of pine-apple plants in Europe" (Inglis 2013). This novel accessibility and Negri's shop ushered in production of pineapple confections to Britain, still luxurious but no longer prohibitively expensive. Victorian Britain introduced hot-water heating in 1816 along with sheet glass in 1833, thereby ensuring that British gardeners with means could grow their own pineapples in their glass greenhouses (Lausen-Higgins Lusby 2008). Also brought from Brazil, the pineapple had already been introduced into Hawaii by at least 1813 with commercial plantations there by 1886, with pineapple jam produced by the late nineteenth century.

Mango (*Mangifera indica*) was another luxury fruit that became part of the nineteenth-century status trade. Originally from India, the mango was taken to West Africa, then Mexico and the Philippines in the early nineteenth century as well as Hawaii (1824) and Durban, South Africa (1860), and then Florida, in the late nineteenth century (Davidson 2006: 491). Of mangoes with skin colors ranging from yellow to red when ripe, those with red were generally favored in the West. British and French colonial gardens and British Empire trade made the mango one of the most popular tropical fruits as it was shipped unripe (Sauer 1993: 18). Fortnum and Mason of Piccadilly became one of the steady purveyors in quality mangoes by the mid-nineteenth century.

By the late nineteenth century other tropical luxury fruits like *guava* (*Pisidium guajava*), which had originated from Mexico and Central America, were grown in other places like Hawaii after the eighteenth century and made into jams before 1900 (MacCaughey 1912: 513–24, esp. 513). While long cultivated in the Ancient Near East, *apricots* (*Prunus armeniaca*), originally from China, also became part of the luxury fruit trade of the nineteenth century because they were not easily grown in much of Europe, brought north with citrus, dried fruit, and nuts (Critz, Olmstead, and Rhode 1999: 316–52). The deep and sunny Rhone Valley of Valais in Switzerland, however, was a sheltered place where apricots were possibly grown since the fifteenth century but certainly since 1838 when 'Luizet' was being cultivated there; although grown for its fresh fruit it was also preserved in compotes or as dried fruit or fermented must, which was distilled into a local brandy, *apricotine* or *Crème d'Abricot*.[9] Although Valaisan distillers like Louis Morand and Cie have made the apricot brandy since the late nineteenth century, Paul Garnier was one French distiller near Paris possibly producing *apricotine* around 1856. The *bergamot* (*Citrus* × *limon* Bergamot Group) is a small sour citrus fruit not usually eaten fresh but its candied peel and highly fragrant oil are used in confections, perfumes, and pharmaceuticals respectively (Riley 2007: 52, 57–8). Calabria and Sicily were the best sources in the nineteenth century and remain so. A Piemontese Italian biscuit, *pazientino*, is produced from it mixed with egg white, sugar, and flour; bergamot perfume has been sold in Sicily for centuries and its essence is one of the distinctive scents of Earl Grey Tea since around 1820 (Kiple and Ornelas 2000: 2:1731).

Syrups and **sweeteners** in the nineteenth-century luxury trade include fruit syrups, and one of the oldest syrups was *grenadine* which had originally been processed from boiling pomegranate fruit (*Punica granatum*) in Anatolia; subsequently the Dutch Bols Family grenadine syrup liqueur included extracts of raspberry and strawberry syrups as well, although Bols now makes a pomegranate liqueur. Pomegranates were planted in California in the eighteenth century, and New York produced a grenadine syrup with pomegranate by 1872, according to a lawsuit (Rillet vs. Carlier 1872), but grenadine syrup had already been part of a Punch "cocktail" mix at the famous Tun Tavern in Philadelphia before 1775 (Barbour 1872: 435; Grasse 2016: 94–6).

Elderflower syrup cordial was a popular nineteenth-century drink, especially in Victorian Britain; made from the European elderberry (*Sambucus nigra*) and produced in a syrup with sugar after the flowers were boiled, it was prepared even by the Romans, more recently the syrup being infused into many drinks such as lemonade in urban Germany or in tonic water. *Cherry syrup* was produced from the wild black cherry (*Prunus cerasus*), cooked with sugar. In France, the cherry syrup family company Teisseire Père et Fils in 1820 was eventually managed by Grenoble distillers Peyraud and Ferrouillat in the second half of the nineteenth century. *Griottes au sirop* was a French confectionary using sour cherries in a liqueur cordial, eventually becoming a candy with chocolate casing without alcohol. In Italy cherry syrup was *Amarena* or *Sciroppo di ciliegia,* usually made around Bologna and Modena in Emilia Romagna. Fabric Amarena: Frutto e Sciroppo of Bologna (founded 1905) with their famous blue and white *faenza* jars—but without liqueur—followed other local non-commercial cherry syrup producers who had made wild cherry cordials and syrups since the Renaissance. A Portuguese cherry syrup variant called *xarope de cereja* also existed both as a cherry liqueur named *ginjina* from Morello cherries or just the cherry syrup that might have also been called *cereja* in the nineteenth century. Other cherry syrups have been made from the sweet cherry (*Prunus avium*), cultivated mainly for fresh fruit.

Other syrups produced for centuries include Canadian and New England *maple syrup*. Maple syrup from sugar maple (*Acer saccharum*), black and red maple (*A. nigrum* and *A. rubrum*), and others was a product of Native Americans who lived in the hardwood "Maple Belt." An engraving by William de la Montagne Carey (1842–1922) shows "Sugar-

FIGURE 2.3 The Comfort Sugar Maple Tree in Autumn, approximately five hundred years old. The oldest maple tree in Canada. Pelham, Ontario. Courtesy of Design Pics Inc/Alamy Photo.

Making Among the Indians in the North" in the *Canadian Illustrated News* (1883).[10] As one of the oldest continuing agricultural practices in North America, the maple sap collection season runs from February through April in the spring sap-rising (Willits 1965: 134). By 1858 in Quebec and Vermont, technology advanced with the invention of metal evaporators—basically large covered frying pans with fire beneath—although the boiling process for reduction was mostly maintained. Sugar maple trees can live to two hundred years old; one Niagara region sugar maple tree in North Pelham, Ontario, Canada, is called the "Comfort Tree" and is estimated to be over five hundred years old, with a circumference of over six metres (twenty feet). The Comfort family acquired the land in 1816 and the recorded tree was well established and surrounded by Native American (First Nations in Canada) lore and artifacts, now under the preservation of the Niagara Peninsula Conservation Authority (Lamb 2008). Old Sturbridge Village in Massachusetts, now a working museum community, has been making maple syrup since 1830 in the region. Most of the nineteenth-century maple syrup producers were also rural individual farm families who collected syrup to augment earnings and were thus too small to operate independently, so after 1860 they likely joined local cooperatives for regional commercial distribution to larger cities and globally (Kelleher 2017).

Molasses, called *treacle* in most Anglophone countries, has been a primary sugar cane syrup, boiled at least twice, often three times in the case of blackstrap molasses, for reduction to higher concentration (Smith 2012: 205). The history of treacle in nineteenth-century Britain was via the Caribbean sugar trade; Dutch treacle was named *stroop* ("syrup") as black treacle. The heavy blackstrap (from black *stroop*) molasses became popular from at least 1875. *Honey* itself—as derived from plant nectar—and its derivatives were natural sweeteners and should never be overlooked as a prized food. In Britain, William Spence (1783–1860) wrote *Introduction to Entomology* (1826) promoting "more 'rational' honey production through the building of more productive beehives" partly as the result of empire (Endersby 2017). Petro Prokopovych (1775–1850) of Ukraine was the likely "father" of commercial beekeeping and honey production, inventing a hive frame and early queen bee separator (Haydak 1957: 474–5). Lorenzo Langstroth (1810–95) was a New England pioneer in apiary modernization and the movable honeycomb hive, as documented in his book *The Hive and Honey-bee* (1853); similarly Charles Dadant (1817–1902) was also an apiary pioneer in France and the United States, as both Langstroth and Dadant hives became commercial successes in the nineteenth century (Horn 2005: 65–80ff.). But the impact of sugar beet and related products like jams on the traditional honey industry noted above has meant that honey production could not really compete.

Biscuits of the nineteenth century—mostly sweet—include many that developed earlier, and the following is by no means comprehensive. The *Abernethy* biscuit was created in Scotland around 1750 as a digestive by a baker connected to surgeon Dr. John Abernethy (1764–1831). It was a form of hardtack (mainly flour and water without yeast but often brined in salt) but the Abernethy had added sugar and caraway seeds and was made with ammonium bicarbonate as a raising agent; it has remained popular since and was promoted for digestion in the nineteenth century. Similarly the *Bath Oliver* was another popular digestive biscuit made around 1750 by Dr. William Oliver (1695–1764) of Bath, Somerset, England. *Amaretti di Saronno* were also from the eighteenth century, and originated in Saronno, Lombardy, Italy, made by a young confectioner couple from almonds, sugar, egg whites, and amaretto liqueur for a visiting bishop, who blessed both the confection and the couple. Lazzaroni and Company have been producing *Amaretti di Saronno* since the eighteenth century and have been incorporated since 1888. *Gingerbread* biscuits of many varieties are another item that saw escalated trade in the nineteenth century with the rise in long-distance shipping. Some gingerbreads were seen as aphrodisiacs for lovesick women, and gingerbread men were effigies for the intended man (Goldstein 2015: 21). So many types of gingerbread existed before and during the nineteenth century, including *Printen* in Aachen (Germany), *Lebkuchen* (Germany), *pepperkaker* and *brunkager* (Scandinavia) with different spellings, *biber* (Switzerland), *priyanik* (Russia), *piernik* or glazed *torun* (Poland), and *turta dolce* (Romania), among many others, with a seasonal emphasis as holiday specialties. Another Swiss spice biscuit famous in Basel is the *Basler Leckerli* (also different spellings) with almond, honey, candied peel, and kirsch liqueur made since the eighteenth century by the Samuel Gassner family along with the Bachofen, Bischof, Burckhardt, and Heussler families, although Basel's Confiserie Schiesser has been operating since 1870. Another sweet Swiss biscuit is the *Berner Haselnusslebkuchen*, flavored mostly with hazelnut and produced in Bern since 1835. *Treacle tarts* were popular from seventeenth-century Britain, based almost entirely on molasses with flour dough casings, changing very little into the nineteenth century although they were no longer luxury items. *Vanilla biscuits* in nineteenth-century France include the famous Biscuit Rose de Reims, a sweet pink biscuit with vanilla flavoring made by Maison Fossier in Reims since 1845. German vanilla biscuits in a style brought to the United States by 1835 include the Berger Cookie, vanilla shortbread topped with chocolate, produced first in Baltimore. In Italy, *biscotti*—derived from the Tuscan *cantucci* from Prato with sliced almonds inside—were mostly hard because they were twice-baked and were either made with sliced almonds or simply almond flavored and intended to be dipped in Vin Santo. The likely original commercial *Prato cantucci* or *biscotti* of Antonio

Mattei of Firenze won acclaim in the Exposition Universelle in Paris in 1867; they were made with pine nuts and almonds as well as flour, eggs, and sugar but without yeast or fat.

In Alsace, *bredele* biscuits or *pain d'epices* used honey, egg white, and sometimes anise, clove, nutmeg, and cinnamon, especially in the *anisbredela*. Another nineteenth-century sweet biscuit was the *florentine*, considered Italian but likely only named after Firenze (Florence), since the sweet caramelized biscuit, often with honey, butter, and nuts, and covered with chocolate, is difficult to trace through Europe but may ultimately be from Austria. *Garibaldi biscuits* were produced in Britain from around 1861 by the Bermondsey biscuit makers James Peek and George Frean (1824–1903) after an earlier visit by Giuseppe Garibaldi (1807–82), the "unifier" of Italy. Essentially consisting of raisins or currants crushed and baked between thin dough, it resembles *Eccles cake*, itself from around 1793, only the biscuit is flatter and not a pastry.

Shortbreads have always been very buttery sweet biscuits, often with almost as much butter as flour, and traditional European varieties include the Dutch *jodenkoek*, not necessarily created by or for Jews but possibly a baker in Alkmaar by the 1870s; Greek versions called *koulourakia* and *kourabiedes* with sesame and almond respectively; as well as the Italian *krumiri*, also with vanilla; the soft Spanish shortbread *Polvorón*; and the French *sablé* of Sarthe, the latter sometimes with cheese and even pepper. One of the most popular shortbreads in the British Empire was McVities (not to be confused with the later digestive but the same company), whose commercial shortbread production began in Edinburgh in 1830.

Macaroons were made of egg white with ground almond or hazelnut and eventually coconut across nineteenth-century Europe; their name is cognate with *macaroni*, which borrowed into Italian from Greek originally meant "Blessed" for a religious holiday pastry or confection. In France, the coconut macaroon was also known as a *congolais*; in Sicily a *pignolo* biscuit was made from almond paste and covered with pine nuts. In Siena, Italy, *ricciarelli* were soft almond cookies. Another traditional Italian biscuit was the *Nocciolini do Canzo*, like amaretto but made of hazelnut in Canzo, Lombardy, by confectioners since medieval times. *Cavalucci* were Tuscan anise biscuits with almonds, candied fruit, coriander, and Tuscan millefiori honey. *Panellets* are traditional Spanish biscuits made of almond paste. *Speculaas* (other name variations exist) have long been a northern European (Flemish, Dutch, German) spicy flat holiday biscuit with cinnamon, cloves, nutmeg, ginger, and even cardamom and pepper. *Springerle* have been traditional hard German anise-flavored biscuits for the holidays. *Tirggel* biscuits have also been traditionally made in Swiss villages for centuries from flour and honey as small, hard Christmas sweets. In Germany, *Wibele* were tiny, sweet biscuits made of egg white, vanilla, a little flour, and sugar icing. Albanian *ballokume* derive from Turkish traditions and are made with corn flour, butter, and vanilla. All of these biscuits have traditions stretching through the nineteenth century but were mostly family-made rather than commercial products until almost the end of the century.

One of the most important fillings in traditional European sweets of the nineteenth century was *marzipan*, almond paste that had long been produced from crushed almonds and sugar or honey. The name may hark back to pre-medieval Sicily, where it was *frutta martorana* for November 2, shaped as colored fruit as *marzapane* in Palermo with the old Norman church of La Martorana. Other Mediterranean cultures had nuns in convents making it for religious festivals, including Easter, since the almond was associated with Resurrection as the first tree to blossom in spring, as seen in Maltese *figolla* treats. The traditional Swedish Princess cake is topped with mostly marzipan icing tinted pale green.

Baltic coastal cities have been associated with marzipan specialties since the time of the Hanseatic League, including the famous Lübeck confectionery house of Johann Georg Niederegger (1777–1856), making small marzipan "loaves" covered with dark chocolate since 1806, always with 67 percent almonds.

Other global confections or biscuits include *ma'amoul* shortbread from the Arabian world, often filled with dates, pistachios, walnuts, or even figs; fried *reshtesh khoshkar* from Iran with sugar, walnut, cardamom, cinnamon, and ginger; *kleika* from Iraq, with dates and cardamon; Turkish *kurabiye* shortbread with almond or nut toppings or the *badem ezmesi* Turkish almond confection as a precursor to marzipan; *ghoriba*, another Middle Eastern shortbread long associated with North Africa's Maghreb; Jewish *hamantash* from poppyseed and fruit made for Purim often with villain Haman head, originating in Persia; *nankhatai* shortbreads from Pakistan; the shortbread *countess* from French Guiana; the sweet hard *tareco* with egg white and sugar from Brazil; Mexican *coyotas* with brown sugar coating; Tibetan *khapse* with butter, eggs, and sugar; Mongolian sweet *boortsag* (or *bawrsak*) deep-fried in yak butter and covered with honey; and Philippine *apas* were sugar-covered wafers; others include generic Macao (China) almond biscuits from hybrid Chinese and Portuguese colonial traditions, as in Goa (India) where *mazpon* is a holiday treat but replaces almond with cashew. Last but not least, *Turkish delight* originated as *rahat lokum*, a "sweet that is throat soothing," or just *lokum* in the late eighteenth century in Istanbul in the shop of Haci Bekir near the spice bazaar. Originally made from honey and corn flour with rosewater and lemon juice in gelatin with fruits and nuts, Turkish delight or *lokum* quickly spread to Europe from the Ottoman sultan palaces. Traditional flavors like rose, lemon, pistachio were favorites in the nineteenth century as *lokum* became a Victorian sensation, remaining popular ever since.

LUXURY PLANT-DERIVED CONDIMENTS AND SPICES

Condiments of the nineteenth century include luxury herbs, spices, and patented concoctions or sauces. One of the most historic is *horseradish* (*Armoracia rusticana*), possibly indigenous to the Balkans and Anatolia, which was used as a European meat condiment from at least the medieval period. In the United States, Henry John Heinz (1844–1919) likely introduced a commercial horseradish sauce around 1869 as one of his first products of Heinz, Noble and Company. *Peppers* in Chinese or Indian *chili oil* derived from New World peppers (*Capsicum* spp.) carried by Portuguese traders in the late fifteenth century to India. Even Hungarian *paprika* was derived from some of the New World capsicum chili pepper products. From India, the British Empire reintroduced chili pepper products into northern European cuisine, although they were not as readily appreciated as *curry* spices (from the Tamil word *kari*, for "sauce") which became a British India favorite, although it has long been a generic term referring to Indian spices and may or may not contain South Asian fenugreek (*Trigonella foenum-graecum*) along with cumin, coriander, and turmeric, and possibly some form of chili pepper. Due to the British East India Company's forays into India by the mid-eighteenth century, the Hindostanee Coffee House began in London around 1809 where Indian curry concoctions were ever present in the recipes of Sheikh Sake Dean Muhammad (1759–1851). The Hindostanee continued until 1833 and curry became very popular by 1852, suffering a temporary taste setback with the Sepoy Rebellion in 1857 but reviving by 1870. Similar South Asian spice mixes such as Punjab *masala* created concoctions of garlic, ginger, and onions taking the form of pastes and dried spices.

Vanilla, derived from the fruits of an orchid (*Vanilla planifolia*) native in Central American was brought back from Mexico with the Conquistadors. Hugh Morgan (1539–1613), apothecary to Queen Elizabeth I (1533–1603), created vanilla confections—expensive to anyone but royalty—that became some of her favorites (Bradford 1939: 30; Rupp 2014). Mexico was the main producer until the nineteenth century, when Edmond Albius (1829–80), a bright slave boy on the island of Réunion (then Bourbon) found a simple way to artificially pollinate vanilla in 1841, thus revolutionizing vanilla production.

FIGURE 2.4 A standing portrait of Edmond Albius, a former slave, who invented an efficient method of vanilla pollination, standing next to vanilla crops, 1863. From the New York Public Library. Photo via Smith Collection/Gado/Getty Images.

Ginger is a tropical Asian plant (*Zingiber officinale*) brought into the Roman world by Red Sea trade that increased greatly as a commodity spice through the Dutch East India Company in the seventeenth century and the British East India Company in the next. Used as a tonic, ginger was a warming spice that sped up metabolic rates. Ginger tea was also a nineteenth-century staple with other teas in general. Ginger was the basis of a crystallized sugar candy and wine from at least the eighteenth century in Europe; ginger beer was a popular British Empire drink, especially in Asian colonies, although Yorkshire and Edinburgh breweries made it in Britain. John Crabbie (1806–91) was one of the ginger-beer pioneers in Edinburgh by the mid-nineteenth century. Ginger liqueur such as Lucas Bols Ginevers from Amsterdam became standard setters for European gin in the nineteenth century.

Turmeric (*Curcuma longa*) is a relative of ginger, the commodity bright yellow and used for many health purposes in India, as well as a textile dye. Turmeric production in 1869–70 was 209 tons (3,729 cwt) of powder—valued *in toto* at over 14,000 rupees—which were shipped from Madras (Ravindran, Babu, and Sivaraman 2007: 11, 17). It became one of the important spices in Indian curries and East Asian cuisine in general. Nutmeg (*Myristica fragrans*), another tropical spice, was fairly expensive in the nineteenth century because it was imported from the East Indies, as were cloves (*Syzygium aromaticum*) (Bender 2009: 383). Willoughby M. McCormick (1864–1932) was a spice importer trading from around 1889 to buy spices wholesale and create an American market niche from Baltimore and Philadelphia.

Although, as the modern authority Paul Freedman has stated, "all spices were luxuries" in the Middle Ages, some became more readily available over time: for example, in 1355, 500 lbs of black pepper (*Piper nigrum*) were sold for 163 gold dinars (around $26,000) in Alexandria, but by the mid-nineteenth century, the price had fallen to 6 cents a kilogram, due to the breaking up of monopolies and increased production along with the abovementioned accelerating volume of shipping and sophistication of industrial technology (2012: 324–40, esp. 336). Vanilla was also very expensive before Albius' pollination breakthrough of 1841, but saffron (*Crocus sativus*) has nearly always been prohibitively expensive, even in the nineteenth century and since, at ten to fifteen times the price of black pepper in the Middle Ages; even today the lowest quality saffron begins at $3,000 a kilogram to at least $30,000 for the purest and highest grade. One kilogram of saffron comprises around 150,000 individual crocus stigmas and the price is so high due to the painstaking hand-gathering, virtually unaffected by the Industrial Revolution that made so many other spices more affordable.[11]

Worcestershire Sauce was one of the original nineteenth-century sauces still produced, that of Lea & Perrins, made by John Wheeley Lea (1791–1874) and William Henry Perrins (1793–1867) in 1835 in Worcester, England. It is a mixture of vinegar, molasses, sugar, salt, anchovies, tamarind extract, onions, garlic, and other spices such as cloves, peppers, soy sauce, and lemon. The relish *Piccalilli* or *India pickle* (containing chopped pickled onions, gherkins, cauliflower, and similar pickled vegetables with mustard and turmeric, possibly with picked red pepper) was made at least since the eighteenth century, and Crosse and Blackwell introduced their Piccalilli before 1867. Dutch Surinam had its own spicier version called *sambal* (an Asian word) mixing British piccalilli with pickled garlic and peppers.

A Mexican company made a sauce using Tabasco pepper (*Capsicum frutescens*), calling it "Salsa Mexicana"; Edmund McIlhenny (1815–90) purchased the company and moved it to Louisiana, growing peppers on subtropical Avery Island, which also produced salt from a local mine, recreating *Tabasco Sauce* with vinegar and the salt in 1868. Similarly, A-1 Steak Sauce was created by Henderson William Brand in 1824–31 while he was a chef to King George IV (1762–1830) in London, soon sold under the Brand and Company label as a condiment for fish, meat, and poultry; its original nineteenth-century ingredients were tomatoes, raisin paste, malt vinegar, molasses, marmalade, garlic, and onions in combination with other herbs and spices.

Satay condiments *sambal kacang* or *bumbu kacang,* traditional in the Dutch East Indies (now Indonesia), are spice and roasted peanut-based sauces produced at least since the peanut (*Arachis hypogaea*) was introduced via the Philippines in the sixteenth century. Other main ingredients have included coconut milk, tamarind, galangal (*Alpinia* spp., ginger relatives), garlic, chili, and cumin. Commercial production in the nineteenth century of satay sauces, which have also been traditional dipping sauces, is difficult to trace in Asia, but was certainly largely home-based. It is likely, however, that the extant Raffles Hotel in Singapore (1887 onwards) and its predecessor Emerson's (from 1878) had satay sauce cuisine for colonial expatriate guests.

Hickory sauce and smoked flavoring in American barbecue sauces of the nineteenth century began to use shagbark hickory (*Carya ovata*), an indispensable hardwood product of the American Southeast. Virginia colonists of the seventeenth century used hickory-wood smoke to flavor and preserve meat. In the nineteenth century hickory became an American barbecue essential. *Ketchup* (catsup) was a mostly tomato-based pourable sauce made and marketed by Henry J. Heinz beginning in 1876 in Pittsburgh, Pennsylvania. Early ingredients in the nineteenth century included egg white, for emulsifying, with vinegar and minor amounts of mushrooms, shellfish, onions, and flavors from garlic, cloves and allspice. H.J. Heinz Company dominated the nineteenth-century market share with its "57 Varieties" products that soon grew to a much larger number by the end of the century, and "ketchup" became the most popular Heinz product worldwide.

Pickling in salt brine or vinegar eliminates most oxidation and has been a food preservative process for thousands of years, extending "shelf life" when the germane plants are out of season. Brining also keeps plant structure by making the natural pectins stronger, creating crispier pickled food, while salt slows down fermentation. The firmer the vegetable or the stronger the fibers, the more ideal they are for pickling. While certain vegetables like broccoli and cauliflower can be pickled, as seen in Italian mix *giardiniera*, they often yield unpleasant sulfurous oxides. German sauerkraut-making and related traditions have pickled cabbage for centuries. The German merchant Carl Ernst Wilhelm Kühne (1841–88) introduced Schützenbachian quick vinegar production in 1832 and within a generation his Kühne Company KG began producing commercial sauerkraut and pickles ranging from gherkins to dills.

The nineteenth century vastly improved pickling technology, especially the non-commercial home industry. In 1850, Scottish chemist James Young (1811–83) created paraffin from coal and oil shale, ideal for jar-sealing wax, and the American tinsmith John Landis Mason (1832–1902) invented the heavy glass Mason Jar in 1858, able to withstand high heat for canning, with rubber seals. Immigrant Jews arriving in New York City from eastern Europe in the latter half of the nineteenth century brought many kosher dill pickling styles and recipes. In Chicago, farmer Claus Claussen started the company C.F. Claussen for pickles in 1870. Shipping pickles and pickled food over ocean lanes to far markets after 1860 was easier than for many other food products due to their already preserved state.

FERMENTED PLANT-DERIVED FOODS

Fermented foods of the nineteenth century besides wine, beer, and spirits discussed above, will still be familiar in some sense as they have continued as staples of modern cuisine, but the concept of the greater the distance or more exotic the culture, the greater the cost has remained true. Understanding of microbial transformation began with Louis Pasteur (1822–95), but many cultures have long utilized fermentation. Cider is a well-known apple fermentation product and British, French, and American companies marketing cider in the nineteenth century include Gaymer's in Norfolk (since at least 1800), Taunton's in Somerset (1805), Hunt's Devon Cider (since 1805), H.P. Bulmer in Hereford (1887), Manoir de Grandouet in Normandy (sixteenth century), Pierre Huet in Normandy (1865). Pommeau is another type of longstanding French cider mixed with Calvados liquor. Laird and Company of New Jersey (1780) is one of the oldest Applejack producers, an American hard cider that flourished in the Hudson River Valley in the colonial period through the nineteenth century.

The original "Marmite" derived from brewer's yeast was made by Justus von Liebig (1803–73), a German agricultural chemist in Giessen; he concentrated and bottled it from waste yeast before 1873. Some of the other more familiar fermented foods are cabbage made into sauerkraut in Germany and spiced *kimchi* of Korea—almost two-hundred-plus varieties from brined cabbage, radishes, pimento, pepper, oysters, members of the onion and garlic family, among others (Bender 2009: 307). Fermented food also include the ancient *kombucha* tea of Tibet, and *miso* salted soy sauce of Asia as a condiment (Aidoo 1992). Ghana in Africa has long produced *ahai* drink from maize as well as cassava tuber making fermented *kokonte* in savanna Africa. *Masato* was a cassava tuber beer of the Amazon, like the corn *chicha* beverage and dried potato *chuno* food of the Andes, other fermented products (Hesseltine 1965: 149–97). Fermentation via lactic acid is also a major form of pickling, as mentioned, with salt brining.

LUXURY SEEDS AND NUTS

Seeds and nuts in the nineteenth century were increasingly available as tropical products, or as European comestibles, as were edible fungi such as porcini or truffles. Some of the luxury seeds and nuts included kola nuts (*Cola acuminata* and *C. nitida*) from Africa that contain caffeine (utilized by pharmacist John Pemberton in the 1860s for his French Wine Cola that became Coca-Cola); macadamia or Queensland nuts (*Macadamia* species) from Australia, high in fat and protein, named by German botanist Baron Ferdinand von Mueller (1825–96) in 1857 and commercially produced by Charles Staff at Rous Mill near Lismore, NSW, in 1882; high fat and protein Brazil nuts (*Bertholletia excelsa*), from South America, harvested from the wild and a considerable trade distance from world markets; also pine nuts (*pignola* or *pinoccera* in Italy) the best from the stone pine (*Pinus pinea*) and even almonds (*Prunus dulcis*) from the Ottoman Empire (Rosengarten [1984] 2004). Other than almonds, pistachios (*Pistacia vera*) and pine nuts, which were luxury seeds or nuts but less expensive in Europe due to greater proximity to Istanbul's bazaars, the price for these exotic tropical nuts was gradually reduced as shipping and trade increased. California orchardists began extensively planting almonds in the Central Valley after 1850 to reduce dependency on the more expensive Ottoman products. Greek

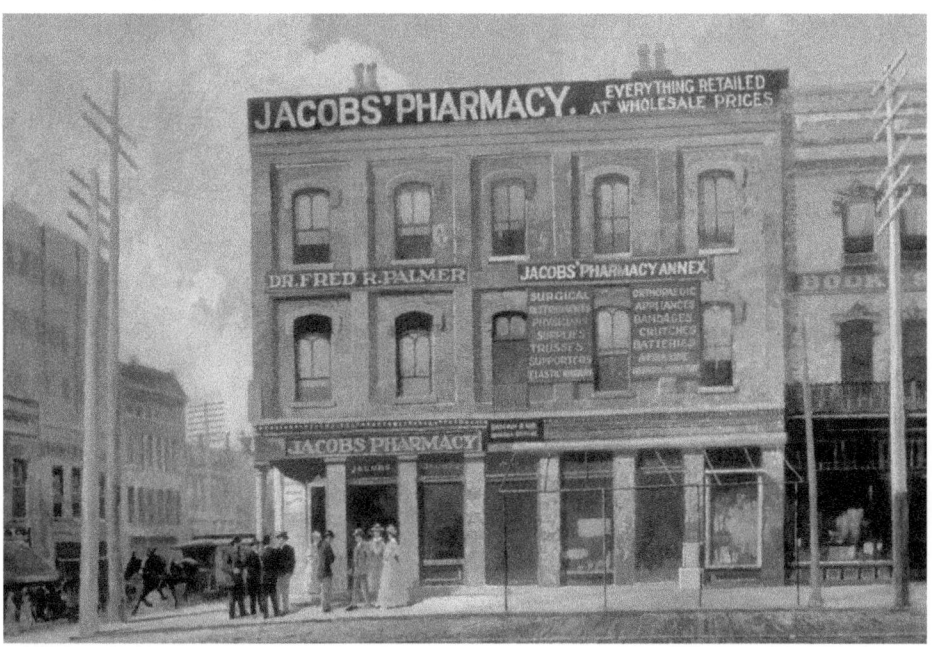

FIGURE 2.5 The first place where Coca-Cola could be bought was at the soda fountain in Jacobs' Pharmacy, 1886, in Atlanta, Georgia. Photo by John van Hasselt/Sygma via Getty Images.

production of pistachios flourished on the island of Aegina and other Aegean islands in the 1830s after independence from Turkey.

Coconut (*Cocos nucifera*) or copra, the dried endosperm of the seed, became a highly desirable product from South Asia and the Pacific by the 1860s. The seeds are able to travel by sea hundreds of miles while retaining viability. Florida's Palm Beach coast may derive from a shipwrecked vessel, *Providence*, that ran aground in 1878 with a coconut cargo (Ganeshram 2017: 27–9). Even the Caribbean produced coconuts as byproducts of colonization and the burgeoning slave trade by the early nineteenth century, as coconut prices in Europe became less dependent on South Pacific sourcing. Dried tropical fruits like coconut were sold in Fortnum and Mason from the mid-nineteenth century.

RITUAL OR DRUG PLANTS

Ritual or drug plants have at times been romanticized or at least strangely glorified even when addictive and illicit, and the nineteenth century was no exception. There was commercial exploitation of addiction in the nineteenth century, as seen in the British Empire, with its Opium Wars as well as in the fledgling tobacco industry. These drugs and ritual plants included opium, tobacco, hashish, San Pedro Cactus, *peyotl,* and cocaine from coca (*Erythroxylum coca*)—the latter two among indigenous Central and South Americans respectively. Some had millennia of medicinal applications, as in the case of opium and coca for pain relief. Some had addicting or habituating behavioral associations, in the case of opium and nicotine, or escapism in forms of cannabis or other psychoactive agents.

By the nineteenth century *opium*, from the poppy (*Papaver somniferum*) was often commonly prepared in *laudanum* as a tincture, even if only 10 percent opium, or as *paregoric* (Hodgson 2001). Opium was an infamous ingredient in many Victorian remedies sold without a physician's prescription, and administered to children to quieten them or quell pain: Kendal Black Drop, Gee's Linctus, Godfrey's Cordial (called Mother's Friend), Dover's Powder, Dalby's Carminative, Collis Browne's Chlorodyne, McMunn's Elixir, Batley's Sedative Solution, and Mother Bailey's Quieting Syrup (Hodgson 2001). Even Bayer produced its Glyco-Heroin Cough Syrup in 1898. Crude opium exports to the United States rose from 4 lbs per 1,000 persons around 1850 to almost 9 lbs per person around 1895, indicative of the lack of both awareness of its dangers and regulation (Musto 1991: 40–7). *Hashish* in Arabic was the Middle Eastern derivative via Egypt from *Cannabis sativa* (hemp). W. B. O'Shaughnessy (1809–89), physician at the Medical College of Calcutta, in 1839 promoted hemp's use as an analgesic, but it was in the official *U.S. Pharmacopoeia* by the 1820s (Grimspoon 2005: 2–3). For further discussion of opium, as well as for discussion of heroin, morphine, cocaine, and other plant-based drugs see Chapter 5 ("Plants and Medicine") of this volume.

Chewing *tobacco* (*Nicotiana tabacum*) was as prevalent as pipe-smoking in the first half of the nineteenth century. In the previous century increasingly intensive plantation cultivation as a lucrative cash crop helped promote slavery in the United States, since indentured servitude was insufficient to provide a labor force, an abrasive social problem that helped alienate the agricultural South from the industrial North and bring about the Civil War (1860–5) (Horn 2011: 327–31). By 1879–80 the US Internal Revenue Service appears to show tobacco tax of $38.9 million out of the total $116.8 million collected, an enormous percentage.[12] James Bonsack (1859–1924) created a tobacco paper-rolling machine in 1881, which greatly sped up commercial production (Burns 2007: 194ff.). Commercial tobacco expanded greatly in 1890 when James Buchanan Duke (1856–1925) began the American Tobacco Company monopoly—parent predecessor of many twentieth-century producers—by merging tobacco manufacturers John Allen and Lewis Ginter of Virginia and E. Goodwin and Company of New York. Duke's company became one of the original dozen Dow Jones Industrial Average members in 1896 and wielded enormous influence on American society and commerce in the southern US economy.

Peyotl is a Nahuatl Aztec word for a regional desert cactus (*Lophophora williamsii*) that had a religious cult in the Pre-Columbian New World, using its psychoactive drug mescaline for ritualized visionary experiences but, in the nineteenth century, was dwindling among indigenous societies with increasing urbanization and United States Indian reservation policy (La Barre 1960: 45–60). The San Pedro Cactus (*Echinopsis pachanoi*) native to the Andes of Peru was another ancient psychoactive plant for traditional medicine and ritual divination for at least three thousand years; producing visions and accompanied by excessive mucus production in its users, it also contains mescaline. Its use was severely waning in the nineteenth century, except in remote areas where shaman *curanderos* had greater sway, due to colonial policy, as the Catholic Church discouraged its use (Cloudsley 1999: 63–78).

CONCLUSION

In conclusion, luxury plant food products of the nineteenth century saw overall trending with reduced prices for many of the long-distance products, thanks to higher volumes of shipping, the move from sail to steam, better navigation, and increased mechanization with advances in engineering and agronomy besides industrialization. In 1869, the Suez Canal cut out a quarter of the distance from Asia to Europe. The nineteenth century made hundreds of food products accessible to more people and, while the idea of luxury persisted, it was an exciting time for a general population now able to afford the luxury foods, diversifying and enriching a diet hitherto very dull. Many of these new luxury products became such available staples that the modern world hardly considers them luxuries at all.

CHAPTER THREE

Trade and Exploration

MARK NESBITT

Plant exchange has a long history, accelerating in the fifteenth to eighteenth centuries—the period of the Atlantic world, during which slavery and the cultivation of sugar cane (*Saccharum officinarum*) in the Americas drove transatlantic exchanges (Bleichmar 2008; Carney 2001; Schiebinger 2004; Schiebinger et al. 2007). However, as this chapter shows, the nineteenth century saw a significant increase in the scale of plant exchange, and in the degree of conscious design applied to it. Colonies of Britain, Germany, France, and the Netherlands became subject to stronger central control, a change in style of government also reflected in the increased capability of federal institutions in the United States. Scientific infrastructure, in the form of museums and botanical gardens, entered its modern form, benefiting too from the widespread use of Linnaean nomenclature for plants, and the development of new analytical techniques in genetics and plant chemistry. Improved transport links, particularly the widespread use of steamships from the 1860s, and the use of the Wardian case to protect plants during transit, meant that few parts of the world were closed to plant explorers. Plant exchange was also affected by wider political changes, above all the abolition of slavery—as early as 1834 in Great Britain's colonies, though as late as 1888 in Brazil—leading to fundamental changes in the organization of plantation agriculture.

Commerce and industry were the main reasons for plant introduction. Plants formed the basis not only of food and clothing, as they do today, but of materials such as wood, rubber and gutta percha, and medicines (Drayton 2000). While economic historians disagree on the underlying drivers of change—free trade, population increase, increase in consumption, banking, state regulation, technological developments, and transport are among the many factors cited—they agree that the nineteenth century was a period of exceptional growth in industry and trade, and thus of consumption of plant raw materials (Federico and Junguito 2016; Lee 2006). Commerce thus enlarged the market for plantation crops such as rubber or cinchona, often introduced to regions where they were not previously grown, and often for reasons of imperial preference.

In retrospect, global plant-exchange in the nineteenth century can appear to be the result of a well-organized system (Brockway 1979; Hobhouse 1985). However, a more nuanced study of the motivations of those involved, and of the operation of the infrastructure available to them, points instead to the importance of context and contingency. Plant introductions often—perhaps even usually—failed. The history of plant introduction, as written, has tended to focus on the success stories. In this chapter plant exchange is examined in its wider context, largely using examples from Britain and its empire, but drawing attention to comparable developments elsewhere.

INFRASTRUCTURE

Plant exchange depends on infrastructure and logistics: botanical gardens to act as staging posts and places of acclimatization; herbaria and libraries as centers for plant identification, and museums of economic botany to demonstrate the connection between plant raw materials and plant products. Recent research has found complex flows of knowledge and plants between these, emphasizing frequent interactions between commerce, in the form of plant nurseries, or industries based on plant products, and government institutes and departments (Pawson 2008; cf. Coote et al. 2017). Examples such as the transplantation of rubber and cinchona are well documented cases in which much of the necessary science was carried out far away from London, in Singapore and India respectively. Even the great international exhibitions were often preceded by local exhibitions, sometimes very elaborate, of the specimens gathered together for a national display (Modest 2018). Understanding the processes of trade and exploration in the nineteenth century demands looking beyond the metropolitan centers of Europe and North America.

The Botanical Garden

The first botanical gardens arose as medicinal-plant gardens, associated with medical schools, in the sixteenth century. However, as Donal McCracken (1997) argues, the modern form of garden, usually much larger, and acting as a base for plant exploration and research, is an Enlightenment creation, becoming prominent in France and Britain in the early to mid-eighteenth century, at the Jardin du Roi, Paris and the Royal Botanic Gardens, Kew. In the forty years up to 1837, some twenty-two botanical gardens had been founded (and in some cases closed) in the British Empire; by comparison, some one hundred operated there during the Victorian era (1837–1901).

In Britain the Royal Botanic Gardens, Kew (known for most of the nineteenth century simply as the Royal Gardens, Kew) dominated plant exchange involving Britain from the founding of its "Physic Garden" in 1759, but in two very different periods. Georgian Kew was a royal garden, though one run effectively as a national botanical garden from 1782 to 1820 under the control of King George III and Sir Joseph Banks (1743–1820). After their deaths in 1820 the Gardens went into decline, being re-founded as a government botanical garden in 1840, and funded by the Ministry of Works. A large part of the rationale for its new funding was "the promotion of Botanical Science throughout the Empire," in particular through coordinating the work of colonial gardens (Cornish 2013, 2015; Desmond 2007; Lindley 1840). Large gardens such as those at Kew, in Paris, or Berlin, were closely linked to botanical gardens in overseas parts of the empire, some equally large, such as Calcutta (now Kolkata), many small. These overseas gardens were locally funded, usually by the local legislature or governor, but in the case of those founded by the British, maintained their close relationship to Kew through their staff, usually recruited from there. The overseas garden gained a trained and tested member of staff; Kew gained an excellent conduit for plants. Botanical gardens were nodes in a global system of plant circulation, from the 1840s via the Wardian case, but were also centers of acclimatization, in which newly introduced plants could be tested for their adaptation to local conditions.

Throughout the world, botanical gardens had similar facilities. Most gardens had ornamental parts, open to the public, and featuring lakes, lawns, and avenues of trees. Glasshouses protected warm-temperate and tropical plants, offering many people their first and only opportunity to experience tropical nature (Arnold 2006; Driver and Martins 2005). Behind the scenes, plant nurseries raised plants for sale—an important route for newly arrived plants to reach a wider public. A remarkable 174,230 plants were distributed by the botanical gardens in Jamaica in 1884, for example (Taylor, D. 2017). There was often tension between the ambitions of the garden staff, and those of the local government. The annual reports of botanical gardens often give a very frank view of these: for example, in Jamaica funding was inadequate until the imposition of government control by London in 1865 following the Morant Bay rebellion. Plantation owners pressed unsuccessfully for greater support for the sugar estates, ruined by the collapse in sugar prices of the 1820s, while the Botanical Department instead focused on the smallholding class of farmers who had formerly been enslaved. In Jamaica, as elsewhere, staff found themselves torn between their desire to carry out scientific work, such as collecting herbarium specimens and cataloging the wild plants of the country, and the requirement to meet immediate needs for farmers and gardeners (Nesbitt 2018). Botanical gardens also played an important role as a base for visiting plant collectors. It should not be assumed that bioprospecting—the hunt for new plants for practical purposes—was the only or even the main driver for plant collecting in the nineteenth century; botanists' travels to the tropics and to islands were in part driven by, and had a profound effect on, the understanding of evolution and biogeography, exemplified by the work of Alexander von Humboldt, Robert Brown, Charles Darwin, Alfred Wallace, and Joseph Hooker (Browne 1997; Driver and Martins 2005; Sutter 2014; Wulff 2015).

Toward the end of the nineteenth century, government agricultural departments established specialist seed-storage facilities, the forerunners of today's seed banks. The two largest were those of the United States Office of Foreign Seed and Plant Introduction founded in Washington, DC, in 1898, and the Bureau of Applied Botany in St. Petersburg,

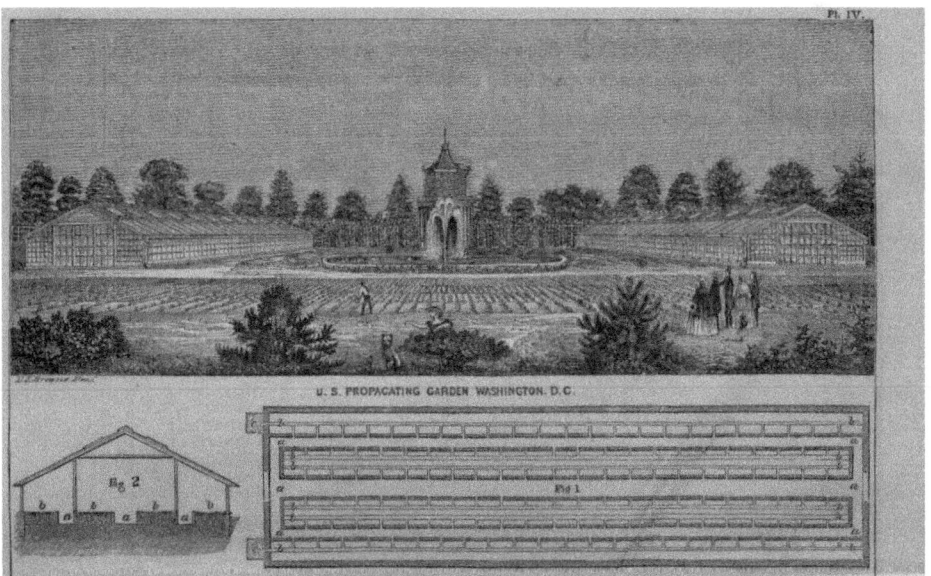

FIGURE 3.1 Propagating Garden, Washington, DC. Prepared in 1858 to receive Robert Fortune's tea plants from China. Courtesy of the Smithsonian Institution (*Report of the Commissioner of Patents for the year 1858. Agriculture*. Washington, DC).

Russia, founded in 1894. These closely resembled botanical gardens in the range of facilities, including nurseries and herbaria, but focused on crops and their wild relatives, and on distribution of seeds rather than plants (Dierig et al. 2014). They evolved in a different direction from botanical gardens, becoming in the twentieth century the modern germplasm repository with its banks of frozen seeds.

Societies played an important role in representing growers interested in plant exchange, whether of ornamental or economic plants. These included the Horticultural Society of London (founded 1804), the Agricultural and Horticultural Society in Kolkata (1820), the Société Zoologique d'Acclimatation in Paris and Algiers (1854), and the Acclimatisation Society of Victoria in Melbourne, Australia (1861) (Anderson 1992; Gillbank 1986; Osborne 1991, 2000). Such societies worked closely with government, but were able to act independently, often in reaction to what was characterized as colonial government's insufficient emphasis on applied science. Many ran their own botanical garden, and their publications reached homes and farms.

The Commercial Nursery

In the nineteenth century there was far greater exchange of plants between botanical gardens and commercial nurseries than there is today; now the mission of botanical gardens has moved away from ornamental plants, and the Convention on Biological Diversity imposes restrictions on commercialization of collections (McCook 2016). By contrast, in the nineteenth century Kew exchanged plants with commercial nurseries all over Britain, including the famous ones of Backhouse and Sons (1816–1950) in York, and the Loddiges and Veitch families in London (Alcorn 2020; Gorer 1976). These operated on a substantial scale, sending their own plant collectors overseas, and also following scientific practices, as in the collection of herbarium specimens in Australia by James Backhouse (1794–1869), and economic botany specimens in Japan by John Veitch (1839–70), collections both now deposited at Kew. Similar close relationships continued into the late nineteenth century with plant explorers such as Ernest H. Wilson (1876–1930) collecting plants for the Arnold Arboretum in Massachusetts, while at the same time for commercial nursery firms in Boston and New York.

The Herbarium

Like the botanical garden, the herbarium—a repository for pressed, dried plant specimens—originated in Italy in the sixteenth century. Working herbaria often remained the property of the collector, as in the case of the herbaria of Sir Hans Sloane (1660–1753) and Sir Joseph Banks, both of which collections were deposited in the British Museum only after their owners' deaths. Sir William Hooker's herbarium and library remained his private property even after he became director of Kew, only becoming part of Kew's collection after his death in 1865. During the nineteenth century very large institutional herbaria began to form, requiring purpose-built housing. This was sometimes in botanical gardens, as in Paris (1835) or Kew (1876), sometimes in a city center museum, as at the Natural History Museum, London (1881). Only a proportion of specimens in such a herbarium were collected by its staff; most came by exchange and donation with global networks of collectors. The historian Jim Endersby (2008) has studied the maintenance of these networks during Sir Joseph Hooker's tenure as Kew's Director (1865–85). There was a delicate balance between encouraging the inward flow of specimens, while at the same time maintaining control of specimen naming, essential to avoid "splitting" of local variants into new species. In addition to supplying names for specimens, Hooker carried out many favors for donors, including the supply of books and equipment. Some botanists were paid to collect on Kew's behalf. When Richard Spruce (1817–93) went to the Brazilian Amazon, George Bentham (1800–84) acted as his agent, selling specimens on his behalf to herbaria. Spruce was able to support himself for fourteen years collecting in South America from 1849 to 1864 (Hemming 2015).

Museums and Exhibitions

In 1847 Sir William Hooker (1785–1865) opened the Kew Museum of Vegetable Products, in 1852 renamed the Museum of Economic Botany. The museum was distinctive in its emphasis on plant use ("economic" in this mid-nineteenth century context meant practical, or useful), and on the processes that attend transformation from living plant to finished product (Cornish 2015, 2017). It made much use of the "illustrative series" of specimens that started with raw material, showing the tools and half-made object, and ending with the commercial product. The Museum was aimed both at the general public and, more specifically, at British manufacturers seeking advice on plant raw materials. The supply chains from distant lands such as the Andes or the Far East were such that important products such as cinchona bark (*Cinchona calisaya*) or rice paper pith (*Tetrapanax papyrifer*) were still of uncertain botanical identity at this time. Kew acted as a source of reliable information, in close collaboration with the brokers of the City of London, who could advise on price and industrial quality.

Like the herbarium, the museum played a vital role in naming and authenticating plant material. It also acted as a center of circulation in its own right: the museum at Kew had accumulated about seventy-five thousand specimens by 1914, but had also distributed fifty thousand specimens to other museums, receiving many in exchange (Cornish and Driver 2019). The Kew Museum was much imitated, and over forty galleries or free-standing museums of economic botany have been identified in Great Britain alone; among the many surviving international museums were those in Adelaide (1864), Berlin (1878), Harvard (1858), New York (1891), and Missouri Botanical Garden (1860), mostly founded in the period 1850–90 (Cornish and Nesbitt 2014). It was Sir William Hooker's view that a botanical garden should comprise gardens, a herbarium, and a museum, the latter showing "all kinds of useful and curious Vegetable Products, which neither the living plants of the Garden, nor the specimens in the Herbarium could exhibit" (1855).

Museums of economic botany (Mackenzie 2009; Sheets-Pyenson 1988) were connected to a much wider museum network, including natural history museums, newly established museums in the colonies, and international exhibitions (World's Fairs). Although each of the last was shortlived, the sequence of international exhibitions that begins with London's Great Exhibition of 1851 and perhaps ends (in terms of significant botanical exhibits) with the British Empire Exhibition (1924–5) was important in disseminating knowledge of plant products in trade. Such exhibitions usually featured large-scale displays of plant products, acting as both a trade fair connecting exporters and importers, and promoting products direct to the public (Hoffenberg 2001). Visitor numbers were large: London's Colonial and Indian Exhibition of 1886 was attended by 5.5 million people. The impact of such exhibitions was increased by the dispersal of their exhibits to museums and, thus, permanent display after the exhibition closed: for example, the founding economic botany collections of the Field Museum in Chicago derive from the World's Columbian Exposition of 1893.

FIGURE 3.2 Santos Museum of Economic Botany, Adelaide Botanic Garden. Courtesy of Mark Nesbitt.

TRANSPORT

Improvements in transport played a crucial role in expanding geographical reach. Sea travel became more reliable after the end of the Napoleonic Wars (Daws et al. 2007), and an extensive network of ships regularly sailed directly to British ports, for example from Manaus on the Amazon to Liverpool. Steamships came into use in the 1830s on mail routes, and were widespread by the 1860s. Previously, the sea journey from England to India took five to eight months by the Cape route; from the 1850s a combination of rail and steamship, crossing Egypt, took a month, reduced still further after the Suez Canal opened in 1869. As well as speeding the transport of living plants, steamships had a major impact on the cost of plant products (Headrick 1981). From the 1860s cheap cereals from Russia, eastern Europe, and the United States reached Britain, leading to a halving of wheat prices over a thirty-year period, with severe effects on British agriculture (Cain and Hopkins 2016; Offer 1999). Building of railways often went hand in hand with the development of new agricultural areas, but these routes were usually less useful to the plant explorer.

Fieldwork continued to rely on the hire of porters and horses or mules. Plant collecting still took botanists into extremely difficult terrain, often only accessible with assistance from local guides. Once fresh plant material had been collected, it needed to be transported to a botanical garden for cultivation and propagation. In the case of seeds that retain viability after drying, this was relatively straightforward. However, the seeds of many species, especially those from the tropics, are "recalcitrant," a term used by seed scientists to describe seeds that die when dried. And in any case, the newly encountered plant might not be bearing mature fruits. However, transport of living plants by sea is far more challenging, with death through salt-water spray, lack of fresh water, extremes of weather, and the depredations of animals onboard ship. Many plants—and sometimes entire collections—were lost in the eighteenth century (Keogh 2017; Parsons and Murphy 2012; Rigby 1998). Only well-funded journeys, such as those sponsored by Sir Joseph Banks, could afford a greenhouse on deck. These problems were resolved in the 1830s, when Dr Nathaniel Ward (1791–1868), a physician and naturalist in London, developed the Wardian case, a highly portable wood and glass greenhouse (Keogh 2019, 2020; McCook 2016). It was designed in collaboration with the commercial nursery of Loddiges and Sons, and was first tested by sending two cases of Loddiges' plants to Australia in 1833, and bringing Australian plants back in 1834. The Wardian case was rapidly adopted, for example in the successful transfer of the Dwarf Cavendish banana (a cultivar of *Musa acuminata*) from the Duke of Devonshire's garden at Chatsworth, England to Samoa in 1838. The plants survived the seven-month journey, and their cultivation spread widely through the Pacific (Simmonds 1959).

GOVERNMENT AND EMPIRE

In broad terms, nineteenth-century trade in plants was driven by the requirement of large-scale industry, in developed countries, for cheap, reliable supplies of raw materials (Drayton 2000). These could be made into more expensive manufactured goods and be exported to world markets (Johnson 2003). However, then, as now, the creation and maintenance of the necessary worldwide networks was not straightforward. Two factors helped: the easy availability of capital after about 1800 in the financial center of London, which funded many of the global developments in transport and agriculture, and the large-scale migration of Europeans, to be farmers themselves, or to supervise plantation agriculture. Over six hundred thousand people left Great Britain between 1815 and 1850 (Cain 1999).

The mutual influence of trade and empire is complex and much debated by historians (Cain and Hopkins 2016; Porter, A. 1999). Globally, the Spanish empire was in decline, but the British, Dutch, and French maintained substantial colonies, joined by Belgium and Germany after the Treaty of Berlin of 1884–5 divided up Africa between the European powers. In Britain the nineteenth century saw the expansion of empire despite the loss of the American colonies in 1776, with a reorientation from the West Indies to Africa, Southeast Asia, and India. The same period also saw increasingly centralized control of empire; in Britain the first ministry for War and the Colonies was established in 1801, in 1854 divided into the Colonial Office and the War Office. British India had been administered by a commercial enterprise, the East India Company, but from 1858 by the British government's India Office. The large volume of correspondence with these departments in the archives of the Royal Botanic Gardens, Kew, shows how closely botany and empire were sometimes connected. In the cases of tea, cinchona, and rubber, the East India Company or India Office played crucial roles in their transfer and introduction.

However, empire was not an exclusive frame for plant trade and introduction. Throughout the nineteenth century the empire accounted for about a third of overall exports from Britain, and just under a quarter of imports of primary products such as grain, cotton, timber, sugar, tea, and indigo (Cain 1999). From the 1840s onwards, a strong impetus for free trade in Britain led to reduced tariffs and freedom to choose the cheapest supplier. This reduced sugar imports from the West Indies and timber imports from Canada. In large parts of the world—China, Thailand, Latin America—British trade and influence, often backed up by the threat of military force, was so extensive that historians have termed them "informal empire" (Lynn 1999; Osterhammel 1999). Although the concept of informal empire is much debated, these regions often had similar advantages to the formal empire in terms of contacts and access for the traveler and trader.

While the commercial impetus was predominant, humanitarian concerns were increasingly visible and sometimes effective (Porter, A. 1999). Britain abolished the slave trade in 1807 and emancipated its enslaved in 1834, with a severe impact on the profitability of the West Indian sugar plantations, as freemen left plantations for their own small-holdings. However, Britain replaced West Indian sugar with imports from Cuba and Brazil, both of which still depended on slavery. "Legitimate commerce" sought to replace slavery, where it survived, by trade. David Livingstone's Zambezi expedition (1858–64) to modern-day Mozambique was unsuccessful in this aim, but other missionary and commercial endeavors, in West Africa, led to the establishment of a thriving oil palm (*Elaeis guineensis*) industry by the 1850s (Dritsas 2006, 2010; Olabimtan 2013).

Plantation agriculture of rubber, coffee, tea, and cinchona in Southeast Asia depended on slavery's replacement, indentured labor (Johnson 2003). Concerns for the rights of workers and indigenous peoples were strongly expressed in European countries, and sometimes supported by colonial governments, but had limited effect in remote rural areas. Two notorious examples are associated with rubber: Julio César Arana's Peruvian Amazon Company, responsible for the deaths of some thirty thousand rubber collectors on the Putumayo river of the Amazon region between 1908 and 1913, and King Leopold II's Congo Free State, in central Africa, which saw the death of well over a million rubber collectors in central Africa between 1885 and 1908 (Loadman 2005; Stanfield 1998; Tully 2011). In both cases prolonged pressure in Europe was necessary to end these practices.

GARDEN PLANTS

The nineteenth century saw major changes that increased the consumption of introduced ornamentals: technological changes such as easily available plate glass for greenhouses, and social and economic changes, which meant there were more of the rich and middle class who could afford exotic plants. Furthermore, there was simply more space for garden plants, with ambitious planting schemes in public parks, cemeteries, and hospitals as well as homes (Billston 2008; Elliott 1986; Wimmer, Chapter 7 of this volume). An increasingly professional class of gardeners had excellent access to information through journals such as the *Gardeners' Chronicle*. Two groups of plants illuminate the story of ornamentals. First, orchids for their extraordinary and much publicized stories and, second, trees for their visual impact. Both illustrate the importance of commercial nurseries in trade and introduction of garden plants.

Orchidmania

More than every other plant, orchids exemplify the intersection of infrastructure, commerce, and desire that drove—and drives—the acquisition of new garden plants. Although the competitive acquisition of large numbers of exotic plants was well established in eighteenth-century gardens, such as that of the Marquess of Bute at Whitton, orchids

FIGURE 3.3 Flowers of *Cattleya labiata gaskelliana*. Illustration from *Reichenbachia: Orchids illustrated and described*, by the nurseryman Frederick Sander (1888–94). Courtesy of New York Public Library.

came late to the scene. The first tropical orchid flowered in Britain in 1728 and, as Jim Endersby has argued, orchids could only become popular with the wide availability of greenhouses after 1800, aided by advances in shipping, and—crucially—in the numbers of very wealthy patrons (Endersby 2016; Reinikka 1995). Successful cultivation also depended on improved scientific knowledge of the orchid's original environment, and on horticultural experimentation to replicate these. By the 1840s both the technology and knowledge was in place for successful imports by the Wardian case, and successful cultivation of tropical orchids outside the tropics.

The main source of tropical orchids in Britain was commercial nurseries, such as the Royal Exotic Nurseries, run by the Veitch family, and those of Conrad Loddiges, and Benjamin Samuel Williams of the Victoria and Paradise Nurseries, all in London, and Frederick Sander of St Albans, Hertfordshire. Arrangements for collecting orchids in tropical America and Asia, and then ensuring their transport to Europe was complicated. Some collectors were employed directly by nurseries, while consignments of orchids were also imported on a speculative basis for auction, often at Messrs Stevens of King Street, Covent Garden in London. Orchids could be extremely expensive at first importation: a single orchid is said to have been sold at auction for 650 guineas. Endersby notes that the first example of the Indian orchid, *Cypripedium* (i.e. *Paphiopedilum*) *spicerianum* sold in London in the 1870s for over £250; within a year plants were selling for two shillings (ten pence), a tiny fraction of the original price. Other notable introductions included *Cattleya labiata*, first sent to Britain in 1818 by William Swainson (1789–1855), who had found it in northeastern Brazil, and reintroduced (after some confusion over its area of origin) by Sander in 1891, and *Dendrobium phalaenopsis* (i.e. *D. bigibbum*), also collected in 1891, in New Guinea by the prolific commercial collector Wilhelm Micholitz (1854–1932). While plants could now be grown in Europe, propagation was much more difficult, and very large numbers of orchids were therefore collected from the wild to supply the sizeable middle-class market. Benedict Roezl (1823–85), a Czech plant hunter, collected many thousands of orchids in the Americas for Sander, and "discovered" eight hundred new species in a forty-year career. One Roezl shipment alone, in 1873, was of a hundred thousand orchids; another was of 10 tons of cacti. However, as the nineteenth century came to an end, there were fewer new territories to explore for new orchids, and nursery cultivation became more advanced, leading to a decline in the mass harvesting of wild orchids.

Trees and Shrubs

The idea of the arboretum arose in the eighteenth century but became most popular from the 1830s, inspired in part by Loudon's *Arboretum et Fruticetum Britannicum* (1838) and the garden designs of Humphry Repton (1752–1818). Well-known arboreta in Britain, such as Derby Arboretum (1840) and Westonbirt (1828), existed alongside ambitious arboreta overseas, such as the Segrez Arboretum near Paris (1856), with 4,267 species of tree, the Spath Arboretum near Berlin (1876), and the Arnold Arboretum near Boston, Massachusetts (1872). Arboreta were also established in many colonial gardens. Often arranged taxonomically, with the trees of the same genus or family planted in proximity, the scientific organization of many arboreta in the British Isles sat within planting that reflected the "cultural status of trees in British myth, culture and society, estate economy and changing fashions in landscape gardening" (Elliott et al. 2007, 2011).

As with orchids, commercial nurseries were the most important source of trees for planting. This is reflected in the presence of an arboretum at Loddiges' nursery in north London, begun in about 1820. Display specimens could be seen alongside young plants for sale. The range of plants and reliability of identification made Loddiges the most important supplier of garden trees in nineteenth-century Britain. The rise of the arboretum in the nineteenth century was based on the huge increase in the number of new species, and selected cultivars, available. In the eighteenth century about three hundred new trees and shrubs were introduced to the British Isles, giving a total of 733 woody species in cultivation; by 1900 the figure was 1,911 species (Bean and Taylor 1970; Jarvis 1979). For climatic reasons few new introductions came from the tropics. North America had been the main source in the preceding two centuries, and in the nineteenth it remained important, as new territories were opened up by colonization in the west. The best-known collector in western North America was David Douglas (1798–1834), sent there in 1814 by the Horticultural Society of London; among many trees sent back was the Douglas fir, *Pseudotsuga menziesii*. Douglas' death—fatally gored in a wild bull trap in Hawaii in 1834—is a reminder of the dangers facing plant collectors at the time. Chile was also an important source of conifers, with the most notable collector being William Lobb (1809–64), sent by Veitch in 1840 (Noble 2009). Lobb was the first to collect the monkey-puzzle tree (*Araucaria araucana*) in quantity; later he also collected for Veitch in California and Oregon, introducing the Wellingtonia (*Sequoiadendron giganteum*).

However, a distinctive feature of nineteenth-century plant collecting is the eastwards turn to Asia. In the case of trees and shrubs, China and Japan were the most similar in climate to Europe and North America, but access for foreigners had long been restricted to coastal traders. Wider access to China was the result of the Opium Wars. Notoriously, the East India Company encouraged the production of opium (the morphine-rich latex of *Papaver somniferum*) in India, exchanging it for Chinese tea via intermediary traders so as to reduce loss of silver currency (Emdad-ul 2000; Trocki 1999). By 1830 illegal exports of Indian opium to China were 2.5 million lbs (1,133 tonnes). In 1839, further efforts by the Chinese Imperial government to control opium importation led to the destruction of large stocks of imported opium at Guangzhou, but military action by British forces in the Opium Wars of 1840–2 ended with the Treaty of Nanking, ceding Hong Kong and giving trading rights in five ports including Guangzhou and Shanghai; the second Opium War of 1856–60 further opened the country, allowing the work of French missionaries such as Père Armand David (1826–1900), who introduced *Davidia involucrata*, the handkerchief or dove tree, via the Jardin des plantes, Paris (Fan 2004; Kilpatrick 2014; Mueggler 2011). British trade with China never met expectations, and opium imports peaked in 1879, thereafter being replaced by cultivation of opium poppy within China itself.

USEFUL PLANTS

Those plants that dominate international trade are but a tiny proportion of those used by humans. The nineteenth century saw a well-developed scientific infrastructure applied to the commercial development of thousands of locally used plants, with varying degrees of success. The appropriation and transformation of indigenous knowledge is under-investigated; Abena Dove Osseo-Asare's work on the arrow poison and heart medicine, from *Strophanthus* species, has demonstrated that careful reading of sources can expand on both indigenous uses and the process of transfer (Osseo-Asare 2008, 2014).

Many plants were commercialized, but with limited impact; for example, New Zealand flax or harakeke (*Phormium tenax*) was introduced to St. Helena and formed the basis of an export industry, but this only survived from 1907–66. The New Zealand flax industry in New Zealand itself began in the 1860s and lasted to the 1980s, but with government subsidies for the last fifty years of its production (Cruthers et al. 2009). Here are examined in some detail three groups of plants, each with complex stories, and two of which—tea and rubber—with histories that have attracted much mythology.

Rubber

There are many parallels between the transplantation of rubber and cinchona (discussed with other medicinal plants by Simmonds in Chapter 5 of this volume), including South American origin, obstacles to introduction, and continuing controversy over ethical and legal aspects. However, in contrast to cinchona—the only significant source of quinine alkaloids—many different plants produce latex with the long-chain polymers that give it the elasticity and strength of rubber. These include the rubber plant of ornamental horticulture (*Ficus elastica*) in India, *Castilloa* species in Central and South America, and *Landolphia* species in the Congo basin. However, it was the Brazilian rubber tree or "Pará," *Hevea brasiliensis*, that entered plantation agriculture in the 1870s and has dominated world production of natural rubber up until today. Its successful transfer from the Amazon rainforest, where it is native, to Malaysia, depended on careful planning. It is likely that lessons were learnt from the introduction of cinchona a decade before.

The impetus for transplantation came from Clements Markham (1830–1916), by 1870 the leading figure in the Geographical Department of the India Office. He had recognized that both Amazonian and Indian rubber trees were threatened by overharvesting. In addition, rubber was in high demand by industry since Charles Goodyear's invention of vulcanization in 1839, which led to a material with more controllable properties. A survey of rubber species identified Pará both as the source of high-quality rubber, and suitable

FIGURE 3.4 The Amazonian rubber tree (*Hevea brasiliensis*), its flower and fruit segments bordered by six scenes illustrating its use by humans. Colored lithograph, *c.* 1840. Courtesy of Hamza Khan/Alamy Photo.

for cultivation in Britain's colonies in the wet tropics of Asia. Robert Cross (1836–1911) was dispatched by Kew to collect seeds in the Brazilian Amazon, while Henry Wickham (1846–1928), a British trader in Santarém on the Amazon, was also commissioned to collect them (Dean 1987; Loadman 2005; Nugent 2017).

Rubber seeds are "recalcitrant"; in other words, they die if allowed to dry out. This had already put paid to two earlier batches of seeds sent to Markham. Without using Wardian cases, Wickham sent seventy thousand seeds which arrived at Kew in June 1876. Of these, for which he was paid £70 per hundred, only 2,397 germinated. In November 1876, Cross returned to Kew with a further 1,080 seedlings. About 2,200 seedlings were sent to Sri Lanka, and in 1877 twenty-two of these were sent to Singapore, and nine or ten of these to Perak in Malaysia. Although a large rubber industry developed in Sri Lanka in the early twentieth century, production in Malaysia came to dominate. However, this was a slow process, depending on innovations in rubber-tapping and cultivation by Henry Ridley (1855–1956) at Singapore's botanical garden. As fungal rust affected Malaysian coffee plantations in the 1890s, planters switched to Pará rubber, and production increased from 2,000 acres in 1898 to 540,000 acres in 1910.

Although the export of rubber seeds from Brazil remained legal until 1914, it is often cited as an early example of biopiracy. However, the reality is more complex, and complicated by the unreliability of the principal witness, Henry Wickham. Those details that can be checked of the shipment of Wickham's rubber seeds from Santarem to Liverpool undermine his romanticized account of chartering a boat. Wickham's subsequent knighthood and generous pension from the rubber industry are testimony to his ability for self-promotion, but it remains unclear whether it was Wickham's seeds, or those of the little-known Robert Cross, that were sent to Malaysia. Looking at the bigger picture, the collapse of the Brazilian rubber industry from about 1920 led to the ending of the associated system of slavery required to harvest wild rubber from far-scattered trees; however, even plantation rubber harvesting and processing is labor intensive, leading British planters to bring in indentured laborers from China, and Tamils from southern India, to Malaysia. Both in its cultivation and in its manufacturing, rubber has a worldwide association with poor labor practices (Tully 2011).

Fiber Plants

In the era before artificial (oil-based) fibers, and the full suite of techniques for modifying natural fibers, there was a strong demand from industry for novel fibers, particularly those with the strength and sheen of silk (Lane 2007). Fibers were also popular with farmers and colonial botanists looking for new export crops that had the potential to be high value, yet also easy to ship on long voyages. Many fiber plants were circulated for trials, but typically failed against one of two criteria. Either the fibers were too difficult to extract to a high standard, particularly from the tough leaves of plants such as sisal (*Agave sisalana*), or prices were too low on the European market to cover production costs and long-distance travel. Nathaniel Wilson (1809–74), superintendent of the botanical garden in Jamaica from 1846 to 1872, experimented with a wide range of fiber plants, and sent many specimens of extracted fiber to international exhibitions (Wilson 1855). Renewed efforts in the 1880s, under Daniel Morris (1844–1933), failed for the lack of the correct machinery for fiber extraction (Morris 1884). Local fiber plants such as lace-bark (*Lagetta lagetto*) were the basis of an important souvenir industry but resisted commercialization (Brennan et al. 2013).

Historians have naturally concentrated on those fibers that were economically dominant, such as sisal, ramie (*Boehmeria nivea*), and jute (*Corchorus* spp.). In some cases the plant did not travel, but its manufacture did. Jute cultivation remained concentrated in Bengal throughout the nineteenth century, initially encouraged by the East India Company as an export crop, but mechanized manufacturing was developed in Dundee, Scotland, which became jute capital of the world—"Juteopolis"—between the 1830s and 1880s. Cotton was by far the most important of these fibers (Beckert 2014; Riello 2013). The early years of the nineteenth century saw New World cotton production shift from the Caribbean, where soils had become exhausted, to the south of the United States. Here production depended on slavery until its abolition in 1865. Concerns over continuity of supply—amply justified in the American Civil War—led British mills to look to India for their supplies. The East India Company introduced long-staple American Upland cotton (*Gossypium hirsutum*) to India, as its longer fibers were better suited to machine spinning. American cotton farmers were even persuaded to settle in the Bombay presidency although they did not stay long. There were substantial obstacles to the initial introduction of American cottons, including resistance from local farmers who found New Orleans cotton poorly adapted to local soils, and the long fibers difficult to separate from the seeds (Hazareesingh 2012). The outbreak of the American Civil War in 1861 spurred on British companies to provide suitable machinery for processing, and to create purchasing arrangements through agents that made the crop worthwhile for farmers.

Tea

By the mid-eighteenth century tea was a popular drink in Europe and North America, appreciated for its flavor and for the stimulant effect of caffeine (Mair and Hoh 2009). Tea was imported from China, by the British East India Company and other European traders, and a steep reduction in tax on tea in 1784 made it a mass-market product in Britain (Moxham 2003). By 1860 tea was widely cultivated in northeast India, then Sri Lanka, and, eventually Kenya, and tea had become widely affordable in Europe. This transformation was a complex process, still incompletely understood, and an example of the transfer of knowledge being arguably more important than the transfer of plants.

Tea plants from China had been traveling the world since the seventeenth century, including to the botanical garden in Kolkata, but there was little incentive for the East India Company to find another source for tea: it had a monopoly on British trade with China, and from the mid-eighteenth century onwards, a system of exchanging Indian opium for Chinese tea that made the trade even more profitable. However, in 1834 the Company lost its monopoly on trade with China, making cultivation in India a more attractive possibility. It established a Tea Committee, which took on the joint roles of procuring seeds and expertise from China, and investigating claims that tea was native to India. Assam tea had first been encountered in 1815, but it was only in the 1830s that Assam became easily accessible to British officials. In 1834 tea, in the form of *Camellia sinensis* var. *assamica*, was recognized by the Tea Committee as indigenous to the Assam region in northeastern India, where its leaves were chewed as a stimulant. Experts were divided as to whether Assam tea (var. *assamica*) or Chinese tea (var. *sinensis*) would grow best, and over the next decade both were tried. George James Gordon, Secretary of the Tea Committee, sent eighty thousand seeds from China to a botanical garden in Kolkata in 1834, with many more seeds later. In 1839, the first auction of Assam tea took place in London, but with only a few Chinese artisans taking up the opportunity to work in India, expertise in the harvesting and processing of tea was scarce (Mair and Hoh 2009).

The need for know-how is demonstrated by the fact that in the 1830s Western botanists were still unsure whether black tea and green tea were different species of *Camellia*, rather than the actuality, a single species processed in different ways. Access to tea plants was relatively easy; access to Chinese tea planters was difficult because Europeans were forbidden to travel to China's interior, and in particular to the Bohea Hills (Wuyishan), which produced the finest black teas. The bold travels of Robert Fortune (1812–80), a Scottish-born plant collector, were to change this. Taking advantage of the improved access for Europeans to China after the First Opium War, Fortune was commissioned by the Horticultural Society in London to collect ornamental plants. On this first trip, in 1843–5, eighteen Wardian cases of garden plants were sent back to Britain from Hong Kong. On this and a further four journeys to China and Japan, Fortune collected over 250 different kinds of garden plant, mostly already grown by the Chinese, including the windmill palm (*Trachycarpus fortunei*) and tree peonies (*Paeonia* spp.). Fortune's second expedition, from 1848 to 1851, was commissioned by the East India Company, at a generous annual salary of £500. By disguising himself in Chinese dress, and traveling with Chinese servants, Fortune was able to circumvent the ban on inland travel, reaching the Bohea Hills in 1849.

Fortune had no difficulty collecting large numbers of tea seeds, which were sown in soil at the bottom of Wardian cases and sent to India in several shipments during the expedition. Most of these went to the province of Uttarakhand in northern India, where

FIGURE 3.5 Nine scenes showing tea cultivation and preparation on an Indian plantation. Engraving by T. Brown, *c.* 1850, after J. L. Williams. Courtesy of the Wellcome Collection, London.

the East India Company was promoting plantations that were to prove unsuccessful in the long run; however, it is likely that Fortune's teas contributed to the Chinese stock that still forms the basis of tea cultivation in Darjeeling today. The explosion in tea cultivation of the 1860s was based on var. *assamica*, cultivated in Assam. Fortune's most important contribution was in being one of several botanists to show in the 1840s that the difference between green and black tea was in the processing, with black teas allowed a much fuller fermentation. On his third expedition (1853–6) Fortune was able to recruit seventeen specialist workers from China, with their equipment, to come to India and train the staff of the new plantations. As Fortune's biographer Alistair Watt (2017) suggests, this, and similar initiatives by others, may have enabled the breakthrough in tea production, rather than the movement of plants.

Fortune's fourth expedition to China, in 1858–9, was on behalf of the American government, and resulted in thirty-five thousand plants being grown in a glasshouse in Washington, DC (Gardener 1971). Much delayed by the American Civil War, subsequent attempts at cultivation foundered on the high labor costs of harvesting, an issue resolved in Sri Lanka through the employment of Tamil laborers from South India. The British had established a successful coffee industry in the hills around Kandy in the 1820s, but in the decade from 1869 the coffee rust disease destroyed the plantations. Cinchona at first appeared a savior, but prices collapsed in the mid-1880s. In 1867 tea plants were ordered from the botanical garden in Kolkata, and by 1885 there were over 100,000 acres of tea, of both Indian and Chinese varieties.

IMPACT

The scale of plant introductions to the Europe was huge; British historian Keith Thomas has estimated that in 1500 there were two hundred species of cultivated plants in England; by 1839, about eighteen thousand (Jarvis 1979). We do not have equally precise figures for tropical countries, but there is good evidence for massive shifts in production; for example, the virtual cessation of wild rubber harvesting in Brazil, and its replacement by the planting of 433,000 acres (175,228 ha) of rubber trees in Malaysia by 1920. This nineteenth-century combination of plant transplantation and large-scale plantation, combined with advances in transport, meant that correspondingly large populations in Europe could be supplied with food and industrial raw materials from around the world. In the years 1834–6 the British cotton industry imported £14.5 million worth of cotton, most of it from the United States—by far Britain's most valuable import. The impact of this easy access to raw materials on European economies then, and subsequently, is much debated by economic historians (Cain 1999; Peers 2004).

Transoceanic plant introductions had a major impact on the environment soon after Columbus' voyages of the late fifteenth century. What Alfred Crosby (1986) termed "ecological imperialism" saw crops and animals such as wheat and sheep spread from Europe into the temperate regions of North and South America, southern Africa, and Australia and New Zealand, while crops such as maize and potatoes moved in the other direction. However, the nineteenth century saw a step-change in the ability of transport to take plantation crops to market, and to bring labor forces to crops. Forests were extensively cleared for plantations of cinchona, rubber, and oil palm, although those sometimes replaced other crops. It is estimated that about 1.5 million square km of tropical forests were converted to crops or grasslands between 1850 and 1920 (Ross 2017; Williams 2003). Large-scale felling of trees occurred in North America and

Australia, but in other cases trees were introduced to treeless lands, as in the case of naturalized cinchona in the Galapagos Islands (Macdonald et al. 1988), or eucalyptus trees planted in Italian marshes (Doughty 2000). These cases are a reminder that plant introduction is often complex in its effects (Brock 2014). Similar impacts occurred in other ecosystems including grasslands, with wheat cultivation in the Argentinian pampas growing from 12,000 square kilometers (4,600 square miles) to 62,000 square kilometers (24,000 square miles) between 1890 and 1914, and the cultivated area of the Russian steppe increasing fivefold to 340,000 square kilometers (131,000 square miles) during the eighteenth and nineteenth centuries: it was not only the tropics that were affected by the expansion in crop monocultures (Isenberg 2014). Many of these regions were to become dustbowls in the twentieth century.

Richard Grove's pioneering study *Green Imperialism* traces the origins of modern environmentalism to the colonial world, initially in the eighteenth century in islands such as St. Helena and Mauritius, in which the effects of deforestation on climate and soils were worked out, and then in the nineteenth century, to a more structured body of knowledge that saw large-scale application to forests in India (Grove 1995; Ross 2017). Although initially marked by a purely pragmatic approach to maintaining timber supplies and climatic stability, by the 1840s some were also raising concerns about the extinction of plants and animals. Tensions between the economic and environmental roles of protected areas, and the rights of local communities within protected areas, arose early on and continue to trouble nature conservation today. By the late nineteenth century orchid collectors were becoming concerned at the effects of the plunder of tropical orchids, all the more foolish as so many plants died en route to their eventual markets (Endersby 2016).

Plant invasions were another consequence of plant introductions. The greatest impact of these has been in island ecosystems, where the effects of introduced animals such as rats and goats on native vegetation paved the way for introduced plants. For example, in the South Atlantic island of St. Helena, first visited by Europeans in 1502, goats browsed the saplings of trees such as gumwood (*Commidendrum robustum*), leading to the replacement of forest by imported grasses such as Bermuda grass (*Cynodon dactylon*). The effects were compounded by the nineteenth-century introduction of New Zealand flax as a crop; although no longer cultivated, it is widely naturalized and forms an "impenetrable monoculture" (Cronk 1989). Ornamental horticulture was an important source of invasive plants too. The best-known example in Britain is *Rhododendron ponticum*, introduced in about 1763 from Spain, selected for hardiness, and proving to grow fast and set abundant seed, allowing naturalization (Dehnen-Schmutz and Williamson 2006). Invasive in woodland since the nineteenth century, it was not recognized as a problem until the mid-twentieth century. Other nineteenth-century introductions of problematic ornamentals to Europe include Japanese knotweed (*Reynoutria japonica*), giant hogweed (*Heracleum mantegazzianum*) from the Caucasus, and Himalayan balsam (*Impatiens glandulifera*). About a quarter of ornamental plants for sale in nineteenth-century Britain escaped into the wild; of these about a third became established there (Dehnen-Schmutz et al. 2007a, 2007b). Although the historical ecology of invasive plants is not yet well resolved on a global scale, these examples show that the nineteenth century was a period of exceptional impact.

Plant trade and introductions have had a long legacy in human affairs. Viewed from the West, new plants brightened homes and gardens as ornamentals, enabled the importation of cheap food, and provided raw materials for major industries that employed and supplied many. Viewed from the perspective of source communities, the impact of nineteenth-century extraction of natural resources, migration, and forced labor remains deeply felt, and is visible in contemporary inequalities (Ross 2017). Past practices of plant translocation, although at the time often legal and in the context of global exchange of plants, took place within highly unequal power structures of which indigenous peoples remain keenly aware (Mt. Pleasant 2014). Unpicking the motives and practices behind nineteenth-century plant exchange is an enterprise of high contemporary relevance.

CHAPTER FOUR

Plant Technology and Science

ANNE OSBOURN

INTRODUCTION

This chapter focuses on key advances in plant science and technology during the period 1800–1920. This was an exciting time during which the foundations of future plant biotechnology and plant breeding were laid. It was also a time of important technical developments—for example, improvements in microscopy and the culture of isolated plant cells. During the nineteenth century the focus of botanical research moved from observing to investigation of fundamental processes such as cell division, natural transformation of plants by the crown gall bacterium, photosynthesis, heredity, and evolutionary adaptation. By the 1920s plant science was poised for the development of modern agriculture and the new emerging field of plant biotechnology.

CELLS, CELL CULTURE, AND PLANT TRANSFORMATION

Cell Theory

In the early part of the nineteenth century two German scientists, Schleiden (1838) and Schwann (1839), identified the cell as the primary unit of all living organisms. Matthias Schleiden (1804–81) was a prominent botanist, while Theodor Schwann (1810–82) was a professor of animal physiology. Recognizing the importance of discovery and naming of the cell nucleus in the orchid, *Cypripedium reginae* by Scottish botanist Robert Brown (1773–1858; Brown 1831; see Stephenson 1932, Mabberley 1985a: ch. 16), Schleiden commented: "every plant and animal is 'an aggregate of fully individualized, independent separate beings', that is, cells" (1838: 231). Schwann noted:

> Now, if we find that some of these elementary particles, not differing from the others, are capable of separating themselves from the organism, and pursuing an independent growth, we may thence conclude that each of the other elementary parts, each cell, is already possessed of power to take up fresh molecules and grow; and that, therefore, every elementary part possesses a power of its own, an independent life, ... The failure of growth in the case of any particular cell, when separated from an organized body, is

as slight an objection to this theory, as it is an objection against the independent vitality of a bee, that it cannot continue long in existence after being separated from its swarm.

(1847: 192–3)

Schleiden and Schwann's extensive use of microscopy led them to recognize that all plants and animals comprise recognizable units or cells (Vasil 2008). The term "cell" was first used by Robert Hooke (1635–1703) in the mid-seventeenth century to describe rectangular structures that he observed when studying cork under the microscope (Hooke 1665). Others subsequently saw structures that would later be recognized as cells in animals but did not make a connection between these structures and plant cells. The collective findings of Schleiden and Schwann led to the recognition that cells were the elementary parts of all living organisms, and hence to the universal cell theory.

Although cell theory represented a major advance in the thinking at the time, this was only the beginning of efforts to understand how cells function and how these apparently simple structures are able to give rise to whole new multicellular organisms. Schleiden wrote that "Botany, as an inductive science, comprehends the study of laws and forms of the Vegetable Kingdom." He also said, however, that "As an experimental science, it takes a very low position; and at present, embraces but a very narrow circle of actually established facts, few indications of natural laws, and no fundamental principles and ideas by which it might be developed" (1849: 1). In doing so, he challenged the descriptive approaches that had dominated the study of plants, which had focused primarily on observation and classification. The advent of microscopy enabled organisms to be studied at higher resolution in more detail but was still descriptive in nature. Schleiden noted that "The objects of Botany are actual existences—natural bodies. These must be examined in all possible ways; and to this many aids are necessary, for even the parts invisible to the naked eye must be investigated" (Schleiden 1849: 1). Robert Brown (see above) had previously described nuclei in plants and Schleiden and Schwann recognized similar structures in animal cells. Schleiden was of the opinion that plant cells formed around nuclei, with cell membranes growing out of nuclear structures: "So soon as the cytoblasts (nuclei) have attained their full size, a delicate transparent membrane, the young cell, rises upon their surface … It gradually expands, becomes more consistent, and at length so large, that the cytoblast appears only as a small body inclosed in one of the side walls" (Schwann 1847: 3–4). In contrast, French botanist Charles François Brisseau de Mirbel (1776–1854) had proposed (in 1815) that formation of new cells took place in the intercellular canals and on the surface of the plant in a thallose liverwort, a species of *Marchantia*. Schwann also noted that Schleiden had suggested a further different mode of formation of new cells by division of existing ones through the formation of partition walls, "if, as Schleiden supposes, this may not be an illusion …" (1847: 4); indeed, in 1833 Robert Brown described just that, in what would later be called meiosis, in the anthers of *Tradescantia virginiana* (Mabberley 1985b: 297).

Totipotency and Cell Culture

The concept of totipotency—the ability of a single cell to divide and produce a whole organism—was inherent in the cell theory but was not empirically tested until later in the nineteenth century. In 1858 the pathologist Ludwig Karl Virchow (1821–1902) challenged the prevailing view of spontaneous generation by stating that "all cells arise from cells" (Vasil 2008). Twenty years later German botanist Hermann Vöchting (1847–

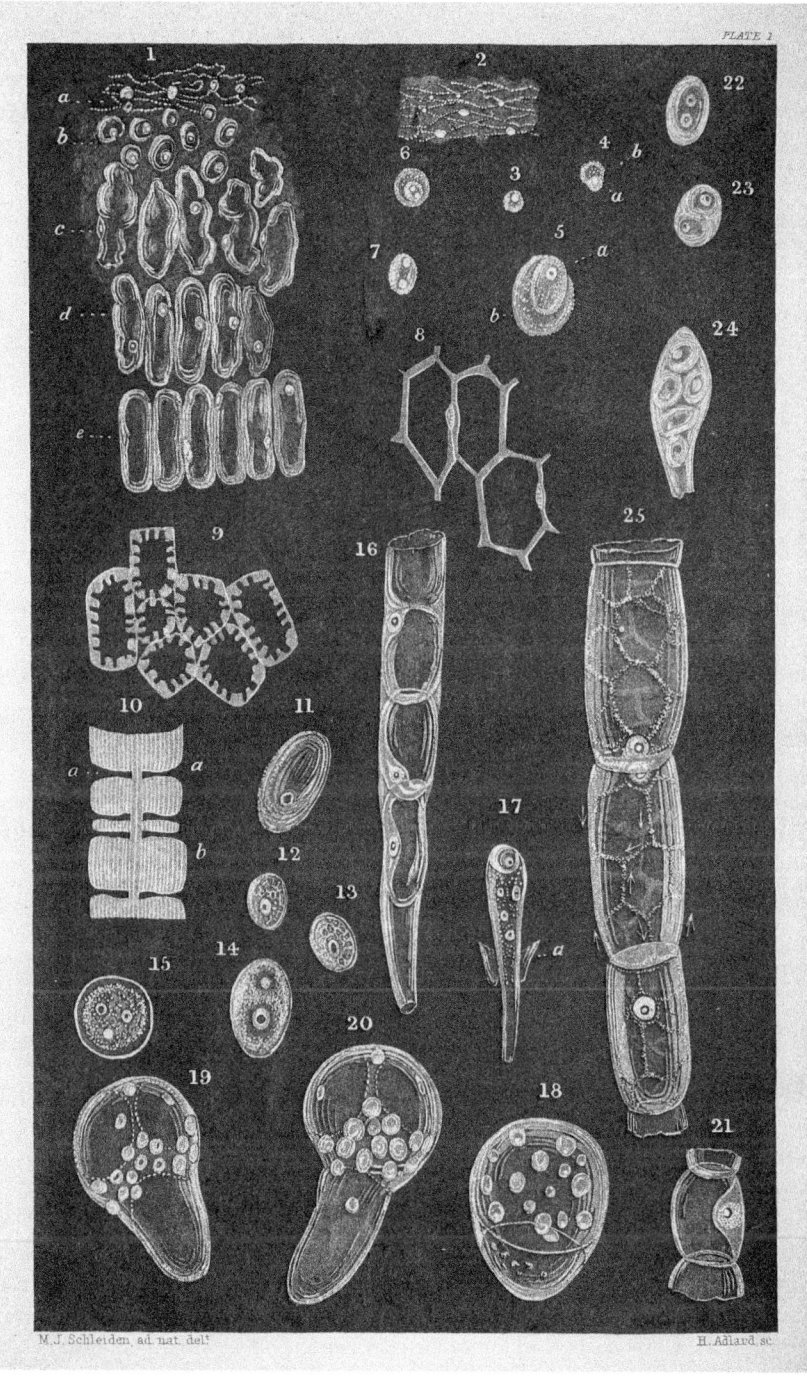

FIGURE 4.1 Schleiden's drawings of cells and embryonic structures from certain plant species. A single free cytoblast is shown (3), with a cytoblast with a cell forming on it, at low and higher resolution (4,5). From Schwann, *Microscopical researches into the accordance in the structure and growth of plants* (1847)—Plate I. Image from the John Innes Archives, courtesy of the John Innes Foundation.

1907) dissected plants and grew plant tissues, his experiments leading him to conclude that, "In every fragment, be it ever so small, of the organs of the plant body, rest the elements from which, by isolating the fragment, under proper external conditions, the whole body can be built up" (Vasil 2008). The demonstration that tiny fragments of plants contain the information necessary to regenerate into whole plants was a pivotal step in the subsequent foundation of the field of plant biotechnology.

The Austrian botanist Gottlieb Haberlandt (1854–1945) was the first to culture isolated somatic cells of higher plants in vitro:

> To my knowledge, no systematically organized attempts to culture isolated vegetative cells from higher plants in simple nutrient solutions have been made. Yet the results of such culture experiments should give some interesting insights to the properties and potentialities which the cell as an elementary organism possesses. Moreover, it would provide information about the inter-relationships and complementary influences to which cells within a multicellular whole organism are exposed.
>
> (Haberlandt 1902, as translated by Krikorian and Berquam 1969)

Although Haberlandt was successful in growing isolated plant cells, he was not able to induce cell division in any of the cells that he cultured. Nonetheless, he is recognized as the founder of plant cell culture because of the innovative methods that he proposed (Sussex 2008).

Haberlandt first attempted to culture green photosynthetic cells (palisade cells) from the bracts of red dead-nettle (*Lamium purpureum*). To do this he used Knop's solution, a simple nutrient solution containing the seven inorganic elements identified by Knop (1865) as being sufficient for water culture of higher plants, variously supplemented with different concentrations of sucrose, dextrose, glycerine, asparagine, and peptone. He carefully monitored the appearance, size, and physiological status of the cells using microscopy, in combination with methods that enabled him to detect oxygen release (based on chemotactic attraction of aerobic bacteria to photosynthesizing cells—"Engelmann's bacteria method"; Engelmann 1882), starch-grains (using iodine staining), cell-wall thickening, and turgor pressure (by plasmolysis). He also observed the size and position of the nucleus in the cultured cells, as well as the changes in protoplasm just before cell death. He further investigated the effects of maintaining the cells in the light or dark. Although he saw cell growth under some conditions, he never observed cell-division.

Other cell types that he experimented with included photosynthetic cells from water hyacinth (*Eichhornia crassipes*), glandular cells lacking chlorophyll from the leaves of "*Pulmonaria mollissima*" (i.e. *P. mollis*), the "stinging hairs" of nettle (*Urtica dioica*), staminal filament hairs from "*Tradescantia virginica*" (i.e. *T. virginiana*), and stomatal cells of *Ornithogalum umbellatum*, *Erythronium dens-canis*, and *Fuchsia magellanica*, again failing to observe cell division (reviewed in Krikorian and Berquam 1969). Haberlandt's experiments were thorough and rigorous. His failure to observe dividing cells was likely to be due to inadequate control of microbial contamination, culture media that lacked the hormones and growth factors that are now known to be essential, and his use of highly differentiated mature cells.

Haberlandt's technical innovations, including use of hanging-drop culture methods (which allow cell growth that would otherwise be restricted by the flat plane of culture dishes, and also minimize evaporation) and use of micropipettes to manipulate single

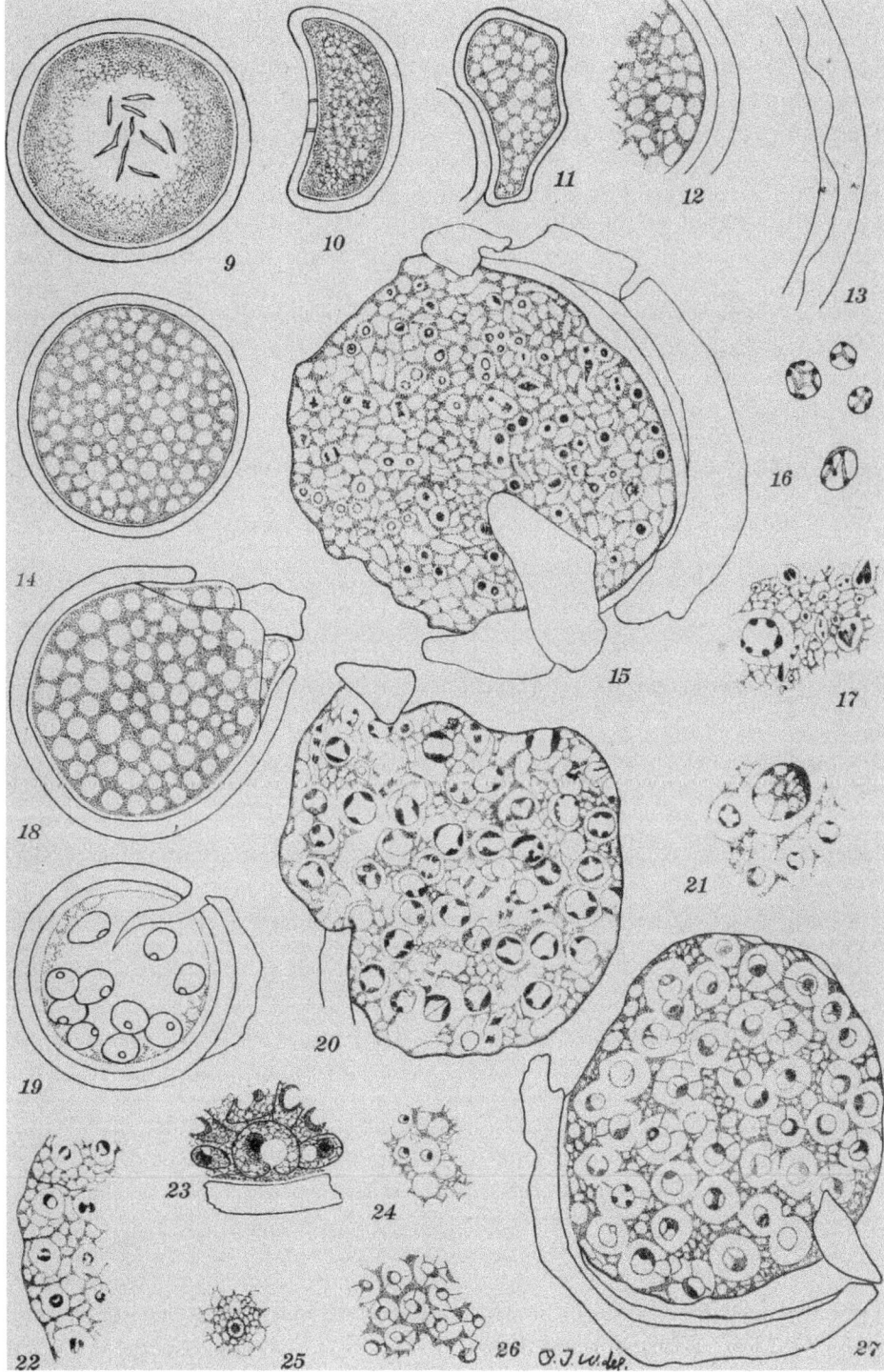

FIGURE 4.2 Crown gall of alfalfa [*Medicago sativa*]. From O. T. Wilson in *Botanical Gazette* (1920).

cells, made immense contributions to plant and animal cell culture studies (Sussex 2008). He also made some far-sighted predictions based on his experiences. He advocated co-cultivation of vegetative cells with pollen-tubes, which were known to produce chemical stimuli that induced growth of orchid ovules, a proposal that foreshadowed nurse culture technology, and he predicted that embryo sac fluids (such as coconut milk) might be used to induce the division of isolated vegetative cells (a method that was later used widely and successfully in tissue culture studies; Sussex 2008; Vasil 2008). Although Haberlandt went on to focus his energy on his other scientific interests ("sensory physiological" investigations), he was optimistic about the future of the cell culture field, concluding his seminal 1902 paper by saying, "Without permitting myself to pose further questions, I believe, in conclusion, that I am not making too bold a prediction if I point to the possibility that, in this way, one could successfully cultivate artificial embryos from vegetative cells" (Haberlandt 1902 in the translation by Krikorian and Berquam 1969).

Charles Darwin's observation that seedlings grow towards the light and that this effect could be prevented by covering the tip of the coleoptile (Darwin 1880: 418–92), along with the demonstration that complete removal of the tip abolished the seedling response (Boysen-Jensen 1910) suggested the involvement of a "material substance." This substance, later to be called "*wuchstoff*" (hormone; Went 1928) by Dutch scientist Frits Went (1903–90), would subsequently be identified as auxin (indole acetic acid, or IAA). The discovery of auxin and other plant growth hormones, in combination with new completely defined nutrient media and improved aseptic culture methods, paved the way for industrial micropropagation and regeneration of plants from cultured tissues, as well as for somatic embryogenesis, the preferred method for the regeneration of most of the commercially cultivated biotech crops (Vasil 2008).

Agrobacterium: The Natural Genetic Engineer

Crown gall disease is a disease of higher plants in which tumors are formed on roots and at the base or "crown" of diverse plant species. It was described as early as biblical times, on trees and grapevines, as galls and nodules. Research on the cause of crown gall disease would ultimately have completely unforeseen consequences for plant biotechnology. First, some background is required on the state of the field of microbiology in the nineteenth century. Previously, inspired by Robert Hooke's microscopic observations in his *Micrographia* (Hooke 1665), the Dutch scientist Antonie van Leeuwenhoek (1632–1723) built a simple microscope in 1671. His skills and passion for lens grinding and his superior light-adjusting techniques enabled him to make microscopes that could magnify up to five hundred times (van Leeuwenhoek and Hoole 1800). In 1676 he claimed to have observed microscopic one-celled organisms. He was the first person to see bacteria. The term "bacterium" was introduced in 1828 by the German naturalist, geologist, and microscopist Christian Gottfried Ehrenberg (1795–1876; Murray and Holt 2015: 1–13). In 1859, Louis Pasteur (1822–95) demonstrated that the growth of micro-organisms causes the fermentation process, and that this growth is not due to spontaneous generation (Pasteur 1909–14). Previously, the idea that disease was caused by "bad air"—the miasma theory (Galen 129–216 CE)—had predominated.

The consensus was now moving in favor of the germ theory of disease, which posited that diseases may be caused by micro-organisms (Pasteur 1909–14). Germ theory was finally proved by German scientist Robert Koch (1843–1910), who had been studying infectious diseases such as tuberculosis, cholera, and anthrax. Koch set out criteria to test whether an organism is the causal agent of a disease—criteria (Koch's postulates) that are still used today, and for which he won the Nobel Prize in 1905 (Blevins and Bronze 2010; Kaufmann and Schnaible 2005). His criteria were: i) the suspected pathogen must be consistently associated with the same symptoms; ii) the organism should be isolated into culture, away from the host: this precludes the possibility that the disease may be due to malignant tissues or disorders of the host itself; iii) the organism should then be re-inoculated into a healthy host; iv) symptoms which are identical to those observed in the original outbreak of the disease should then develop; v) the causal agent should be re-isolated from the test host in a pure culture and should be shown to be identical with the organisms originally isolated (Lucas 1988: 24).

The first scientific description of galls on grapevines was printed in France (Fabre and Dunal 1853). The causal agent was isolated from such galls at the Real Orto Botanico, the Royal Botanic Garden of Naples, Italy, by Fridiano Cavara (1857–1929), who showed that it caused the tumor disease, which he called *"tubercolosi della vite"* (Cavara 1897a, 1897b). George C. Hedgcock subsequently reported the isolation of a causal bacterium from grapevine galls in the United States (Hedgcock 1905). In 1907 two other American plant pathologists, Erwin Smith and Charles Townsend, published a paper in *Science* reporting that a bacterium caused plant tumors or crown galls on a variety of plants, including marguerite daisy, chrysanthemum, tobacco, tomato, potato, sugar beet, and peach (Smith and Townsend 1907). They systematically followed Koch's postulates and named the causal agent *Bacterium tumefaciens*. By the 1920s, there were numerous reports of crown gall on fruit trees. The organism was later named *Pseudomonas tumefaciens* (1909), then *Phytomonas tumefaciens* (1923) and finally *Agrobacterium tumefaciens* (Conn 1942). As Smith and Townsend (1907: 671–3) comment:

> The number of vegetable galls known positively, i.e. by exact experiment, to be due to bacteria, is not very great. The discovery of a new one of undoubtedly bacterial origin is, therefore, of considerable interest to plant pathologists, and may be of some interest to animal pathologists, especially to those interested in determining the origin of cancerous growths.

This report was some twenty years after the discovery of the first plant pathogenic bacterium.

Smith observed secondary tumors that developed on the leaves or stems of plants infected with *A. tumefaciens* at some distance from the primary gall and concluded that these secondary tumors were outgrowths from the primary one. The work of Armin Braun (1911–86) at the Rockefeller Institute would subsequently fail to find histological support for this claim and suggested that the mechanism of formation of secondary tumors may differ from that of primary ones. His demonstration that secondary tumors were bacteria-free led to the proposal that a "tumor inducing principle" (TIP) was responsible for tumor development (Braun 1951; Braun and Mandle 1948), and ultimately to the demonstration that *A. tumefaciens* is able to transfer DNA into plant nuclei and that these introduced DNA sequences are inherited as stable Mendelian loci (Lacroix and Citovsky 2013; Nester 2008). Thus, research that began as a classical study in plant pathology led to a serendipitous discovery that has greatly enabled the transformation and genetic engineering of plants, both for fundamental research purposes and for crop improvement.

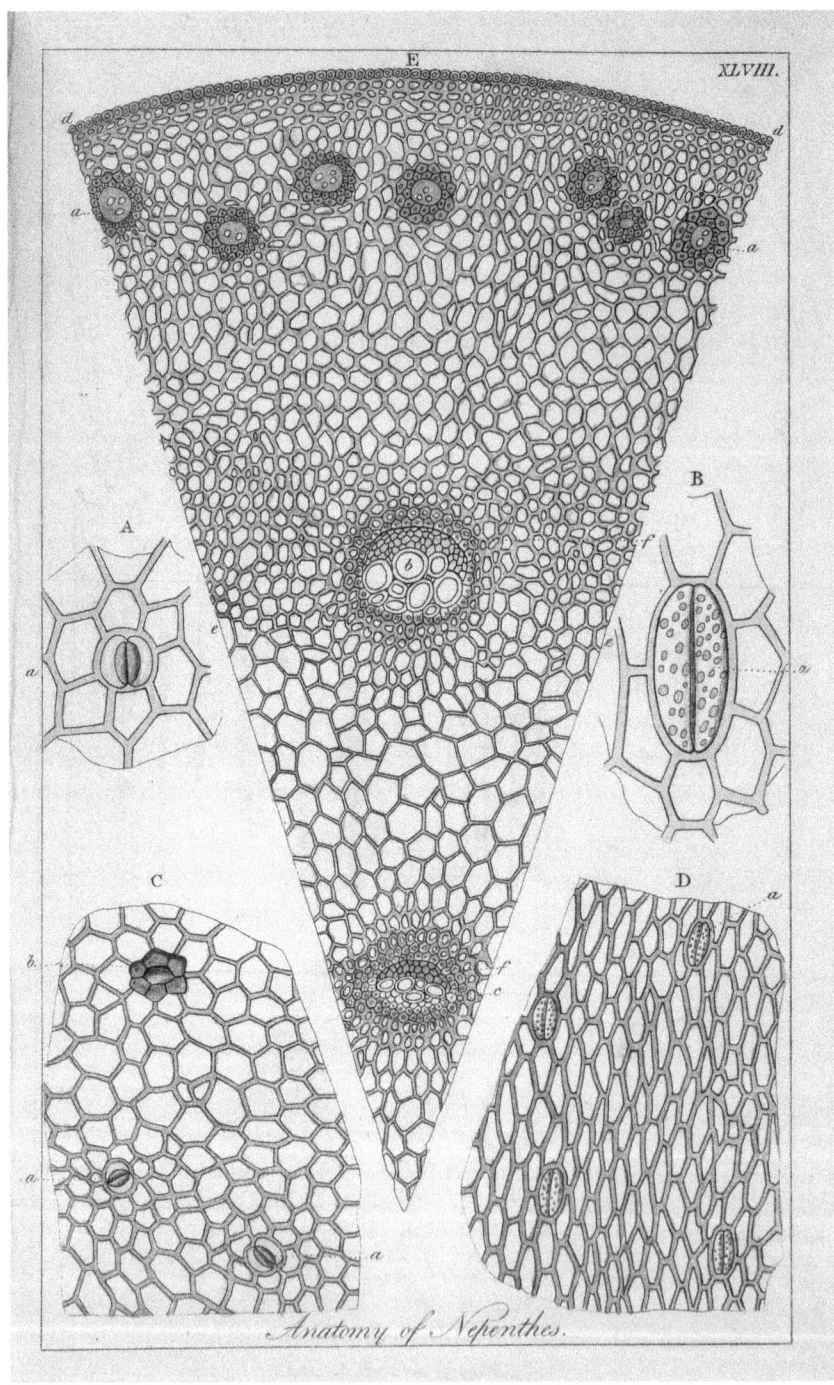

FIGURE 4.3 Anatomy of *Nepenthes distillatoria* showing vascular tissue and stomata. From *Ladies' botany, or a familiar Introduction to the study of the natural system of botany* by John Lindley (1840, vol. 2, 2nd edn., London: Ridgway)—a section of Plate XLVIII. Image from the John Innes Historical Collections, courtesy of the John Innes Foundation.

PHOTOSYNTHESIS

In the eighteenth century, the English clergyman and naturalist, Stephen Hales (1677–1761), made careful measurements of water vapor given off by plants, and observed a considerable decrease (approx. 15 percent) in the volume of air above the surface of water when he grew a plant in a closed chamber:

> The farther researches we make into this admirable scene of things, the more beauty and harmony we see in them: And the stronger and clearer convictions they give us, of the being, power and wisdom of the divine Architect, who has made all things to concur with a wonderful conformity, in carrying on, by various and innumerable combinations of matter, such a circulation of causes and effects, as was necessary to the great ends of nature. And since we are assured that the all-wise Creator has observed the most exact proportions, *of number, weight and measure*, in the make of all things; the most likely way therefore, to get any insight into the nature of those parts of the creation, which come within our observation, must in all reason be to number, weigh and measure.

He concluded that "air was being imbibed into the substance of the plant" (1738: 1). Hales also suggested that plants derive nourishment from the atmosphere through the leaves, and that light may contribute to "enobling principles of vegetation" (Govindjee and Krogmann 2004). The future discovery and elucidation of the photochemical process of photosynthesis would depend on the integration of discoveries and thinking across the disciplines of biology, chemistry, biochemistry, and physics.

Other seminal advances in the eighteenth century provided key insights that would pave the way for ensuing research on understanding the interactions between plants and air. Yorkshireman Joseph Priestley (1733–1804) demonstrated that air that had been depleted by burning a candle or keeping mice in a closed container could be restored by adding a living sprig of mint to the container (Priestley 1772). French chemist Antoine Laurent Lavoisier (1743–94) subsequently identified Priestley's active component of air as oxygen. Lavoisier also developed the concepts of oxidation and respiration and showed that "fixed air" is composed of carbon and oxygen (Govindjee and Krogmann 2004). Dutch physician Jan Ingen-Housz (1733–99) demonstrated that the ability of plants to purify common air in Priestley's experiment was dependent on sunlight reaching its green parts, and that this phenomenon specifically needed the sun's light, not its heat (1779: 1–19; 1796: 1–399). He noted: "*La production de l'air déphlogistiqué des feuilles, ne peut pas être attribuée à la chaleur du soleil, mais principalement à la lumière*" (1780: 36). Ingen-Housz also proposed that carbon dioxide was the source of carbon in plants. Swiss scientist Jean Senebier (1742–1809) established that "fixed air" (carbon dioxide) was essential for photosynthesis and showed that while carbon dioxide is absorbed by plants from the air, oxygen was released (1788: 417). In 1832 French physiologist René Joachim Henri Dutrochet (1776–1847) showed that gas exchange in plants took place via minute openings (stomata) on the surface of leaves and through their deep underlying cavities (Dutrochet 1832).

Nineteenth-century research focused on understanding the chemical processes by which carbon is fixed to form carbohydrates. German botanist Julius von Sachs (1832–97) showed that starch-grains are the first visible product of photosynthetic activity in leaves and suggested that they were formed from carbon dioxide (Sachs 1862, 1864). In 1818, two French scientists, Pierre Joseph Pelletier (1788–1842) and Joseph Bienaimé

Caventou (1795–1877), named the green pigment found in plants "chlorophyll," i.e. "green leaf" (Pelletier and Caventou 1818); in 1837 the German botanist Hugo von Mohl (1805–72) provided the first definitive description of "Chlorophyllkörnern" (chlorophyll granules) in green plant cells using light microscopy (Staehelin 2003). Sachs suggested that chlorophyll catalyzed photosynthetic reactions in the presence of light. Much later (1914) Hans Molisch (1856–1937) made pictures of starch inside a leaf by illuminating through a photographic negative (Gest 1991; Molisch 1914).

In 1845, the German physician Julius Robert Mayer (1814–78) described the plant as a reservoir of "solar force," which the plant absorbs and transforms into "chemical potential" for use in growth and metabolism. This phenomenon was referred to as "carbon assimilation"—"Assimilation in the botanical sense of the word is the formation of carbohydrates from carbonic acid and water while oxygen is released. This process is

FIGURE 4.4 Starch picture of Jan Ingen-Housz on a geranium (*Pelargonium*) leaf, made using a simplified version of Molisch's method (Molisch 1914). From Gest (1991).

limited to the chlorophyllous assimilation system (assimilation tissue) and requires the involvement of sunlight" (Nickelsen 2015: 16). Carbon dioxide (in its dissolved form, carbonic acid) and water were taken to be the starting materials of the process and, in the green chlorophyll-containing parts of plants, were converted in the presence of light into carbohydrates and oxygen. Assimilation was known to take place in living cells but not in dead tissue. However, the process was considered undoubtedly complex and it was not known how these factors interacted with one other (Govindjee and Krogmann 2004; Nickelsen 2015: 15–25).

In 1860 French chemist Jean Baptiste Boussingault (1801–77) made the first accurate measurements of gas exchange and found that the volume of oxygen evolved relative to that of carbon dioxide used up during assimilation was close to unity (Boussingault 1864). A central question concerned the course of events by which the absorbed carbon dioxide was incorporated into sugars. The German organic chemist Justus von Liebig (1803–73) proposed that organic acids were the intermediates in the gradual reduction of carbon dioxide to carbohydrates. An alternative pathway was put forward by another German organic chemist, Adolf von Bayer (1835–1907), who proposed that formaldehyde was the product of photosynthesis, and that several formaldehyde molecules were condensed to form sugars. Both ideas ultimately proved to be incorrect. In 1893 H.T. Brown and J.H. Morris suggested that most leaves contain glucose, presumably as a result of photosynthesis. Forty years later it was established that the major products of photosynthesis are disaccharides (sucrose) (Govindjee and Krogmann 2004; Nickelsen 2015).

Russian physiologist Climent Arkad'evitch Timiryazev (1843–1920) established the red maximum of the absorption spectrum of chlorophyll (Timiriazeff 1877), while Jacques-Louis Soret (1827–90) discovered an intense absorption band in the blue region of the spectrum of porphyrins (the "Soret" band; Soret 1883). German botanist Theodor W. Engelmann (1843–1909) illuminated cells of an unnamed species of the filamentous alga *Spirogyra* with light that had passed through a prism, thus exposing different segments of the alga to different wavelengths of light. He then used aerobic bacteria that seek oxygen to establish which parts of the alga were releasing the most oxygen and demonstrated that the bacteria congregated around the parts illuminated with red and blue light (Engelmann 1884). The development of the technique of chromatography by Russian botanist Mikhail Tsvet (or Tswett, 1872–1919) enabled plant pigments (chlorophylls and carotenoids) to be separated for the first time (Tswett 1906). The first detailed chemical investigations of chlorophyll including its chemical structure were published by Richard Willstätter (1915), an achievement for which he won a Nobel Prize in Chemistry.

The term "assimilation," as applied to light-dependent assimilation of carbon in plants was changed to a new term, "photosyntax," at the suggestion of American botanist Charles R. Barnes (1858–1910) to avoid confusion with processes in animals (Barnes 1893). Conway MacMillan suggested the alternative term "photosynthesis," a term that was not initially favored but became accepted by 1898 (Gest 2002). An important advance was the proposal by English plant physiologist Frederick Frost Blackman (1866–1947), based on quantitative experiments on the rates of photosynthesis under different light intensities, temperatures, and carbon dioxide concentrations in a species of the aquatic plant *Elodea candensis*, of "the law of the limiting factor"—the premise that the slowest step or factor in shortest supply limits the overall rates of photosynthesis (Blackman 1905; Blackman and Matthaei 1905). At low light intensity and high carbon dioxide concentrations there was no temperature effect. However, in strong light and limiting carbon dioxide concentrations, increasing the temperature increased the rate of photosynthesis. This led to the concept of light-limited and dark-limited photosynthesis. It would not be until 1924 that Otto Warburg (1883–1970) and T. Uyesugi would explain these results by showing that photosynthesis has two types of reactions—light and dark reactions (Warburg and Uyesugi 1924). With the aim of enhancing crop productivity, the mission to overcome the barriers that limit photosynthesis continues to be a focus of major international research efforts today.

FORM AND FUNCTION, CHROMOSOMES AND HEREDITY, AND THE NEW FIELD OF GENETICS

The nineteenth century was a period during which the key foundations of modern light microscopy were developed. Although lenses and subsequently light microscopes were in use well before the nineteenth century, it was Leeuwenhoek's extensive observations of many different biological samples that first brought the power of light microscopy to the fore (Wollman et al. 2015). He created microscopes with the greatest magnification available at the time, pioneering research into many and diverse areas of biology. Joseph Jackson Lister (1786–1869) developed the English compound microscope in 1826:

> Among the many ingenious novelties enumerated in his published papers we find the graduated lengthening of the body-tube of the microscope; a stage-fitting for clamping and rotating the object; a subsidiary stage; a dark-well, and a large disc to incline and rotate opaque objects; a ground-glass light moderator; a live-box with levelled flat-glass plate; an erector-eye-piece; an adapter for using Wollaston's camera lucida for microscopical drawing; and, above all, a combination of lenses to act as a condenser under the object (evidently the first approach to the present achromatic substage condenser).
>
> (Hogg 1911: 81)

Key theoretical and technical advances included diffraction-limit theory, aberration-corrected lenses, and optimized illumination (Köhler illumination). This resulted in microscopes with improved focus, reduced chromatin aberration and brighter field of view ("brightfield") (Wollman 2015).

There was increasing interest in the relationship between form and function in plants. Gottlieb Haberlandt classified different plant tissues as twelve different functional types (absorptive, mechanical, photosynthetic, etc.; Haberlandt 1884). Other publications on comparative plant anatomy included *The anatomy of woody plants* (Jeffrey 1917) and *Water plants: a study of aquatic angiosperms* (Arber 1920). It had also been shown that ovules of plants could not develop into seeds from the female style and ovary without pollination (Reed 1942: 165–6). In the nineteenth century, investigations of the reproductive mechanisms of plants, including ferns and bryophytes, especially by the largely self-taught Wilhelm Hofmeister (1824–77), whose work on the "alternation of generations" appeared when he was just twenty-seven, demonstrated that the process of sexual reproduction involves an alternation of generations between sporophytes and gametophytes (Reed 1942: 127–35).

In 1865 Mendel presented his work on plant hybridization and speculated that cells contained a factor that carried trait-determining information from one generation to the next (Mendel 1866). In 1868 Darwin published his hypothesis on pangenesis, proposing that traits could be passed down by "gemmules," hypothetical particles that were dispersed throughout whole organisms and then became stored in the sexual organs (Darwin 1868: 2:374, 378–81). These first attempts to explain heredity lacked scientific support, but nevertheless provided the foundation for future research. Thirty years after the observations of Schleiden and Schwann that concluded that the cell was the fundamental unit of living organisms (between 1874 and 1876), the German-Polish botanist Eduard Strasburger (1844–1912) demonstrated that new nuclei can only arise in cells from the division of other pre-existing nuclei (Strasburger 1876) though in fact this had been accurately described in 1831 by Brown, using a simple microscope (see above).

The physical nature of the hereditary material was still unknown. As noted by cell biologist Clarence McClung (1870–1946), "The value of the results obtained from any microscopical study is directly dependent upon the quality of the technique employed in preparing the material" (1929: vii). German anatomist Walther Flemming (1843–1905) used innovative methods to fix and stain cells to enhance the resolution and contrast of intracellular structures and explored the fibrous network within the nucleus which he named "chromatin" (stainable material). Through painstaking observation and with great patience he observed cells at different stages of division and noted that the chromatin formed thread-like structures (later to be known as chromosomes). Importantly, he correctly deduced the sequence of chromosome movements during mitosis and noticed that chromosomes split during mitosis; he hypothesized that the split halves were partitioned into different daughter cells at the end of mitosis (Flemming 1882; Paweletz 2001) as had in effect been shown long before, for meiosis, by Brown.

Work in animals in the early twentieth century collectively led to the establishment of the chromosome theory of inheritance. The term "genetics" was first used by William Bateson (1905). Strasburger and others continued to use "pangene," a term first coined by Hugo de Vries (1889: 187–212) to refer to the fundamental unit of heredity, until in 1909 the Danish botanist Wilhelm Johannsen (1857–1927) abbreviated this name to "gene" (Roll-Hansen 2014). The stage was then set for the development of the field of genetics and its application to plant breeding.

TAXONOMY, ECOLOGY, AND EVOLUTION

Matthias Schleiden's "modern" textbook *Grundzüge der Wissenchaftlichen Botanik*, published in English as *Principles of scientific botany* (Schleiden 1849) did much to stimulate interest in botany. Linnaeus, in the eighteenth century, had led modern biological systematics through his standardization of binomial nomenclature (Koerner 1999; Linnaeus 1753). His classification system based on reproductive organs ("sexual system") was reformed by Bernard de Jussieu and his nephew Antoine-Laurent de Jussieu by returning to a "natural system" (see Volume 4 of this series), based on all characteristics, with some one hundred orders, most of them currently recognized as families (Jussieu 1789). This "natural" taxonomic framework and understanding of the relatedness of different groups of plants provided a critical foundation for future scientific research into the form and function of plants. Augustin Pyramus de Candolle (1778–1841) and his son Alphonse (1806–93, later with his son Anne Casimir, 1836–1918) published the seventeen-volume treatise *Prodromus systematis naturalis regni vegetabilis*, which covered the taxonomy and distribution of a large number of angiosperms (Candolle and Candolle 1824–73).

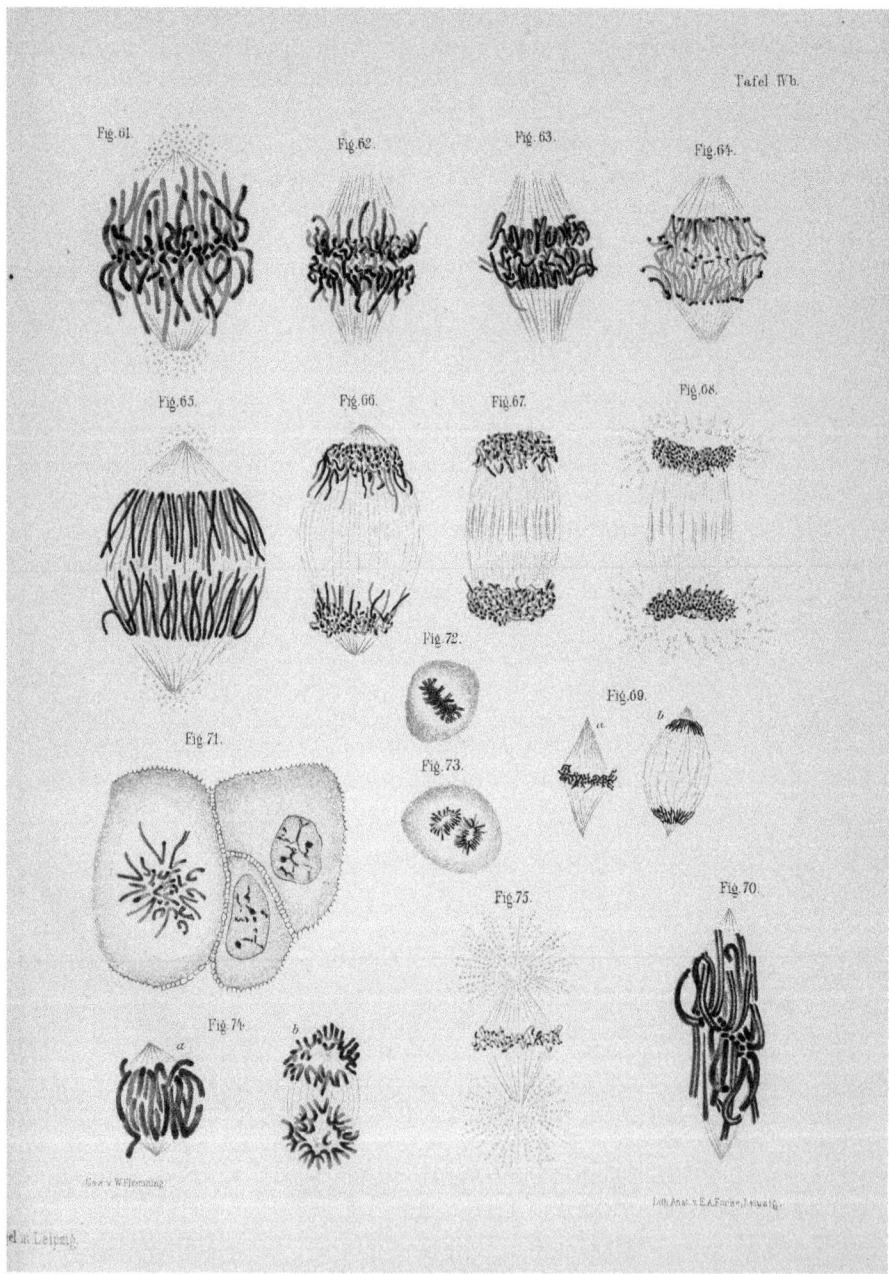

FIGURE 4.5 Drawing of mitosis by Walther Flemming, from Flemming (1882). Image from the John Innes Historical Collections, courtesy of the John Innes Foundation.

Linnaeus had been of the opinion that tea and other such tropical plants could be made to grow in his home country of Sweden by gradually accustoming them to the cold climate, but his endeavors were inevitably unsuccessful (Fara 2003: 28; Koerner 1999: 150–2). The nineteenth century saw an increasing awareness of the relationship between plants and the environment through the work of botanists such as Carl Willdenow,

Alexander von Humboldt, Aimé Bonpland, Robert Brown, Joakim Schouw, Alphonse de Candolle, Joseph Hooker, August Grisebach, and Asa Gray, who traveled extensively or studied plants from all over the world (Morton 1981: 314, 432–3). This interest in the influence of the environment on plants gave rise to the new field of ecology, pioneered by the Danish professor Eugen Warming (Warming 1895; Warming and Vahl 1909; see Mabberley 1992: 204 and the Introduction in this volume). Warming's work had wide-reaching impact. British botanist Arthur Tansley was drawn to ecology after reading *Plantesamfund* (Warming 1895), commenting: "I well remember working through it with enthusiasm in 1898 and going out into the field to see how far one could match the plant communities Warming had described for Denmark in the English countryside …." Tansley would later state that: "Though the organisms may claim our prime interest, when we are trying to think fundamentally, we cannot separate them from their special environments, with which they form one physical system" (1935).

The publication of *On the origin of species* by Charles Darwin introduced the scientific theory that populations evolve over the course of generations through a process of natural selection, replacing the assumption of constancy of species with the concept of descent with modification (Darwin 1859). The mutability of traits intrigued plant breeders, who began working with chemical solutions and radioisotopes with the aim of generating heritable variation "at will" in order to improve crop traits (mutation breeding) (Curry 2016).

FIGURE 4.6 Rowland Biffen (1874–1949), one of the great pioneers of modern plant breeding, examines a giant ear of wheat furnished by the University of Cambridge agricultural students to convey playfully aspirations for the future of wheat breeding. Image from the John Innes Archives, courtesy of the John Innes Foundation.

PLANTS AND INDUSTRY

Improving Agricultural Efficiency

Agriculture moved from hand-production methods to mechanization during the Industrial Revolution and afterwards, with the development of threshing machines, reapers, binders, and combine-harvesters (Grantham, 1984; see also Chapter 1 of this volume). Interest in the potential of fertilizers for increasing agricultural productivity also grew during this time. In 1836, French chemist Jean-Baptiste Boussingault (see above) recognized that the effectiveness of fertilizers was correlated with their nitrogen content (Smil 2001). In 1840, Justus von Liebig concluded that plants only needed a limited number of nutrients present in the soil to grow, but that these nutrients may be used up and so needed to be supplemented to guarantee the success of the next crop (von Liebig 1840).

Rothamsted Experimental Station (now Rothamsted Research) was founded in England in 1843 by John Bennet Lawes (1814–1900). Lawes, working with chemist Joseph Henry Gilbert, initiated long-term field experiments at Rothamsted to measure systematically for the first time the effects on crop yields of inorganic compounds containing elements known to occur in considerable amounts in crops and farmyard manure. This work led to the development of superphosphate, the first commercial mineral fertilizer. Lawes and Gilbert also found that crops with added nitrogen (applied as ammonium sulphate) gave substantially enhanced yields, confirming the belief that nitrogen is important for plant growth (Dyke 1993: 46–7; Galloway et al. 2017). In 1909 German chemist Fritz Haber (1868–1934) fixed atmospheric nitrogen in the laboratory, paving the way for industrial-scale nitrogen-fixation through the Haber-Bosch process in 1913 (Galloway et al. 2017; Smil 2001). The industrial production of nitrogen would be critical both for production of explosives for military purposes and for making fertilizers for agricultural use. In addition to Rothamsted Experimental Station, the longest-running agricultural research institution in the world, numerous other organizations were founded in Europe, North America, and the colonies during the nineteenth century with the remit of increasing crop yields and agricultural efficiency (Grantham 1984).

However, as the New and Old Worlds increased trade and cultivated crops in the absence of quarantine regulations, pests, and pathogens were inadvertently exchanged, resulting in catastrophic outbreaks of disease, the best known being the devastating late blight epidemic that caused the Irish potato famine of the 1840s (see Chapter 1 of this volume). This disease, which appeared simultaneously in the British Isles and much of Continental Europe, came from America (center of origin probably Mexico), where it had been recognized as early as 1843. More than twenty years before Pasteur's germ theory, Carl von Martius (1794–1868) and Miles Berkeley (1803–89) suggested that the causal agent was a fungus (Bourke 1991; Judelson 1997). The pathogen (an oomycete), originally named *Botrytis infestans* by Montagne in 1845 (Berkeley 1846), is now *Phytophthora infestans*, the name coined by DeBary (1876). The lack of genetic variability in the potato crop presented a susceptible host population to the organism. Darwin was involved in early research to find resistance to the disease and personally funded a breeding program in Ireland (Ristaino and Pfister 2016).

In 1869, the first-recorded epidemic of coffee rust disease (caused by the fungus *Hemileia vastatrix*) broke out in Ceylon (now Sri Lanka), and then spread to other coffee-growing regions around the world. The disease was first described and named in a November 1869 issue of the *Gardeners' Chronicle* (Berkeley 1869). Coffee growers experimented with newly developed fungicides such as Bordeaux mixture and, in India, identified some cultivars of coffee that seemed to show resistance to the rust. However, these endeavors were ineffective with the result that many coffee estates failed or switched to other crops such as tea (McCook and Vandermeer 2015).

Grapevines in Europe also suffered heavy losses, firstly due to mildew outbreaks and then to phylloxera infestation. Phylloxera, caused by the grapevine louse, *Daktulosphaera vitifoliae*, native in eastern North America, would ultimately be controlled by grafting scions of desired cultivars on resistant American vine rootstocks (Granett et al. 2001).

The possibilities of biological pest control were pointed out as early as 1800 by Erasmus Darwin (1731–1802), the grandfather of Charles Darwin (Darwin 1800: 356–8; Riley 1893), and a number of articles appeared during the first half of the nineteenth century on the beneficial effects of entomophagous (insect-eating) insects for controlling various parasites. The first international shipment of an insect as a biological control agent was made in 1873 by Charles Valentine Riley (1843–95), who sent predatory mites to France in an ultimately unsuccessful attempt to fight phylloxera. The term "biological control" was first used by Harry Scott Smith (1883–1957) in the early twentieth century (Smith 1919).

Plants as Sources of Commodity and High-value Chemicals

The nineteenth century saw improvements in wood technology including the production of tar and turpentine and eventually the synthesis of viscose, besides advances in pesticides and fungicides. A new crop was sugar beet (*Beta vulgaris*). In 1705, French agronomists had shown that beet roots contain sugar, but the discovery was not exploited until 1799 when a method for extracting sugar from them was developed, leading to the establishment of a small sugar beet factory at Cunern in Silesia, funded by Frederick William III (Austin 1928: 10–14). The importance of this industry increased during the Napoleonic wars, when commercial restrictions imposed by England's embargo cut off cane sugar imports from the West Indies (see Introduction in this volume). France in particular suffered from a shortage of sugar when the imports from its colonies ceased. Napoleon issued an edict compelling French farmers to devote at least 90,000 acres (approximately 365 square km)

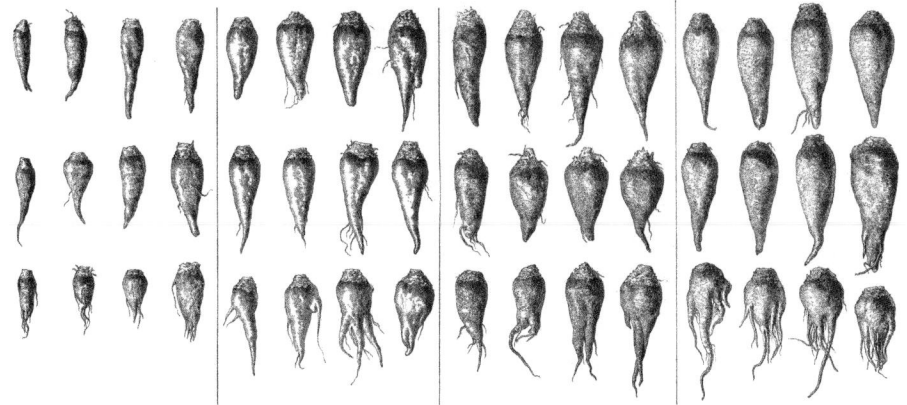

FIGURE 4.7 Sugar beet experiments, nineteenth century. Illustration of sugar beet grown under different conditions and with different fertilizers. Artwork originally from Turgan (1860). Courtesy of the Science Photo Library.

of land to the production of sugar beet and provided one million francs from the public treasury to support the building of sugar beet factories. Within two years, 334 small factories were established in France. Napoleon's action marked the beginning of a new and important epoch in the history of the sugar industry. Other European countries also began to encourage the establishment and development of the industry by the granting of bounties and subsidies, so that by the outbreak of the First World War there were in Europe over 1,200 large sugar beet factories, producing annually over 9 million tons of sugar, about a half of the world's sugar supply (Austin 1928: 12–13).

Advances in analytical and structural chemistry during the nineteenth century opened the way for the discovery and use of pharmacologically active natural products from plants (Atanasov et al. 2015; Ji et al. 2009). Morphine was isolated from opium at the beginning of the century by the German pharmacist Friedrich Wilhelm Sertürner (1783–1841), and became the first pure naturally derived medicine (Sertürner 1817). It was also the first naturally derived pure drug to be commercialized (by Merck, in 1826; Kaiser 2008). During the following decades, many bioactive natural products were purified from plants and their structures determined. The first natural product to be produced by chemical synthesis was salicylic acid in 1853 (Kaiser 2008). It would not be until the twenty-first century that the possibility of commercial *bio*synthesis of natural products by expression of plant pathways in heterologous production species would begin to be realized, opening up new routes to harness the vast chemical diversity of the Plant Kingdom (Paddon et al. 2013).

FIGURE 4.8 A laboratory at the Merck chemicals factory in Darmstadt around 1900. Image reproduced with permission of Merck: Merk-Archiv, Y01/ea-2122.

CONCLUSION

In the early part of the nineteenth century scientific ideas were exchanged through oral communications, letters, and descriptive works, primarily by authoritative individuals, often wealthy and influential gentlemen. This style of communication was superseded by the publication of research papers and the regular meetings of scientists, beginning with those in Germany in the 1820s. Laboratories were established as foci for plant science where previously there had only been herbaria and other collections. Building on the eighteenth-century legacy of classification, nineteenth-century research generated the cell theory, cell cultures, and *Agrobacterium tumefaciens*, setting the scene for plant tissue-culture, transformation, and biotech crops; it provided mechanistic insights into photosynthesis and the factors that limit this; and it gave rise to the new fields of genetics, plant pathology, ecology and evolutionary biology, which would inform and drive plant breeding. It also saw the establishment of sugar beet as an alternative feedstock for sugar production, and the beginnings of the pharmaceutical industry.

ACKNOWLEDGEMENTS

I would like to thank Sarah Wilmot (John Innes Historical Collections) and Helen Ghirardello for their advice and assistance in the preparation of this chapter, and Andrew Davis for photography. I would also like to acknowledge funding support from the Biotechnological and Biological Sciences Research Council Institute Strategic Programme Grant "Molecules from Nature—Products and Pathways" (BBS/E/J/000PR9790) and the John Innes Foundation.

CHAPTER FIVE

Plants and Medicine

MONIQUE S. J. SIMMONDS

INTRODUCTION

The nineteenth century saw the birth of pharmacy and the movement from herbals to the identification and use of single-molecule drugs. The century was marked by scientific advances in our understanding of human physiology and the spread of diseases, especially bacterial infections. These advances should be placed in the context of the significant political and economic changes taking place in the world, including the development of the United States of America, the founding of the French Republic, the decline in the status of China, and the industrial revolution in Britain (Porter, R. 1999: 353).

During the nineteenth century the care of the sick in hospitals also started to change. For example, there were changes in the management of hospitals, which historically were often run by religious orders for the care of the infirm and dying. These hospitals often paid for access to a local herbalist or a physician. More hospitals now started to be run by professional doctors, a change pioneered by physicians such as Jean Nicholas Corvisart (1755–1821), who was personal physician to Napoleon (Porter, R. 1999: 306). In cities, especially those with universities teaching medicine, such as Paris, the physicians would use the hospitals for training and experimentation on the sick in their quest to develop effective treatments. At the start of the nineteenth century, Paris was considered to be the Mecca for students studying medicine from all parts of the world, as they had access to patients in the public hospitals (Porter, R. 1999: 314). Towards the end of the century a similar trend occurred in Britain with the development of the teaching schools in the University of London (Porter, R. 1999: 317).

To complement the progress in our understanding of human physiology and patient care, there were advances in the field of chemistry that helped to further our knowledge about the compounds in medicinal plants associated with their medicinal or toxic properties. This is especially true when the active compounds in the plants were alkaloids. For example, this was the century for the discovery and isolation of alkaloids from poppies (1806), ipecacuanha (1817), *Strychnos* (1817), and *Cinchona* (1820) (Wink 1998). It is these discoveries that inspired many practitioners to move from the use of herbal-derived extracts and tinctures to the use and development of single-compound drugs—the start of molecular-based pharmacy. As the methods to study the chemistry of plants advanced, other active substances such as phenolics like salicin (1828) (Samuelsson and Bohlin 2009) and cardiac glycosides like digitoxin (1875) (Wade 1986) were discovered. These plant-derived compounds provided substances for the chemists to modify and supported advances in chemical synthesis of man-made drugs that then led to the development of pharmaceutical companies.

Given the diversity of medicinal plants available to the chemists and physicians, it is of interest that they initially focused on trying to understand how the toxic plants worked. This was for medicinal as well as legal reasons, the legal aspects usually being associated with the use of toxins in murders (Bertomeu-Sánchez and Nieto-Galan 2006). One of the first to collate information on toxic medicinal plants was the physician and chemist Mathieu Joseph Bonaventure Orfila (1787–1853), who was born on the island of Minorca and then traveled to Paris, where he held chairs in legal medicine and medical chemistry (Porter, R. 1999: 333). At that time the main toxin used to kill people was arsenic. In his book *Traité des poisons*, Orfila (1814) provided an overview of toxins, including those from plants, with examples from the court records. His work on the detection of toxins was the foundation for the scientific discipline of toxicology and raised the awareness of plant chemistry in the legal system as well as the medical professionals. Robert Christison (1797–1882) was a professor and physician in Edinburgh, who collated information on the toxicity of poisonous seeds. He also undertook some self-experimentation on a diversity of seeds, as he was interested in the dose required to have a beneficial effect rather than death. His findings about toxic substances were published in a *Treatise on Poisons* in 1829 (Christison 1829).

The isolation of active compounds from specific species of plants increased the demand for these plants, especially to support the developing pharmaceutical companies. Thus, there was an increased need to get key species into cultivation. Chemists and pharmacists started to optimize methods to extract the compounds from plants. There was also interest in seeing if the active compounds from some of these plants could be synthesized; this was both an academic challenge but also a practical need, as the actives usually occurred in very low concentrations in the plants. It was this need that was an important driver in the development of synthetic medicinal chemistry and our modern drug discovery programs. However, some of the experimental pharmacists studying the biological activity of the pure compounds were, like many herbalists, aware that the pure compounds often had a narrower range of medicinal activity than the medicinal plant from which they were isolated. It was also true that in many cases the plants were less toxic than the pure compounds. By the end of the nineteenth century, the more holistic approach to the use of whole plant extracts in healing had started to be marginalized in many countries with an increasing emphasis on single-compound drugs: this was the beginning of medicinal chemistry.

Individual scientists in mainland Europe, especially those in France, Germany, Italy, and Switzerland drove these changes. For example, many of these advances were influenced by people like the French physician François Magendie (1783–1855), who was keen to have scientific evidence explaining why plants as well as plant-derived compounds had beneficial medicinal properties (Porter, R. 1999: 327). He also wanted to improve the standards by which plant-derived products were industrially produced and used by medicinal practitioners. He was clearly influenced by the changes in France taking place during the French Revolution as well as major advances in medicine also taking place in France. He had an independent and sharp mind and the energy and frankness to influence the established medical views to drive forward advances and changes (Porter, R. 1999: 336). He was also keen to improve the quality of medicine being produced and prescribed. His work contributed to the development and content of the new generation of pharmacopeias that were a feature of the century.

MEDICINAL PLANTS: ADVANCES IN OUR KNOWLEDGE OF THEIR CHEMISTRY AND USES

Knowledge about the chemistry and medicinal uses of some species of plants advanced greatly in the nineteenth century. The following text provides some examples of the advances made in the knowledge about plants used in herbal remedies and those known to be very toxic.

Aconitum *and Aconitine*

Plants in the genera *Aconitum* and *Delphinium* (family Ranunculaceae) were used for pain control and to treat fevers. They were known to be very toxic so had to be used with care and administered only in very low doses. The German pharmacist Philip Lorenz Geiger (1785–1836) was keen to find the active ingredient in these plants and isolated nor-diterpenoid alkaloids from *Aconitum*, including the toxic aconitine, also known as aconite (Sonnedecker 1986). The chemical structure of aconitine is complex. It contains a nitrogen atom in one of the six-membered ring structures, and this enables the compound to easily form salts and ionize. Aconitine is able to bind with cell membranes and receptors, including the sodium-ion channels in the membranes of tissue such as cardiac and skeletal muscles and the neurons that stimulate these muscles. This ability contributes to the toxicity of the compound (Wink 1998). The first symptoms of aconitine poisoning appear approximately twenty minutes to two hours after oral intake and include paranesthesia, sweating, and nausea. This leads to severe vomiting, diarrhea, intense pain, and then paralysis of the skeletal muscles. It has been used in a few criminal cases because only a small amount is required to kill someone. However, because of its high toxicity physicians were reluctant to use it in treatment, although it was included in a range of products.

Atropa belladonna *and Atropine*

Atropine is a tropane alkaloid which occurs naturally in a number of plants in the Solanaceae, often known as the nightshade family, which includes deadly nightshade (*Atropa belladonna*) and mandrake (species of *Mandragora*). Atropine was first isolated in 1831 by the German pharmacist Heinrich F.G. Mein (1799–1864) and shown to be highly toxic, but also to have a diversity of beneficial properties (Behçet 2014). For example, it was used as an anesthetic, in treating heart disease, to reduce bronchial and salivary secretions, and for the treatment of muscular rheumatism and sciatica. Although atropine is toxic, a range of medicinal plants was found to contain this compound and to remove some of these plants from use would have negatively impacted the range of useful medicinal plants available to treat people. Knowledge was obtained about the levels of atropine associated with toxicity, and the new pharmacopeias provided pharmacists and physicians with advice as to the levels of atropine they should use and also how to make extracts so that the level of atropine in extracts prescribed to patients was low.

Carapichea ipecacuanha *(Ipecacuanha) and Emetine*

The Brazilian plant, *Carapichea ipecacuanha* (family Rubiaceae), formerly known as *Uragoga ipecacuanha,* played a role in the treatment of infections in the nineteenth century. Powdered rhizomes of this plant were used traditionally to invoke vomiting, which links to its local name ipe-cac-unanha, loosely translated from Portuguese as "roadside vomiting plant." The "root-powder" was thought to have come to Europe in the 1600s when it was used to treat dysentery. The use of the root-powder then became popular as an ingredient of Dover's powder, a combination of ipecacuanha, opium (from *Papaver somniferum*), saltpetre, and cream of tartar devised by the English physician Thomas Dover (1660–1742) for the treatment of pain and gout (Morton 1968). The demand for the powder increased, and there were issues associated with the supply and quality of ipecacuanha root-powder as more than one species of plant was traded as ipecacuanha. Therefore, there was a need to identify the best species to use and to find a good reliable source of material. Researchers in Paris started to investigate the chemistry of *Carapichea ipecacuanha*, and, in 1817, Magebdie, working with Pierre-Joseph Pelletier (1788–1842) on plants that cause nausea and vomiting (emetic plants), showed that the vomiting properties of ipecacuanha were caused by a substance they called emetine (Janot 1893). Later in 1894, Benjamin Paul (1827–1917) and Alfred Cownley (1825–1906) isolated both emetine and cephaeline (alkaloids) from *C. ipecacuanha* and also showed why the plant could have traditionally been used to treat dysentery (Greenwood 2008). Just before this, the Russian physician Fyodor Lösch (1840–1903), who was studying the organism associated with dysentery, discovered in 1875 the pathogenic amoeba which was later called *Entamoeba histolytica*. This discovery allowed a distinction to be made between the two main forms of dysentery (amoebic and bacillary). Emetine was shown to be active against the amoebic form of dysentery, but ineffective against that caused by bacteria, which could explain the earlier use of ipecacuanha to treat some forms of dysentery. However, the treatment with emetine was stopped as it could be very toxic. Although the ipecacuanha compounds did not lead to new drugs, they played a pharmacological role in furthering the understanding of the traditional uses of *C. ipecacuanha* and in the type of compounds that could inhibit the growth of pathogenic amoebae (Grollman and Jarkovsky 1975).

Cinchona and Quinine

The nineteenth century was important for the globalization of the use of cinchona bark-based products to treat fevers. In 1852 the British Foreign Office contacted its consular staff in South America to source *Cinchona* (Rubiaceae) seeds that could be planted in British colonies. At that time there were about thirty-five species of trees that were known as "cinchona." To head the exploration team, they appointed Clements Markham (1838–1916) who recruited Richard Spruce (1817–95), later to become a world authority on the plants of the Amazon. At the same time the Dutch government was working with botanists, especially Justus Hassakarl (1811–94), to expand their cinchona plantations. Hassakarl selected seeds from trees in the botanical gardens in Java to cultivate in the Dutch East Indies (today's Indonesia) and other parts of the world (Karch 2003: 5). When the explorers were seeking which material to select, they had little knowledge of the importance of the alkaloid quinine to the anti-fever properties of the bark of the *Cinchona* tree. It took time for news of the work of the two French chemists, Pelletier and Joseph Caventou (1795–1877) isolating quinine in 1820 to reach some of the growers (Achan et al. 2011). Without this knowledge the selection of seeds for establishing plantations was based on the size of the trees. In 1900, when the British started to harvest the bark from trees being grown in parts of Sri Lanka and other British territories, they found they had invested in growing poor-quality trees. Overall, the seeds were from trees with low levels of quinine in the bark, so the British government had wasted money in the cultivation of over a million relatively inactive trees. Aside from this government-sponsored activity, a British trader, Charles Ledger (1818–1905) who was living in Puno, Peru, purchased a 14-lb consignment of *Cinchona* seeds in 1865, which he sent to his brother George in London (Karch 2003: 6). George Ledger contacted the Royal Botanic Gardens, Kew, England, to see if there was interest in buying them. When Kew declined, he contacted the English chemist John Eliot Howard (1807–83), who worked for the family pharmaceutical company, Howard and Co. Ltd in London, making cinchona-based products. Howard realized the commercial significance of the seeds Ledger had collected and suggested that he contact Dutch government representatives, who purchased some of the seeds (Rocco 2003: 246). The others went to the British botanist, William McIvor (1824–76) who took them to the Nilgiris in India, where they failed to germinate. The "Dutch" seeds, which proved to produce trees with high levels of quinine, were sent to Java and were planted there. In consequence Java became, for a time, the world's main source of quinine and an important source of income for the Dutch economy. The trees that contained the higher levels of quinine were later named *Cinchona ledgeriana* (now *C. calisaya*) in recognition of the work undertaken by Charles Ledger. As Britain and the Dutch were ensuring they had a supply of cinchona bark to extract quinine, the chemists were attempting to synthesize it. One such chemist was the Englishman William Perkin (1837–1908) who, although he failed to synthesize quinine, synthesized mauve, one of the first aniline dyes (Brightman 1956). These dyes were shown to have medicinal properties, and his research resulted in the synthesis of the painkiller drug antifebrin in 1886. A breakdown compound from antifebrin was later commercialized as paracetamol.

Coffea arabica *and Caffeine*

Seeds from *Coffea arabica* (Rubiaceae) had a long tradition of being used to make a drink in different parts of Yemen, Ethiopia, and Arab countries and reached Europe early in the sixteenth century. The stimulation effect of coffee soon became well known, and by the nineteenth century its use as a medicine was documented in pharmacopeias. It was considered a tonic and could be given to patients to reduce fevers. The German chemist Friedlieb Runge (1795–1865), like some of the other chemists in Germany and France, was interested in anti-fever plants, and because *Coffea arabica* is in the same plant family as the "fever plant," cinchona, wondered if both contained the same active compound quinine, which he was the first to isolate. This was not the case as the active alkaloid was caffeine isolated from coffee in 1819 and initially called *Kaffebase* (Gilbert 1984). Caffeine was synthesized in 1895 by the German chemist Hermann Fisher (1852–1919), who, in 1902, was awarded the Nobel Prize for his work (Nobelprize.org 2014). Caffeine was later identified as the active compound in a few other plants such as tea used as stimulant drinks and medicines. It was isolated from tea in 1827, although it was initially called "théine." In the nineteenth century coffee was often taken to offset the effects of opium and alcohol and was given to people who had taken either in excess. It was also used to make a thick syrup for the treatment of children with whooping cough; this syrup also contained belladonna and ipecacuanha (King 1875: 274). In another combination it was mixed with citric acid to treat headaches. The diversity of pharmacological properties that can be attributed to caffeine were discovered only in the twentieth century.

FIGURE 5.1 *Cinchona ledgeriana* (*C. calisaya*), Ceylon. Royal Botanic Gardens, Kew. Permission from the Wellcome Collection. Credit: Wellcome Collection. Attribution 4.0 International CC BY 4.0 (https://wellcomecollection.org/works/p3jffpgf).

Colchicum autumnale *and Colchicine*

Autumn crocus or meadow saffron, *Colchicum autumnale* (Colchicaceae) occurs throughout Europe including Britain. The plant is very poisonous but extracts of the bulb have a long tradition in the treatment of joint pain and gout. The alkaloid colchicine was first isolated from the bulbs in 1820 by Pelletier and Caventou (1820). It was shown to relieve pain and to have anti-inflammatory properties such that colchicine-rich extracts started to be used in commercial products to treat gout.

FIGURE 5.2 *Coffea arabica*—fruiting stem. Watercolor, *c.* 1823. Credit: Wellcome Collection. CC BY (https://wellcomecollection.org/works/td95hg54).

Curcuma longa *and Curcumin*

Turmeric (*Curcuma longa,* Zingiberaceae) is a plant from Southeast Asia and was traditionally used in Ayurvedic and traditional Chinese medicine to treat respiratory disorders, colds, and liver ailments. It was imported into Europe and America from the Indian subcontinent, and in the nineteenth century it was often used in mustard plasters and ointments for the topical treatment of coughs and colds, as well as rheumatic and joint pain (Walsh 2015). The polyphenol curcumin was shown to be responsible for the bright yellow color of turmeric and was isolated from rhizomes in Pelletier's laboratory in 1815, although later his isolate was found to be a mixture of curcumin and a resin (Priyadarsini 2014). In the nineteenth century curcumin was used more as a dye than a medicine, and it was not until the mid-twentieth century that the medicinal properties of curcumin were scientifically studied.

Digitalis purpurea *and Digitoxin, Digitaline*

The foxglove, *Digitalis purpurea* (Plantaginaceae), is native to temperate Europe but has become naturalized in many other temperate parts of the world. An extract of this toxic plant was used in mixtures with other plants for the treatment of heart conditions. It was the British physician, chemist, and botanist William Withering (1741–99) who started to study the chemistry of this plant at the end of the eighteenth century. He was interested in its ability to treat those with heart failure (dropsy) (Goldthorp 2009), but was unable to identify a pure compound. In 1875, the German pharmacologist Oswald Schmiedeberg (1838–1921) was able to obtain a pure sample of the cardiac glycoside digitoxin. The therapeutic use of digitoxin for the treatment of heart conditions was researched by the French pharmacist and chemist Claude-Adolphe Nativelle (1812–89). It is of interest that extracts of *Digitalis purpurea* were associated with adverse responses, but not often death (Wade 1986). This is partly due to the fact that the consumption of a small amount of plant material causes nausea and vomiting within minutes and can prevent the patient from taking a toxic dose, whereas the purified digitoxin does not cause nausea and vomiting, so that a toxic dose can actually be consumed. The delays in being able to identify and "characterize" (to probe and measure a material's structure and properties) the cardiac glycosides in *Digitalis* most likely occurred because they can be very difficult to isolate, as they break down to inactive compounds during the standard isolation process.

Ephedra *and Ephedrine*

Species of *Ephedra* (Ephedraceae*)*, especially those used in Chinese traditional medicine, such as *E. sinica,* have a long tradition of being used to treat respiratory problems such as colds. Ephedrine was first isolated in 1885 by the Japanese chemist Nagai Nagayoshi (1844–1929). It then started to be used in the West as a nasal decongestant to relieve cold symptoms and to treat asthma (Lee 2011).

Erythroxylum coca *and Cocaine*

The nineteenth century saw advances in the knowledge and development of coca as a drug as well as in our knowledge about how to cultivate the plant. At that time it would have been difficult to predict that this knowledge would later support foreign drug cartels involved in the cultivation and supply of drugs from this plant. Coca was known to be an important stimulant in parts of South America, and samples started to arrive in different parts of Europe, including Britain. In 1835, the then director of the Royal Botanic Gardens, Kew, Sir William Hooker (1786–1865) made some of the first drawings of coca *Erythroxylum coca* (Erythroxylaceae) to have been seen in Britain (Karch 2003: 6). His work was based on material obtained from Peru. Further material was sent to Kew, including some from Java (where it is not native), these plants having larger leaves than other samples had. Were these the same species? It was concluded that there was a need

FIGURE 5.3 *Erythroxylum coca*—gathering the coca plant in Bolivia. Credit: Wellcome Collection. CC BY (https://wellcomecollection.org/works/t59rv6pq?query=Erythroxylum%20coca).

to acquire good-quality material if plantations of coca were to be developed outside of South America. The solution to this problem was found in the twentieth century.

The supply of quality material for creating the plantations of coca needed to meet market demands has many parallels with that of cinchona. This is partly due to the fact that there are overlaps in the people involved and they faced the same challenges in getting good-quality supplies. The problems the British and Dutch governments had with the cultivation of cinchona impacted how they dealt with coca. Whilst the Dutch (Justus Hassakarl) and the British (Clements Markham) government representatives were trying to identify quality cinchona, they were both investigating the methods and costs involved in the cultivation of coca (Karch 2003: 5). They were also starting to be aware of some of the addictive effects associated with the chewing of coca leaves. Nevertheless, they could see the economic potential of the plant as a crop. Hassakarl petitioned the Dutch government to promote the cultivation of coca, but initially the Dutch were not keen to develop the crop. They were also worried that if farmers in Java started to chew the leaves, they might become addicted to the material and thought that there was no need for another strong stimulant as betel nut (*Areca catechu*) was widely used there already. At the same time, they did not want to rule out the cultivation of coca entirely so decided to establish a small plantation in Buitenzorg, an area in Java near the modern town of Bogor. By contrast, the British government wanted the Royal Botanic Gardens, Kew to continue to work on coca and find quality plants, and it supported the collection of seeds in the Amazon. In 1854, Richard Spruce collected seeds for Kew from the Rio Negro region of the Amazon in Bolivia. However, like the Dutch, the British were cautious and did not invest in large plantations as they had with cinchona (Karch 2003: 5). And so, when the beneficial properties of cocaine as an anesthetic became apparent in 1884, there was not a ready supply of good-quality material from either British or Dutch sources, although there was plenty of poorer material available via the New York market. This is when the Dutch were able to scale up production from their plantations in Java, and the harvested leaves were then sent to the Netherlands for extraction under tight quality-control standards. The coca plants sent by the Royal Botanic Gardens, Kew to Martinique, Trinidad, Jamaica, Sri Lanka, Zanzibar, and Australia met with varying levels of success by the end of the nineteenth century (Karch 2003: 147). They did not match the quality of material being grown by the Dutch.

The stimulatory effects of chewing coca leaves was well known at the start of the nineteenth century, and this use increased during the century. However, the number of people exposed to coca increased during the century when extracts from the leaves started in be used in drinks. For example, in 1863 the French chemist Angelo Mariani (1834–1914) marketed a blend of Bordeaux wine and coca called Vin Mariani that was reported to have analgesic, anesthetic, and carminative properties. When it was approved by the Vatican it gained increased popularity in Europe. A similar cocktail along with a caffeine extract from *Cola nitida* (cola nuts) and *Turnera diffusa* (damiana), were the key ingredients in a product developed by the American chemist John Pemberton (1831–88) in the late 1860s as a patented medicine. He then realized it was easier to sell it as an alcoholic drink which he called Pemberton's French Wine Coca. In 1885, parts of America passed legislation that inhibited some forms of alcohol, so Pemberton removed the alcohol and developed a non-alcoholic version of the drink. This was sold in 1888 to Asa Griggs Candler (1851–1929) who then went on to develop it as Coca-Cola, the product that launched the Coca-Cola company (Pendergrast 2013: 30).

Albert Niemann (1834–61) isolated the stimulating compound cocaine from the coca leaves in 1860 the year before he died (Streatfield 2002: 60). Other scientists, including those working for the chemical company Merck in Darmstadt, Germany, improved the methods that he had used to extract cocaine from the leaves with the result that the compound was available in greater amounts for study by researchers and to incorporate into products. For example, in 1884 Sigmund Freud (1856–1939) experimented on the euphoric and stimulatory properties of cocaine and coca leaves, and between 1883 and 1887 he wrote several articles on the medical uses of cocaine, including its use as an anti-depressant. His student, Karl Koller (1857–1944) saw the potential of cocaine as a local anesthetic, initially for eye surgery. These anesthetic findings supported the earlier work of Karl von Schroff (1802–87), a professor of pathology and pharmacology in Vienna. By the end of the nineteenth century the use of coca and cocaine had spread, and practitioners in many countries were worried about their addictive properties so that countries started to enact legislation to restrict the sale of cocaine. However, by then more surgeons started to see the potential of cocaine as an anesthetic in surgical practises (Markel 2011). The international demand for cocaine increased and the companies had to face how they were going to be supplied with legally grown good-quality material: a matter that was to become a real issue in the twentieth century.

Hydrastis canadensis *and Hydrastine*

Goldenseal, *Hydrastis canadensis* (Ranunculaceae), from North America had many traditional uses in multi-purpose remedies, and in nineteenth-century America it gained some popularity for treating eye problems. In 1851 the alkaloid hydrastine was isolated and physicians started to use it for gastric disorders (Durand 1851).

Papaver somniferum, *Opium, Morphine, and Heroin*

The poppy (*Papaver somniferum,* Papaveraceae) is from Eurasia and the dried latex from poppy seed heads, often referred to as opium, has played a long and varied role in different aspects of health care. However, the nineteenth century witnessed great changes in the use and societal impact of this medicinal plant. The trade and recreational use of opium increased, and by the 1830s the trade of opium from India being sold to China by the East Indian Company was a major source of tax revenue to the British Exchequer. The Chinese authorities, wanting to suppress this trade by seizing consignments of opium entering China, triggered the First Opium War (1839–42), and then when matters were not resolved, the Second Opium War (1856–60). This resulted in the defeat of the Chinese and the legalization and expansion of the opium trade in China (Lovell 2011). The importation of opium to China contributed to high levels of addiction, and, when Chinese workers went to America to help build the railways they took their addiction with them. The problems associated with the use and supply of opium then spread to communities in the United States. Meanwhile, in Europe significant advances were being made in the understanding of opium's chemistry. The French pharmacists Louis (Charles) Derosne (1780–1846) and Armand Sequin (1767–1835) isolated a crystalline substance from opium but they failed to characterize the structure of the compound. It was the German pharmacist Friedrich Serurner (1783–1841) who in 1804 isolated and characterized the alkaloid that he called morphium (morphine), although its structure was not fully confirmed until 1923. As is the case now, this drug was used to treat pain and fueled the scientific interest in the use of opiate-type alkaloids to treat pain (Schmitz 1985). Further work on opium was undertaken by the French chemist Pierre Jean Robiquest (1780–1840), who in 1832 identified another alkaloid from opium which he called codeine (Porter, R. 1999: 334). Codeine was found to be very effective in treating coughs. The interest in the chemistry of morphine continued, and the British chemist Charles Alder Wright (1844–94), while looking for a drug that would be less addictive than morphine, synthesized diamorphine, which became known as heroin, in 1874. This was then developed as a drug in 1898 for use as an analgesic and cough suppressant by the German company Bayer. The highly addictive properties of heroin were realized only later.

Physostigma venenosum *and Eserine (Physostigmine)*

Thomas Fraser (1841–1920) was considered to be one of the "big five" medicinal scientists in Edinburgh in the nineteenth century. He was a professor of pharmacology at the University of Edinburgh and had a fascination with medicinal plants. One of his major studies was on *Physostigma venenosum* (Leguminosae), a liane from tropical Africa. Seeds from the plant are called Calabar beans and were used medicinally as well as a "truth drug" in Africa. As a truth drug, a decoction of the seeds was given to a person accused of a crime. If the accused vomited and survived, the accused was proclaimed

FIGURE 5.4 *Papaver somniferum*. Credit: Wellcome Collection (https://wellcomecollection.org/works/pnu6j3n2).

innocent; death confirmed guilt (Dowling 2001: 134). In the 1870s, Fraser isolated and studied the toxic alkaloid eserine (now known as physostigmine) from the Calabar seeds. There were some discussions as to whether or not eserine could be used for the execution of prisoners in Britain, but this was not pursued. Fraser was able to show that physostigmine could be used as an antidote to exposure to plants containing the toxin atropine. For a time physostigmine was used in eye surgery and for the treatment of glaucoma.

Salix alba *and Salicylic Acid*

Extracts from the bark of the willow trees have a long history of being used to treat pain and fevers. Research on the chemistry of *Salix alba* by the German chemist Johann Andreas Buchner (1783–1852) resulted in the isolation in 1828 of a compound that he called salicina (salicin) (Buchner 1828). The Italian chemist Raffaele Piria (1814–65), while working in Paris, converted salicin into salicylic acid in 1838 and the compound started to be used to treat rheumatism. At the time it was not known that salicylic acid had previously been isolated in 1834 from meadowsweet (*Filipendula ulmaria*—which was previously known as *Spiraea ulmaria*) by a Swiss pharmacist Johann Pagenstecher (1783–1856). It was in 1839 that the German chemist Karl Jacob Löwig (1803–90) realized that the compound isolated from meadowsweet was the same as the salicylic acid from willow bark (Jeffreys 2005: 40, 46). Although salicylic acid had a range of beneficial properties, it also had side effects. Other teams in Europe were looking at how to modify salicylic acid and decrease the side effects and, in the process, synthesized acetylsalicylic acid. Acetylsalicylic acid was then developed by the company Bayer into the pain-relieving drug which, in 1899, they named aspirin. The name "a-spir-in" was derived from the acetylation (a) of salicylic acid, the then current name of the meadowsweet, *Spirea ulmaria*, (-spir), followed by "–in" added to make it easy to say (Jeffreys 2005: 73).

Strychnos *and Strychnine and Curare*

Strychnos nux-vomica (Loganiaceae) is a tree from India with a long history of medicinal uses, but it is also known as a toxic plant used to kill rats and was applied to the tips of arrows. A related species, *Strychnos ignatii* from the Philippines, was used as a nerve tonic as well as a fish toxin and also to treat the tips of arrows. Both species attracted the attention of chemists in France in their search for toxic compounds. *Strychnos ignatii*, also known as Saint-Ignatius' bean, was studied by Caventou and Pelletier who in 1818 isolated the active toxin strychnine (Pelletier and Caventou 1820). They then isolated another alkaloid, a derivative of strychnine called brucine in 1819. Brucine is not as poisonous as strychnine, but it contributes to the toxicity of the plant. François Magendie was professor of anatomy at the College de France and worked on nerve physiology. He was interested in arrow toxins and researched the pharmacology of strychnine and brucine (Rilliet and Barthez 1839: 79). Toward the end of the nineteenth century, strychnine was popularly used at low doses as a performance enhancer for athletes and as a recreational stimulant. However, it could cause convulsions and was therefore withdrawn.

Strychnos toxifera was one of the plants collected by the explorer Alexander von Humboldt (1769–1859) when he traveled in parts of South America in the early 1800s.

FIGURE 5.5 *Strychnos nux-vomica*. Credit: Wellcome Collection (https://wellcomecollection.org/works/s3rawysv/images?id=dsddfju6).

He recorded the use of the resin (curare) from this and other plants as hunting-arrow poisons as well as for the treatment of stomach disorders. He also showed that it was stable when kept in a bamboo tube and gave it the name "tubo curare." The English naturalist and explorer Charles Waterson (1782–1865), while traveling to British Guiana in the 1830s, came across curare which he called *wouralia* and brought some back to Britain. The potential of the resin to be used in anesthesia was realized by the English physiologist and surgeon Sir Benjamin Collins Brodie (1783–1862). However, the active alkaloids in curare, such as tubocurarine, were not discovered until the twentieth century along with their use as muscle relaxants (Lee 2005).

ADVANCES IN MEDICINE THAT IMPACTED ON THE USE OF HERBALS

By the first half of the nineteenth century, the bases of modern pharmacy had been established in Europe. More diseases were identified by symptom, and attacking the symptom was seen as being the key role of many practicing physicians and surgeons by whom herbals were used. However, this was still the age of bloodletting, especially for those in society who had the funds to pay (Greenstone 2010). Physicians would also usually prescribe the use of powerful purgatives and cathartics in an attempt to match the power of the disease. They often considered their treatments to be effective because the symptoms decreased even when the disease was not cured. Maybe this is because it is difficult to sustain a fever or have a high pulse rate when blood pressure decreases due to lack of blood! Patients were often dosed purgative herbal mixtures or mercurous chloride (known as mercury chloride or calomel) which causes sickness and severe diarrhea. At this time the toxicity of mercury was not understood.

A major advance in nineteenth-century medicine was the increased knowledge of the role bacteria play in many diseases. This advance was facilitated by improvements in the lenses for microscopes. Scientists were able to detect bacteria in samples from patients and then link their presence to many infectious diseases. In 1861, the year that the American Civil War began, the Hungarian physician Ignaz Semmelweis (1818–65) published his research on the transmissible nature of childbed fever (Porter, R. 1999: 369). His theories were at first denied by physicians as they would not believe that their unwashed hands could transfer disease from corpses or dying patients to healthy women. This work, along with the pioneering work of Robert Koch (1843–1910) on anthrax, of Joseph Lister (1827–1912) on the use of carbolic acid to sterilize instruments, and of Louis Pasteur (1822–85) on the role of bacteria in different diseases and the importance of the pasteurization of wine and milk, contributed to the "germ theory" of diseases (Brock 1999). The discoveries of Koch, Lister, and Pasteur added proof to the existence and disease-causing abilities of micro-organisms such as bacteria and triggered an increased interest in the search for plants and compounds that could kill these bacteria. Lister's work inspired St. Louis-based American doctor Joseph Lawrence (1836–1909) to develop an alcohol-based formula for a surgical antiseptic that could kill bacteria. This included compounds that had recently been characterized from plants, such as eucalyptol (from species of *Eucalyptus*), menthol (from species of *Mentha*), methyl salicylate (from willow bark), and thymol (from thyme, *Thymus vulgaris*). Lawrence named his antiseptic "Listerine" in honor of Lister. This product was licensed as a surgical antiseptic in the United States in 1881 and was then developed for oral health care in 1895 (Taylor, R.B. 2017: 129).

In 1879, as part of the studies on bacteria, *Bacterium coli* was discovered and was renamed *Escherichia* after its discoverer, Theodor Escherich (1857–1911), in 1919 (Hacker and Blum-Oehler 2007). It became an example of an easily grown bacterium "safe" for laboratory use. Other plants traditionally used to treat inflammation were tested for their antibacterial properties against this bacterium, including members of the Labiatae family such as species of *Lavandula* (lavender), and there followed an increased use of distilled extracts from these plants in hospitals.

IMPACT OF AMERICAN HERBS AND HERBALISTS ON MEDICINE IN BRITAIN AND EUROPE

The nineteenth century saw an increase in the use of North American herbs in Britain and Europe (Marco 2003), for example, plants such as *Polygala senega* (Seneca snake root) to induce vomiting and for getting rid of worms, *Sassafras albidum* (sassafras) to treat fevers and wounds, *Spigelia marilandica* (pink root) to purge people of worms, and *Veratrum viride* (American false hellebore) to induce vomiting (Hooper 1820: 534). The fact that all these species were used as emetics is in line with the trend of the time to purge patients of their ailments. By contrast, *Asclepias tuberosa*, a species used by native Americans to treat respiratory problems, was added to the prescriptions to treat pleurisy in 1861. This was based on the work done by Constantine Rafinesque (1783–1840), eccentric professor of botany at Transylvania University, Kentucky, who published a *Medical Flora: or, Manual of the Medical Botany of the United States of North America* (Rafinesque 1828). The addition of American herbs to some of the mainstream medicines in Britain distinguished British herbal practice from that on the Continent. On the other hand, in America the professional physicians who initially were influenced by the medicinal plants from Europe started to evaluate more of their own plants for medicinal properties (Marco 2003).

American herbalists also impacted the practice of herbal medicine in Britain. One of the key figures in America was Samuel Thomson (1769–1843) who practiced as an herbalist in New Hampshire. He began a form of herbalism that became known as Thomsonianism. He outlined his thoughts in his publication, *New Guide to Health or, botanic family physician* (1822). He wanted to "raise the inward heat," "overpower the cold," and promote "free perspiration." His concept was that loss of internal heat caused a disease, and if this heat was replaced by a "hot" herb, patients would regain their health. He suggested this be done in a domestic setting via the use of steam baths along with a small range of herbs, which included America plants such as Capsicum annuum, Cayenne pepper (Coffin 1845). His products became popular in America and were also promoted in Britain by his student Albert Isaiah Coffin (1791–1866) and others including John Skelton (1805–80) (Denham 2013: 9).

The herbalist John Skelton based many of his treatments on the Thomsonian principles (Denham 2013: 65). An overview of his treatment for respiratory diseases, as outlined in his *Family Medical Adviser* (1852), provides an insight into how he approached treatments. For example, he advocated the use of "heat," but he also gave prominence to the importance of treating inflammation which he considered more than just heat. He was also an advocate of the use of indigenous British herbs alongside those from North America or the tropics (Denham 2013: 16). He published a textbook, *Science and Practice of Herbal Medicine* (1870), for use by students of herbal medicine, setting out his views. Many of his treatments combined herbal medicine with advances being taught by the medical professionals in Britain. Thus, he combined the traditional with the new. However, an issue with many of his treatments was that they were considered by professional physicians to promote the Thomsonian principles rather than modern medicine (Miley and Pickstone 1987). This is most likely due to the fact that Skelton was not afraid to voice concerns about some modern practices. For example, he and co-workers believed that the methods being adopted by medical professionals working in cities and hospitals were killing rather than healing their patients, especially when they were prescribing mercury, opium, alcohol, and bleeding.

In Britain, there was much debate on the part of surgeons and physicians and by others in positions of authority about the value of herbalists. Medical practitioners in Britain, as in most parts of Europe, were medically trained through attending formal university courses. Many did not consider that herbalists received any appropriate formal training for treating patients. In Britain the rules for the prescription (but not the practice) of medicine had already been tightened through the 1815 Apothecaries Act, which specified that apothecaries needed to have a license from the Society of Apothecaries if they were to prescribe medicine (Porter, R. 1999: 316). This meant that those prescribing had to study anatomy, botany, physics, chemistry, and the materia medica. At this time the medical profession was dominated by physicians (as then defined), with surgeons and apothecaries considered as lower orders within the profession. The practitioner now campaigned for statutory regulation of those who could practice medicine so that only those with a formal training as physician or surgeon could practice. This brought the disciplines together, as physicians received training in surgery and could qualify as a general physician and as a surgeon. The need for formal qualifications was so that patients could distinguish between qualified and unqualified practitioners. This would exclude herbalists and others considered by the qualified practitioners to be quacks (Brown 2007). The British Medical Association was established in 1856. This led to the passing of the Medical Act in 1858 and the setting up of the General Medical Council, which required surgeons and physicians in England and Wales to be registered before they could formally practice medicine (Porter, R. 1999: 355). The number of registered physicians and surgeons in England rose from 14,415 in 1861 to 22,698 (of whom 212 were female) in 1901 (Marsh 2016).

At the same time, the serious herbalists wanted to have their trade recognized, and in 1864 they created the National Institute of Medical Herbalists, which provided codes of practice for their members. However, registration often occurred at a regional level and there is no known record of the number of herbalists. It is of interest that Skelton qualified as a registered medical practitioner in 1863 and was then registered with the National Institute of Medical Herbalists in 1864. He spent most of his life as an herbalist treating working-class patients in the manufacturing districts of northern England (Denham 2013: 37). Because of the tightening of the practice of herbalists and the lowering of their status in Britain, many left for North America, where they qualified in botanical or homeopathic medicine. Homeopathy as a form of medicine was initiated by the medically trained Samuel Hahnemann (1755–1833) who promoted the use of small doses of medication and the use of what was termed "similar." For example, the use of hot compresses to treat burns or substances like cinchona that would induce fevers in healthy people for the treatment of those suffering from fevers. He was banned from practicing in Germany, as he questioned the evidence behind many of the new medicines being produced, and so moved to Paris (Porter, R. 1999: 391). He gained an international following among those worried by the increased use of chemical drugs; courses in homeopathic medicine started to be established, especially in North America.

AN OVERVIEW OF THE NEW PHARMACOPEIAS

Advances in the knowledge about the chemistry of plants and the production of plant-derived products by the pharmaceutical industry started to be reflected in the new type of pharmacopeia published in the nineteenth century. These pharmacopeias moved away from the classical Latin-based materia medica that in Europe were based on the first-century Dioscoridean tradition, which provided general information about substances that could be used in healing. Those had had an emphasis on the description of the approved plants. In the new pharmacopeias the plant descriptions were included but often in less detail, and, over time, the number of plants covered decreased. However, they contained more information about the preparation of herbs as drugs (here the term drug relates to plants and other natural substances as well as isolated or synthetically produced chemicals or minerals that are used as medicines). They started to include information about the chemistry of the plants along with information about toxins and how to detect them, as well as more information about how to prepare the drugs and the different drugs that could be prepared from one species of plant.

In 1812, a new pharmacopeia was produced in Austria for use in the Hapsburg Empire, the *Pharmacopoeia Austriaca*. This was a small booklet of 156 pages consisting of detailed instructions of how to collect and process the plant parts and covered 224 substances approved for use in 309 recipes. The information was collated by eight practitioners. Although it was regularly updated during the century to reflect the advances in pharmacy, it was not until the 1855 revised version produced in Latin by the Ministry of Internal Affairs that it became commonly used. The updated version covered 867 substances. Within this version, each part of Austria indicated which substances the apothecaries were to have in stock. It also started to contain information about adulterants (Kletter 2015). In Austria the pharmacist played an important role in making sure that the pharmacopeia reflected the drugs being prescribed, so that new drugs were included and older ones that were infrequently used dropped.

After the Compromise of 1867 between Austria and Hungary, Hungary published its own pharmacopeia (bilingual in Latin and Hungarian). From 1871 it was just published in Hungarian (*Pharmacopoeia Hungarica*). Because Croatia and Slavonia had a political union with Hungary, they also adopted the *Pharmacopoeia Hungarica* of 1871 (Ganzinger 1961). Many countries adopted or used similar pharmacopeias produced by others or prepared them for specific groups of people. For example, the first official pharmacopeia of the Ottoman Empire was published in 1844 under the title *Pharmacopée Militaire Ottomane* (*Pharmacopoeia Castrensis Otomana*). It was written in French and was based on the 1841 version of the *Pharmacopoeia Castrensis Austriaca*. Its main purpose was to provide information about medicines that could be used by the military (Arslan et al. 2017).

The first edition of the Russian Pharmacopeia was published in 1866 and, comprising 906 monographs, was considered a major advance when compared with the previous state pharmacopeias. It covered new drugs from plant-derived compounds such as the alkaloids (like morphine, strychnine, codeine, and atropine). A special feature of the new pharmacopeia was the inclusion of several chemical methods for the analytical control of drugs; in many cases these replaced the older organoleptic methods that relied on evaluating the taste, color, and odor of drugs (Babayan et al. 1978). The first edition of the Japanese Pharmacopeia was published in 1886 (Labbé 2002: 313).

In 1818, the French had produced their first national pharmacopeia (*Pharmacopée Française: Codex Medicamentarius, sive Pharmacopoeia Gallica*), which was still in Latin. This pharmacopeia also provided guidelines for the scope of drugs that could be prescribed in those countries still ruled by or closely linked to France (Bonnemain n.d.). It was published in Latin so that the professional medical workers, wanting to exclude what they considered as the untrained "quacks," would benefit and have access to the drugs. However, many professional workers did not have enough Latin to be able to use the pharmacopeia, and a French version was produced in 1837. In this book 818 herbs were covered and the appendices described how to prepare some of the recently discovered plant-derived drugs such as morphine and emetine.

The Americas also started to publish their pharmacopeias in the nineteenth century. The first edition of the *United States Pharmacopeia* was compiled in 1820. This contained 221 monographs and covered plants used as tonics, laxatives, diuretics, and flavorings. It was updated every ten years. However, there appeared a complete revision in 1890 to reflect advances in plant-derived drugs, general advances in medicinal chemistry, and the need to set clear quality standards for the materials being used in the preparation of drugs (Tirumalai and Long n.d.). In 1846 México published its first pharmacopeia, which contained information on 450 plant-derived products (Schifter and Aceves n.d.). In Argentina the Spanish and then the French pharmacopeias were used in the nineteenth century, but in 1898 the Argentinians published their own pharmacopeia (Viglione n.d.).

The first edition of the Romanian Pharmacopeia was published in 1862. It differed from many in that the contents were written in two columns, one in the local Romanian language and the other in Latin. It comprised 301 monographs, of which 207 were devoted to herbal drugs. It contained many examples of local plants, thus maintaining a close link to their herbal tradition. For example, it covered the therapeutic uses of *Matricaria chamomilla* (chamomile), *Hypericum perforatum* (St John's wort), *Artemisia absinthium* (wormwood), *Arnica montana* (arnica), and *Chelidonium majus* (celandine) (Soroceanu n.d.).

By the end of the eighteenth century the Spanish had already started to refine the legal framework for the management of medicines, and they produced their third pharmacopeia, *Pharmacopoeia Hispana* in 1803. However, it contained many formulae of dubious health value, and the Pharmacopeia went through many changes during the 1800s (González n.d.). In Portugal two pharmacopeias were published in the nineteenth century, the *Codigo Pharmaceutico Lusitano* (1st edition, 1835) and the *Pharmacopoêa Portugueza* (1876). In 1935 the first more complete *Farmacopeia Portuguesa* was published (Pita n.d.). The first German pharmacopeia, *Pharmacopoea Germanica*, was published in 1872 and reflected the different pharmacopeias produced by the different states that were incorporated into modern Germany (Friedrich n.d.).

In the early part of the nineteenth century the Swiss adopted pharmacopeias from neighboring countries. However, between 1840 and 1860 some cantons started to produce their own. The first Swiss one was produced in 1865, although it was not widely used until the end of the 1870s. This was because physicians and pharmacists expressed concerns about the quality of information in the early versions, asserting that it was not addressing the quality of medicines being traded. There was a need to improve the quality of drugs sold in Switzerland, and the pharmacists worked together to produce a much-improved version in 1880 and that became widely used (Ledermann n.d.).

In Britain, the first edition of the *British Pharmacopoeia* was published in 1864. It brought together the pharmacopeias of London, Dublin, and Edinburgh (Dunlop and Denston 1958). It included an authoritative list of all approved herbal drugs, with descriptions of their properties, uses, dosages, and tests of purity (Matthews 1962). The *British Pharmacopoeia* was also adopted as the official approved list of drugs for use in most of the countries that were part of the British Empire, including territories like India (Anderson 2010). Although in India physicians and pharmacists were supposed to use drugs covered by the *British Pharmacopoeia*, they also prescribed locally available medicines which included many adulterated drugs (Bhattacharya 2016). This was because the official drugs were often not available or very costly. The traditional Indian forms of medicine such as Ayurveda and Unani were often used within the family home and products were purchased at local bazaars. Within India the use of these local medicines was suppressed by the British authorities (Mukharji 2009). In 1813, the British surgeon Whitelaw Ainslie (1767–1837) published *Materia medica of Hindoostan*, which was one of the first books to disclose the study of Indian botanical therapeutics to the English-speaking world. This increased scientific interest in the plants used in the different forms of Indian medicine, especially on the part of Dutch and British scientists. Later, Edward Waring published *Remarks on the uses of some of the bazaar medicines and common medicinal plants of India* (1859). The first edition of this book was in Tamil and English, and provided medical practitioners with information about locally used medicines. This book along with other similar work attracted the attention of the authorities in Britain and resulted in the *British Pharmacopoeia* being published in 1898; an annex that covered more of the Indian plant-derived drugs with the aim of improving the quality of products being sold in India that contained these local plants was issued in 1900 (Berger 2008).

The pharmacopeias produced by Britain in the eighteenth century were influential in spreading information about the advances in medicinal products, especially to parts of Europe and North America (Cowen 2001). This was in part because of their quality. They were produced by experienced physicians and surgeons, and the colleges of medicine, especially in London and Edinburgh, who had an oversight of their quality. The compilers were also recording many of the new discoveries in medicine. Their influence declined by the mid-nineteenth century as many countries were producing their own pharmacopeias. The diversity of plants covered in these new pharmacopeias also declined, in part reflecting the move away from herbals. However, because plants were still being used, it was felt that in Britain there was still a need for a botanical materia medica. In 1839, Jonathan Pereira (1804–53), a professor of materia medica at the Pharmaceutical Society of Great Britain, published *Elements of materia medica and therapeutics*. This covered many of the plants entering Britain from the Americas and was considered to be one of the most scholarly books of the time. Because of the advances in pharmaceutical science, other books were written to support the *British Pharmacopoeia*. For example, in 1859 Peter Squire (1798–1884) published his *Companion to the British Pharmacopoeia*, while another significant publication was the *Extra Pharmacopoeia* published by William Martindale (1840–1902), pharmacist, and William Wynn Westcott (1848–1925), coroner and theosophist, in 1883. This work was later to become *Martindale: The Complete Drug Reference*, still being published today.

The Danes published in Latin a series of pharmacopeias during the nineteenth century for those treating people in hospitals as well as the poor and the military. These local pharmacopeias were usually of a high quality and used by local physicians. The Danes published their first pharmacopeia in Danish in 1893. This reflected the fact that by the end of the nineteenth century they considered that Latin was no longer the language of modern medicine (Jensen and Schaeffer 1960).

Before the nineteenth century China had been at the forefront of bringing together and documenting knowledge about medicines, especially medicinal plants, in their equivalent of a materia medica. However, from the Opium Wars of 1839–42, 1856–60, to the mid-1950s, the traditional forms of medicine in China declined as Western forms of medicine increased. The Chinese also used translations of Western pharmacopeias (Xi 2014). For most of the nineteenth century the use of traditional Chinese medicine in China was forced underground by the local warlords.

CONCLUSION

The nineteenth century witnessed major changes in our understanding of diseases as well as the drugs available for use to cure those diseases. The century ended with the development of the first of two synthesized drugs that are linked to plant-derived research: antipyrine in 1883 and aspirin in 1897. This was a century of change and advances: a decrease in herbals and an increase in plant-derived compounds or synthesized chemicals being used to treat people. With these and other improvements, survival rates improved and mortality statistics slowly declined. In Britain, annual death rates fell from 21.6 per thousand in 1841 to 14.6 in 1901. However, the main factors thought to be behind this decline were largely improved public hygiene and better nutrition associated with higher earnings—that is, prevention rather than cure. Nevertheless, the benefits of the advances of the nineteenth century in plant-based drugs and the understanding of diseases on treating and curing patients started to be realized in the twentieth century.

CHAPTER SIX

Plants in Culture

ROY VICKERY

Throughout the nineteenth century people left the countryside; villages became immersed in towns, and many towns expanded to become cities. However, agriculture still employed many workers, and most city-dwellers were only one generation away from their rural roots. The rural poor gathered wild plants for food, fuel, and medicine. The aristocracy and middle classes used hot-house flowers and exotic fruit to display their wealth and their gardeners' expertise. This was the pattern across the Westernized world including its colonies, so here the most powerful imperial nation of the century, what is now the United Kingdom and Republic of Ireland, is examined in some detail, with striking examples from other countries. No doubt parallel patterns are to be found in the history of other such territories.

THE RURAL CYCLE

Large numbers of farm laborers and the paternalistic concerns of many employers ensured the survival of a number of long-established customs throughout much of the British Isles. January 6, Epiphany or Old Christmas Day, was a favorite time for the wassailing of apple trees. This practice, first recorded in 1585 (Hutton 1996: 46), was believed to ensure, or at least encourage, a good crop in the following autumn. Orchards were visited, songs were sung, tree trunks beaten, their roots splashed with cider, and cider-soaked toast placed in their branches, the ceremony concluding with guns being fired through the branches, and, often, a "general hullabaloo" (Simpson and Roud 2000: 380). In the Belgian province of Hainaut, until about 1840, torch-bearing children participated in a custom known as Scouvian, touring orchards on the first Sunday in Lent while shouting: "Bear apples, bear pears, / And cherries all black. To Scouvian!" in the belief that this deterred witches and encouraged an abundant crop (de Cleene and Lejeune 2003: 1, 97). Also practiced at this time of year was burning the ashen faggot—a bundle of young ash (*Fraxinus excelsior*) saplings. According to a nineteenth-century writer this was "an ancient ceremony transmitted to us from the Scandinavians who at their feast of Yuul were accustomed to kindle huge bonfires in honour of Thor" (Poole 1877: 6), but the first record of the custom dates from no earlier than 1795 (Hutton 1996: 40). However, it gained popularity during the early part of the nineteenth century when some towns in Somerset held Ashen Faggot Balls. That held in Taunton on January 2, 1826 was "most respectably attended by the principal families of the town and neighbourhood," and it was still held twenty years later, but by then was losing its appeal (Legg 1986: 54).

Elsewhere the burning of faggots was less decorous:

> When placed on the fire, fun and jollity commence—master and servant are now all at equal footing. Sports begin—jumping in sacks, diving in water for apples and many other innocent games engage the attention of the rustics. Every time the bands [which bind the faggot] crack by reason of the heat of the fire, all present are supposed to drink liberally of cider or egg-hot, a mixture of cider, eggs, etc.
>
> (Poole 1877: 6)

Apparently restricted to Somerset and adjacent Devon and Dorset, ashen faggots continue to be burned in a small number of public houses (Vickery 2010: 149).

As spring arrived crops were sown. Local sayings suggested the correct time for such activities, but it is probable that rather than being rigidly adhered to, these provided reassurance. Throughout the growing and harvesting seasons local events, often parish feast-days, provided assurance that things were being done at about the right time; all was well. In some parts of the country it was observed that two crops needed similar weather conditions to thrive. In north Nottinghamshire "A good wheat year, a fine plum year" was a saying which in 1887 was "heard from many persons this year, the crops of both being very good indeed" (*Notes and Queries*, ser. 7, 4: 485, 1887). It was reassuring to know that barley was being sown, as it should, when the blackthorn (*Prunus spinosa*) was as "white as a sheet" (in full bloom) (Johnston 1853: 57). In Huntingdonshire it was thought that broad beans (*Vicia faba*) should be planted on February 14: "On St Valentine's Day, Beans should be in clay" (*Notes and Queries*, ser. 4, 1: 361, 1868).

Good Friday, which can fall on any date between March 20 and April 23, was a favored time for garden work. This was partly because farm laborers had free time

FIGURE 6.1 "Firing at the apple-tree," in Devonshire, etching, *Illustrated London News*, January 11, 1851: 28. Photo by Roy Vickery.

on that day, but there was also a belief that Christ's blood, dripping from the Cross, sanctified soil. In Devon it was thought that parsley (*Petroselinum crispum*) sown on Good Friday would "thrive all the year round" (Amery 1905: 114). Other superstitions associated with parsley included the widespread belief that its seeds visited the devil before they germinated. The number of visits varied in different parts of the country, with seven recorded in Lancashire, Sussex, and the North Riding of Yorkshire, and nine in Herefordshire (Vickery 1995: 274). Transplanting parsley was thought to lead to illness or death; in Warwickshire a family member would die within a year (*Folk-lore* 24: 240, 1913), while in Hampshire "your father will die within a year" (*Notes and Queries*, ser. 4, 11: 341, 1873). Children in Guernsey were told that babies were "dug out of the parsley beds with golden spades" (Stevens Cox 1971: 7). Elsewhere, too, babies were dug up from parsley-beds, but sometimes it was vexatiously added that the boys were dug up from beneath a gooseberry bush (*Ribes uva-crispa*) bush. Beyond the British Isles it is said that the song "There's a maiden of fifteen, Jean, / As innocent as may be / 'Mongst the parsley she was seen, Jean, / Searching for a baby" was sung round the campfires of Napoleon's army (*Notes and Queries*, ser. 4, 9: 35, 1872).

A common belief was that parsley grew best when sown by the dominant partner in a marriage, or where the wife "wore the trousers." This belief was also applied to other culinary herbs. In Buckinghamshire "the wife rules where sage [*Salvia officinalis*] grows vigorously" (Friend 1884: 8), while in Yorkshire, rosemary (*Salvia rosmarinus*) "would not grow in the garden of a house unless the woman is master" (*Notes and Queries*, ser. 5, 11: 18, 1879).

Poultry keepers were wary of bringing various flowers into their homes until their chicks and goslings had safely hatched. Primroses (*Primula vulgaris*), or at least fewer than thirteen primrose flowers, brought indoors would mean bad luck hatching chicks (*Notes and Queries*, 1 ser. 7: 201, 1853), while few goslings would hatch if flowering goat willow (*Salix caprea*) was brought in (Burne 1883: 250). Fishermen thought plants provided indication of when various fish were in season. In Herefordshire, alder (*Alnus glutinosa*), known locally as aul, was observed, and it was said: "When the bud of the aul is as big as a trout's eye / Then that fish is in season in the River Wye" (Britten and Holland 1886: 19). In Guernsey the flowering of foxgloves (*Digitalis purpurea*) indicated the start of the mackerel season; according to a verse in the local patois: "Quand tu vé epani l'claquet / Met tes leines dans ten bate/ Et t'en vâs au macré" (When you see the foxglove blossoming, put your fishing-tackle into your boat, and go off for mackerel; Marquand 1906: 39).

By mid-July, or earlier, the first fruits of the summer were ready for harvesting. In southwest England gooseberries were welcomed by holding feasts, or "revels." At Stoke-sub-Hamdon, Somerset, the villagers climbed the local hill and enjoyed a Gooseberry Feast, described as a "curious old custom" in 1875 (*Pulman's Weekly News*, July 15, 1975), but apparently not surviving until the end of the century. At Drewsteignton, Devon, gooseberry pasties and cream were traditionally eaten at 'Teignton Fair, held at Trinity Tide—the first Sunday after Whit Sunday. Helston, Cornwall, held its Gooseberry Fair on the third Monday in July, while a similar event was held at Hinton St. George, Somerset, on the first Sunday after Old Midsummer's Day, July 5 (Vickery 1995: 154). At Lichfield, Staffordshire, it was reported in 1830:

> Lamb and gooseberry, it is well-known, are customary dishes at Whitsuntide. In this city, the usage seems to be religiously kept up. The number of lambs killed here on

Friday and Saturday, was 252; and many besides were sold by country butchers who attend the market.

(*Lichfield Mercury*, June 4, 1830)

The eighteenth-century popularity of gooseberry shows continued, peaking in the 1850s by which time there were some 720 cultivars of gooseberries and over 170 shows in Britain. Enthusiasm for these was particularly strong among cottage-based handloom weavers. With the development of power-driven looms, the hand-weavers left their cottages and moved to towns where garden space was often unavailable, so gooseberry shows went into decline and few survived the First World War (Smith 1989: 109). Also popular, particularly in industrial areas in north and midland England, were competitions held in public houses at which relatively humble people, such as weavers and miners, and stocking- and lace-makers, displayed their tulips (*Tulipa* × *gesneriana*), auriculas (*Primula auricula*), pinks, and carnations (*Dianthus caryophyllus*). Copper kettles were frequently awarded as prizes at such shows, and such were sometimes hung outside public houses to advertize the shows (Duthie 1988: 27).

From 1819 onwards parish wardens were empowered to set aside land for letting as allotments. It was usual for half the space in an allotment to be devoted to potatoes (*Solanum tuberosum*), but a wide variety of other vegetables cultivated were also grown by some. Typically, these were worked only by men, but if space was available around their cottages the women would cultivate herbs and flowers (Uglow 2004: ch. 18). In mid-July, usually on St. Swithin's Day, July 15, it was hoped that rain would fall, for it was thought necessary for apples to be christened if they were to crop well and keep throughout the winter. Typically, in Surrey:

> Today ... is St Swithin's Day, brilliant, cloudless, hot. But last night, as the soft white mist rose over the meads, an old dame said to me. "We must have some rain tomorrow, sir, to christen the apples". "What is that?" said I. "Why they always say, if there's no rain on St Swithin's Day, the apples don't get christened, and then they comes to nothing".

(*Notes and Queries*, ser. 6, 4: 67, 1881)

Elsewhere it was hoped that the apple christening would take place on St. Peter's Day (June 29), in Herefordshire, for example, or on St. James' Day in Somerset, Wiltshire, and Cornwall—apparently the feast of St. James the Less (May 1) in the first two counties, and the feast of St. James the Greater (July 25) in the third (Vickery 1995: 9).

The most important crops, however, were cereals—wheat, barley, and oats. Throughout corn-growing areas of the British Isles it was usual for the final patch of corn to be cut with some ceremony and for it to be plaited or twisted into an ornamental shape. In Wales:

> When the corn harvest was reaped one tuft was left uncut in the centre of the last field reaped. When all the reapers had gathered together, each with his sickle, the head-servant would kneel before the tuft, divide it into three parts and plait the parts skilfully together in the same way that he would plait a mare's tail securing the plaited tuft a few inches above ground level. The reapers, six, eight or more would then stand at a distance of at least ten yards from the plaited tuft and, in turn, would hurl their sickles at it, the sickles travelling horizontally just above ground level. The intention was, of course, to cut off the plaited tuft. If this were not accomplished by one of the reapers, the head servant would then himself cut the tuft.

(Peate 1971: 177)

Names given to the final sheaf, or ornament made from it, varied, but were usually female. In the Orkneys, it was known as a *bikko* (bitch) and often placed in a prominent position in the stackyard or on a farm building (Marwick 1975: 69). In England the names baby, dolly, or maiden were frequently used, while in Wales *caseg fedi* or *caseg ben fedi* (harvest mare) were used, and *cailleach* (old woman or hag) was commonly used in Ireland and in Gaelic-speaking parts of Scotland (Vickery 2010: 16). Elsewhere in Europe the use of female names was less general, and although names such as the corn-cow have been recorded, the final sheaf could be spoken of as one of a range of mostly domesticated animals, including bull, cat, goat, pig, and wolf (Peate 1971: 183). In his *Golden Bough*, first published in 1890, James Frazer (1854–1941) devoted much space to what he termed "the Corn Mother" and her representation as a human or an animal (Frazer 1922: chs. 45–50). He believed that the ancestry of nineteenth-century corn ornaments could be traced back to ancient Greek gods; although his theories were accepted and of great influence for some fifty years, they have been rejected by most recent scholars.

Wild Foods

In addition to what they grew in their gardens, country people gathered wild plants for food. Some of these, such as wild strawberries (*Fragaria vesca*) and wood sorrel (*Oxalis acetosella*), were eaten as they were found; others, such as mushrooms (*Agaricus campestris*) and blackberries (brambles, *Rubus fruticosus*) were taken home and cooked, producing welcome seasonal variations to monotonous diets. Bread-and-cheese, a name given to wood sorrel, was shared by a number of other plants, none of which tasted the least like bread or cheese, but which were nibbled by country children. These included common bent (*Agrostis capillaris*), unripe seeds of common mallow (*Malva sylvestris*), and young leaves of hawthorn (*Crataegus monogyna*), which might be added to savory suet roly-poly puddings and dumplings (Vickery 1995). The root tubers of pignut (*Conopodium majus*) were widely dug up, cleaned, and eaten by children, though it was sometimes thought that eating too many of these could lead to an infestation of lice (Braithwaite 2012: 26).

The period from late winter until early summer was particularly difficult for housewives. Few vegetables were available in gardens—mainly root vegetables and various brassicas, all of which were past their prime. Young leaves of common nettle (*Urtica dioica*) were collected at this time for use as a green vegetable or in the preparation of soups. It was claimed that eating three meals of nettles each spring would cleanse the blood and ensure good health throughout the coming year, a belief which undoubtedly could be proved to be efficacious, since they would have provided useful vitamins and minerals at a time when these were in short supply. Walter Scott (1771–1832) in his novel *Rob Roy* (1817) mentioned that at Loch Leven, Perthshire, nettles were grown under glass to provide an early spring vegetable (Vickery 2008: 2). Later in the year nettle beer was an important summer drink. Its alcohol content was low, but probably sufficient to kill any germs present in poor-quality water.

Some wild fruits were collected on a commercial scale, providing rural communities with additional income. Bilberries (*Vaccinium myrtillus*), which abound on heathland, were collected by women and children. In County Tipperary, Ireland, it was claimed in the 1930s that "whorts" (whortleberries, another name for bilberries) had been collected from the slopes of Galtee (or Galty) Mountains for about two hundred years: "pickers, consisting sometimes of the whole family, leave early in the morning for the wood … they often sing songs to lighten their work." The fruit was sent to England, where it was said to be used in the production of dyes and the "cheaper kinds of jam" (Irish Folklore Commission's Schools Scheme 1937–9 [National Folklore Collection, University College Dublin] vol. 575: 382).

In their *Dictionary of English Plant-names*, Britten and Holland (1886) mentioned "French hales" as a fruit sold in Barnstaple, Devon, market "for a half-penny a bunch." They identified these as "*Sorbus intermedia*" (194), but, at that time, the taxonomy of whitebeams was poorly understood, and the fruits were undoubtedly those of Devon whitebeam (*S.* [*Karpatiosorbus*] *devoniensis*). "Hales" is a variant of haws—hawthorn fruits—so French hales can be interpreted as "foreign haws." Also in Devon "there was a large trade done in elderberries [*Sambucus nigra*], to be made into wine. I have known troops of boys, girls and women sent scouring the country for these berries" (Thornton 1907: 172). In Surrey warmed, homemade elderberry wine was villagers' favorite Christmas tipple: "friends and neighbours who called to wish the compliments of the season would be sure to be offered a glass" (Alexander 2006: 113). On Cranborne Chase, Dorset, nutting expeditions were "great events," the collected hazel (*Corylus avellana*) nuts being sold to dealers who sold them on as dessert nuts, or, more usually, for use in dyeing. Families expected to pay a year's rent, if not more, with the proceeds (Dacombe 1951: 44).

Other wild plants collected included the flowers of cowslip (*Primula veris*), coltsfoot (*Tussilago farfara*), and dandelion (*Taraxacum officinale*), which were sold to urban wine makers. According to a letter written in 1885:

> The Moreton Pinkney [Northamptonshire] women and girls go into Buckinghamshire, and as far as Stowe, to gather cowslips and the next day take them to Daventry for sale, sometimes passing through Preston by 4 o'clock in the morning. They sell them picked at from 8d to 10d a gallon for cowslip wine. Now all the cottage windows are full of coltsfoot laid out to dry also for wine; later they will get dandelions.
>
> (quoted in *Plant Life* 74: 44, 2016)

Cowslip Sunday was celebrated in Lambley, Nottinghamshire, but it appears that many who attended the event were little interested in the flowers. In the 1860s there were reports of rowdy crowds and police being needed "to sort things out." This event died out sometime before the 1970s, to be revived in 2009. Current publicity claims it is an ancient festival, but it seems to have started only after the opening of the village railway station in 1852 (Vickery 2013: 52).

Other Wild Resources

An example of the many ways in which a common plant might be used in the mid-nineteenth century is provided by a children's writer's account of broom (*Cytisus scoparius*) in 1863:

> Common broom is a winter-food for sheep; a good thatch for cottages and ricks; good litter for animal; the flower-buds, just before they become yellow, are pickled like capers; the branches are said to be capable of being tanned into leather [i.e. tanning leather], or woven into coarse cloth, and, when tender are mixed with hops

for brewing ... the seeds are emetic; the tender tops boiled in water, form a decoction or soup, good for dropsy.

(Hill 1863: 102)

In the nineteenth century there was a huge demand for wild songbirds which were kept in cages in urban areas. This, in turn, led to a demand for food for these captives, so that early in the 1850s it was calculated that there were a thousand itinerant sellers of groundsel (*Senecio vulgaris*) in London, who between them were calculated to sell 5,616,000 bunches of the plant each year, producing a total annual income of £11,700 (over £1.6 million/US$2.2 million in 2021 money) (Mayhew 1851: 155). Chickweed (*Stellaria media*) was similarly hawked around the capital.

Another plant, also now considered little better than a weed, but which was valued then, was bracken (*Pteridium aquilinum*). In Worcestershire an important local industry was burning bracken for the sake of the alkali in the ashes, which, when made into round balls the size of a cricket ball, "Ess-balls," were sold in the neighboring Black Country for use as soap. Elsewhere bracken was used for bedding for cattle, and, in Ireland, to provide "soft bedding" for humans (Amphlett and Rea 1909: 426). Bracken thatch was said to last for up to thirty years, "at least on the sunny side of the house" (Wyse Jackson 2014: 647).

THE HUMAN LIFECYCLE

While the nature of illness remained poorly understood, a range of plants was used to provide protection and to heal. Nineteenth-century folklore collectors tended to concentrate on the odd, possibly magical, and supposedly "ancient," rather than practices and remedies which might have been of some practical use. Even if things did not work, such remedies provided carers with the opportunity to try and assist, rather than stand by, feeling helpless. Although it is sometimes assumed that people sought out rarities to provide cures, it seems that seeking of rarities was extremely uncommon in domestic "folk" medicine. The plants most frequently used were wayside plants and weeds, or common vegetables, though paid physicians were unlikely to advertise that their cures consisted of old cabbage leaves, for example, and stressed the exotic nature of ingredients they used.

Birth and Childhood

In Victorian England villagers widely believed that an abundance of hazel nuts in the autumn would lead to many births the following year (Vickery 2010: 119). Methods of preventing conception or producing abortion are probably under-recorded. The urban poor in Salford, now in Greater Manchester, valued pennyroyal (*Mentha pulegium*) as an abortifacient (Roberts 1971: 100). In Cornwall, where pennyroyal was known as *organ*, women enjoyed organ tea: "women in the last century were usually well versed in herbal lore ... so were these modestly held gatherings to drink tea really a way of trying to control the size of one's family?"[1]

Newborn babies, particularly before baptism, needed careful protection to ensure that no harm befell them. Children who were sick were suspected of being changelings, sickly fairy babies left in place of healthy human ones. In County Leitrim, Ireland

(Duncan 1896: 163), a test for a changeling was to put three drops of foxglove juice on the baby's tongue and three in each ear: "then place it at the door of the house on a shovel (on which it should be held by someone) and swing it out of doors on the shovel three times, saying: 'If you're a fairy away with you!'" If it was a fairy it would die, if it was not it would "surely mend." The use of foxglove juice in such a way might well prove fatal to a malnourished child, and, in at least one case, from Caernarfonshire (Sir Gaemarfon), Wales, in 1857, a suspected changeling was killed by foxglove poisoning (Harte 2004: 119; Sikes 1880: 57). It seems probable that the use of foxglove juice (and other ordeals to which supposed changelings were subjected) was an acceptable form of infanticide.

Children were subject to many ailments during their early years. Infants who had ruptures were passed through the split stem of a young tree which would then be tightly bound up so that as the two pieces grew together the rupture would heal. Ash (*Fraxinus excelsior*) was usually selected for this, and it is apparent that such cures involved the entire village community, not least the land-owner on whose property the tree stood, who would ensure that it remained unharmed during the child's lifetime (Latham 1878: 40). However, there is one record, from Surrey, of a holly tree (*Ilex aquifolium*) being similarly used: "I don't know that it was any good, but the old women at that time used to hold with it" (*Notes and Queries*, ser. 6, 11: 46, 1885).

Whooping cough was an affliction which was particularly feared and attracted many cures. In West Sussex the gall caused by the wasp *Diplolepis rosae* on dog rose (*Rosa canina*), hung around the neck, was said to be "the finest thing for whooping cough" (Latham 1878: 38). Elsewhere these galls were used to treat, or prevent, toothache in Shropshire (Burne 1883: 194), rheumatism in Wiltshire (Dartnell and Goddard 1894: 23), insomnia in Wales (Trevelyan 1909: 98), and in Northamptonshire they were "placed by boys in their coat cuffs, as a charm to prevent flogging" (Baker 1854: 78). Richard Owen (1804–92), first director of London's Natural History Museum, lived from 1852 at the edge of Richmond Park, Surrey, near what was said to be a shrew-ash—an ash tree in which a common shrew (*Sorex araneus*) had been immured, and which was valued for curing "'bewitched' infants, or ... young children afflicted with whooping cough, decline, and other ailments" (Ffennell 1898: 334). However, it appears that such ash trees were primarily used to cure lameness in cattle, which was said to result "from the running of a shrew-mouse over the part affected" (White 1822: 1:344).

A whooping cough cure which fascinated folklorists was crawling, or being passed, under a bramble (*Rubus fruticosus*) arch. Usually, a number of rituals had to be carefully followed if the cure was to be efficacious; presumably any failure could be explained by not adhering to the required procedure. A Herefordshire woman cured her granddaughter by passing her under a bramble bush rooted at both ends, for nine mornings (Leather 1912: 82):

> She got better ... The bramble bush was supposed to be quite effectual in a recent case at Weobley, but the child was passed under nine times only, and an offering of bread and butter was placed beneath the bramble arch. She left her cough there with the bread and butter.

Similar, but usually less complex, practices were used to produce a wide range of cures: blackheads (comedones) in Cornwall (Deane and Shaw 1975: 135), boils in Dorset (Udall 1922: 255), hernia in Somerset (Palmer 1976: 114), and in Wales to treat rickets and children who were "slow to walk" (Trevelyan 1909: 320).

Courtship and Marriage

Victorian girls seem to have spent a great deal of time trying to determine when and whom they would marry. One widely used form of love divination involved the sowing of hemp (*Cannabis sativa*) seeds. In Guernsey in the 1880s:

> A vision of your future husband can ... be obtained by the sowing of hemp-seed. The young maiden must scatter on the ground some hemp-seed, saying: Hemp-seed I sow, hemp-seed grow, For my true love to come and mow.
>
> Having done this she must immediately run into the house to prevent her legs being cut off by the reaper's sickle, and looking back she will see the long-for lover mowing the hemp, which has grown so rapidly, and so mysteriously.
>
> (Stevens Cox 1971: 10)

This practice, which is a pivotal action in Thomas Hardy's 1887 novel, *The Woodlanders*, had a few variations, the main one being the choice of date. Midsummer's Eve (June 20) appears to have been most popular (Wright 1940: 12), but other nights which were considered appropriate included St. Valentine's Eve (February 13) in Derbyshire and Devon (Wright 1938: 152), St. Mark's Eve (April 24) in parts of East Anglia (Wright 1938: 187), and St. Martin's Eve (November 10) in Norfolk (Baker 1974: 6).

Another plant used to gain information on marriage prospects was yarrow (*Achillea millefolium*), which was employed in a bewildering variety of ways. In Devon, yarrow plucked from a young man's grave at midnight, would, if the right words were recited, and the right rituals followed, ensure that information about a girl's lover would be gained during the night: "He come to her in the night, and he saith, 'I be thee own true love Jan.' And first her married Jan Scoble, and then her married Jan Wakeham" (Morris 1925: 306).

In County Donegal, Ireland, it was recorded in the 1880s that young people who slept with a "square sod in which grows yarrow" under their pillow on May Eve (April 30) would dream of their sweethearts. This custom was said to have been introduced to Ireland by settlers from Scotland (Kinahan 1884: 90). In Suffolk yarrow could be used to find out if a lover was faithful; a leaf was placed in the nose, with the intention of making it bleed, while the following lines were recited:

> Green 'arrow, green 'arrow, you wears a white blow,
> If my love love me, my nose will bleed now;
> If my love don't love me, it on't bleed a drop;
> If my love do love ne, 'twill bleed every drop.
>
> (Britten 1878: 156)

In Dorset, ash leaves that lacked the usual terminal leaflet and thus had an even number of leaflets were used to find out whom a girl would marry. In 1831, the ash leaf was frequently invoked by young girls as a matrimonial oracle in the following way: the girl who wished to divine her future lover or husband to be plucked an even ash leaf, and holding it in her hand, said, "The even ash is in my hand / The first I meet shall be my man." Then putting it into her glove, added, "The even ash leaf in my glove, / The first I meet shall be my love." And lastly, into her bosom, said, "The even ash leaf in my bosom, / The first I meet shall be my husband"—soon after which the future lover or husband would be sure to make his appearance (Udal 1922: 254).

The uncertain joys of courtship might have been complicated by that much-mentioned nineteenth-century invention, the Language of Flowers. This "language" was first made

FIGURE 6.2 "Love makes sweet use of the language of flowers." Postcard printed in France, sent by Harold Carver, "on active service," to his wife in Green End, Hertfordshire, England, September 1919. Photo by Roy Vickery.

popular by the publication in Paris in 1819 of *Le Langage des Fleurs*, by Charlotte de Latour, likely the nom de plume of a Mme Louise Cortambert; it was swiftly followed by a number of rival publications in France and elsewhere (Goody 1993: 235). In 1839, Catherine Harbeson Waterman (1812–97) in her *Flora's Lexicon*, published in Philadelphia, claimed that the Language had "recently attracted so much attention, that an acquaintance with it seems to be deemed, if not an essential part of polite education, at least a graceful accomplishment" (Goody 1993: 264).

Different flowers signified different sentiments, which could be conveyed by giving bouquets containing these flowers. However, as early as the 1820s, different meanings were ascribed to the same plant: acacia (likely false acacia, *Robinia pseudoacacia*) could mean "platonic love" or "anxiety" (Goody 1993: 237). Subsequent volumes derived from de Latour's work continued to give varying meanings to individual flowers, due to mistranslation of the French text and other adaptations to take into account other traditions; a bouquet could be interpreted differently depending on which book one consulted; if this means of communication was attempted, it was perhaps advisable to give the appropriate volume with the flowers the appropriate (Vickery 1995: 210; 2010: 133). Indeed, when the main character in an "erotic tale" set in Paris in the 1870s attracted the unwanted attention of a young ballet-girl, he, "never liking to treat any woman scornfully, ... sent her a huge basket of flowers and a book explaining their meaning. She understood that [his] love was elsewhere" (Wilde 1999: 161). Passion flower (*Passiflora caerulea*) was said by de Latour and some English writers to mean "faith," but other British writers ascribed "religious superstition" to the flower, while *The Catholic Language of Flowers*, published in London in 1861, associated it with

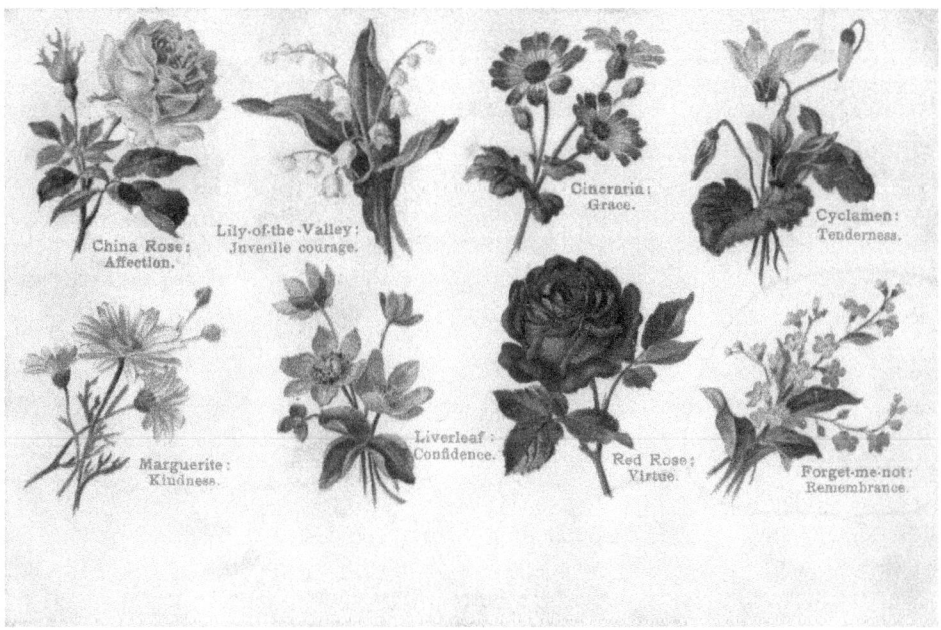

FIGURE 6.3 "Language of Flowers, China Rose for Affection etc." Postcard, produced by H.M. & Co., London, printed in Germany, mailed in Glasgow, Scotland, August 1904. Photo by Roy Vickery.

"meditation" (Vickery 1995: 211). It was also necessary for recipients to have an acute awareness of color: was that poppy red, meaning consolation, or scarlet, meaning fantastic extravagance (Vickery 2010: 128)?

There is little evidence to suggest that the Language was actually much used. One example of its use is in Dublin in May 1868, when Rose La Touche (1848–75) tried to communicate with John Ruskin (1819–1900) by sending him a package of flowers. He was able to work out what was meant by "a rose enfolded in *erbe della Madonna*" (ivy-leaved toadflax, *Cymbalaria muralis*, which he regarded as his personal emblem), but he was unable to interpret the remainder (Mahood 2008: 164). Conversely, according to a writer describing country-house weekends before the First World War, "Saturdays-to-Mondays were a heaven-sent opportunity for sex ... a note left (in collusion with the maid) beside the bottled water on the bedside table, or the placing of a code-laden flower outside a bedroom door ensured that extra-marital sex went on with ease" (Nicolson 2006: 84–5). One wonders what evidence exists to support this assertion.

An extensive series of postcards produced early in the twentieth century depicted couples in appropriate poses, a flower, and a caption, such as "heliotrope [*Heliotropium arborescens*]—devotion" or "chrysanthemum (yellow) [*Chrysanthemum × morifolium*]—slighted love," judging by such surviving cards, provided a popular method of communication between semi-literate couples. Two plants featured on hundreds, if not thousands, of postcard and greetings card designs throughout Europe and North America were forget-me-not (*Myosotis sylvatica* cvs) and four-leaved clover (*Trifolium* cultivars), both of which were apparently appreciated by a wide range of correspondents.

Although the Victorian era is often considered to be one in which marriage was important, evidence suggests otherwise. Henry Mayhew (1812–87), in his mid-century study of poverty in London, estimated that not more 10 percent of costermonger couples were married (Quennell 1984: 57). No doubt country squires and small-town factory owners enforced greater respectability. The wedding bouquet which most brides carried did not appear to have been an essential accessory until the eighteenth century, and it seems that originally a simple posy of whatever was available was considered appropriate. Later there developed a vogue for white flowers, stimulated by Victorian ideas about the purity of the bride, and reinforced by the invention of photography; white flowers photographed well, whereas darker flowers tended to appear as a dark amorphous mass. It is often said that myrtle (*Myrtus communis*) became popular after it was used in Queen Victoria's wedding in 1840 (Baker 1974: 28). However, contemporary reports suggest otherwise, as it seems that the first British royal wedding at which myrtle was used was that of Princess Augusta of Cambridge (1822–1916) to Frederick, Grand Duke of Mecklenburg-Strelitz (1819–1904), in 1843. Augusta's dress, head-dress, and wedding cake were all decorated with myrtle, it "being introduced ... as the emblematic flower of Germany" (*The Times*, June 29, 1843). Another Victorian innovation was the use of orange-blossom by brides, including Victoria (1819–1901) herself. According to some versions of the Language of Flowers such blossoms symbolized chastity (Anon. n.d.: 41); alternatively, as oranges bear both flowers and fruit at the same time, they are said to be associated with fertility (de Cleene and Lejeune 2003: 1:202). Also popularized by Victoria, and frequently used as a wedding flower, was lucky white heather (*Calluna* and *Erica* spp.). The belief that white heather is "lucky" appears to

FIGURE 6.4 "Language of Flowers, Heliotrope for Devotion." "Trichromatic postcard by J. Welch & Sons, Portsmouth, printed at our works in Belgium," sent to Mr. G. Balcombe, Icklesham, Sussex, England, August 1906. Photo by Roy Vickery.

FIGURE 6.5 "Love's Symbols, Forget-me-not." Postcard, printer unknown, mailed in New Hampshire, USA, September 1911. Photo by Roy Vickery.

be a Highland, or perhaps German, belief which was made more widely known by the Queen. In September 1855 she recorded:

> Our dear Victoria [Princess Royal, 1840–1901] was this day engaged to Prince Frederick William of Prussia [1831–88] ... during our ride up Craig-na-Ban this afternoon he picked a piece of white heather (the emblem of "good luck"), which he gave to her; this enabled him to make an allusion to his hopes and wishes.
>
> (Victoria 1868: 154)

In 1862 when Queen Victoria met Princess Alexandra of Denmark (1844–1925), future wife of the Prince of Wales (1841–1910), "her heart warmed towards the exquisite creature ... [she] presented her with a sprig of white heather picked by the Prince at Balmoral, saying she hoped it would bring her luck" (Battiscombe 1969: 36).

For the wealthy, the use of plant materials at weddings could become so extravagant that it might be difficult to see what was happening in the church or to converse across the wedding-breakfast table. In 1890 a London florist's catalog announced:

Church Decorations for Weddings

> We devote special attention to this particular branch of our business, always having at hand a large stock of graceful Palms and Foliage Plants of all sizes specially grown for this purpose up to twenty feet high ... We shall be pleased at any time to meet our patrons at the Church, to receive their commands, make suggestions, and furnish estimates. The price of a Wedding Decoration may range from Five to Fifty Guineas.
>
> (Davies 2000: 144)

FIGURE 6.6 "For Luck! A Sprig o' Real White Heather." Postcard, produced by The Cynicus Publishing Co. Ltd, Tayport, Fife, Scotland, mailed in Elgin, Scotland, 1911. Photo by Roy Vickery.

The accompanying illustrations, showing decorations for a wedding at St. Margaret's Westminster, depict a jungle-like profusion of palms, precariously placed pots of lilies and other flowering plants.

If a marriage offended local opinion there were ways of expressing disapproval. A husband suspected of wife-beating might have chaff, the byproduct of threshing, scattered on his doorstep (*Notes and Queries*, 5 ser. 6: 463, 1876). Early in the nineteenth century in Cheshire, May Birchers would visit houses after dark on May Eve and leave a sprig of a tree on the doorsteps. In the morning householders would discover what people thought of them: "nut for a slut, pear if you're fair, plum if you're glum, bramble if you ramble, alder (pronounced 'owler') for a scowler, gorse [*Ulex europaeus*] for whores" (Simpson 1976: 148).

Death and Mourning

Death omens abounded (see also parsley above). The landed gentry could expect a death in the family if a limb fell from a certain tree, or group of trees, and, it seems, most country-dwellers believed that an apple tree producing flowers and fruit at the same time foretold death (Vickery 2010: 138):

> Remarking an apple blossom a few days ago, month of November, on one of my trees I pointed it out as a curiosity to a Dorset labourer. "Ah! Sir," he said "tis lucky no women folk be here to see that"; and upon my asking the reason he replied "Because they's sure to think somebody were a-going to die".
>
> (*Notes and Queries* ser. 4, 408, 1872)

During the latter half of the century the use of flowers at funerals could be extravagant, but, in country areas, simple, locally grown flowers were still used. In Shropshire, wallflowers (*Erysimum cheiri*), roses, and other blooms were arranged inside the coffins of the poor (Burne 1883: 299). In chapter 6 of her 1847 novel *Wuthering Heights*, Emily Brontë (1818–48) described the mourning practices of Yorkshire gentry: when Catherine Linton died early on Monday morning her body was placed in an open coffin, strewn with flowers and scented leaves, and left until her funeral on Friday. It is noteworthy that, in both of these cases, scented flowers or leaves are mentioned, possibly hinting at the original purpose of funeral flowers, namely to mask the odor of the decomposing body.

Early in the 1880s the Wesleyan minister Hilderic Friend (1852–1940) noted with approval:

> The pretty custom of sending wreaths for the coffins of deceased friends is also growing, and it is certainly a delicate, expressive, and touching method of paying tribute to their memory. The Queen and Royal Family have set us an example again and again in this matter, and it is an example which we have not been slow to imitate.
>
> (Friend 1884: 8)

By the close of the century the art of making floral tributes for funerals had reached its zenith, and florists were adept at producing elaborate designs, such as pearly gates, hearts, broken columns, books, and empty chairs, in addition to a wide range of wreaths and posies. Even the urban poor had abundant flowers at their funerals. In his 1886–1903 investigation into poverty in London, Charles Booth (1840–1916) mentioned a pious Roman Catholic woman who with her two daughters occupied a first-floor room in one of the capital's poorest streets. Following the death of one of the daughters, their neighbors showed their respect by covering the coffin and almost filling the room with vast quantities of flowers (Fried and Elman 1969: 63). Apparently, fish-sellers and cat-meat dealers were particularly fond of elaborate funerals (Fried and Elman 1969: 247). Presumably this use of flowers was the basis for the belief amongst urban children that it was unlucky to pick up flowers which they found on pavements. Fallen food was acceptable: "Waste not, want not, pick it up and eat it." But picking up a dropped flower would lead to the picking up of fever (Gamble 1979: 94).

The flowers at the funeral of Queen Victoria on February 4, 1901, are estimated to have cost £80,000 (almost £10 million/US$13.7 million in 2021 money), producing a display of unparalleled extravagance: "the Queen Regent of Spain sent a wreath seven feet high; the business firms of Queen Victoria Street presented a Royal Standard five feet by nine composed entirely of violets [*Viola odorata*], geraniums [*Pelargonium* cvs] and mimosa [*Acacia dealbata*], Australia's large wreath was of finest orchids," and so on (*The Graphic*, Supplementary Funeral Number, 9 February 1901; quoted in Jones 1967: 122). However, it appears that at country funerals the use of flowers remained rare. In rural Oxfordshire early in the twentieth century: "There warn't no money for wreaths; the coffin were often bare or, at times, strewn with a few wild flowers from the verge" (Stewart 1987: 12). Similarly, in Aberdeenshire, the flowers at a funeral often comprised merely a wreath prepared by the local minister's wife (Fraser 1973: 164).

THE RITUAL YEAR

During the nineteenth century Christmas became established as the pre-eminent festival, but other Christian festivals, including local feast days associated with saints to whom churches were dedicated, continued to punctuate the year. On Old Christmas Eve (January 6) people visited hawthorns believed to be descendants of the Glastonbury, or Holy, Thorn (*Crataegus monogyna* 'Biflora') to watch them produce flowers. Sometimes these visits were encouraged. People who came from miles around to see a Thorn in the garden of Kingstone Grange in Herefordshire were liberally entertained with cakes and cider (Leather 1912: 17). Sometimes they were unruly events, as was the one near Crewkerne, Somerset, in 1878, when the weather was "unfavourable and the visitors were impatient," so that when the tree failed to bloom a quarrel ensued at which stones were thrown, and the owner of the Thorn pulled it up and took it indoors, "receiving a blow on the head from a stone for his pains" (*Pulman's Weekly News*, January 10, 1978).

The use of flowers in churches is a largely nineteenth-century innovation. In 1874 it was observed that "the decoration of churches is now much more popular than it was a few years ago," the writer noting that his earliest memory of plant material being used to decorate consisted of the grumbling old sexton filling the font at Christmas, "with an armful of holly and ivy [*Hedera helix*] and to stick a few sprigs of laurel [?*Prunus laurocerasus*], holly, or box [*Buxus sempervirens*] in gimlet-holes bored in the tops of old-fashioned high-backed pews" (Burbidge 1874: 199). Twenty years later a Peebles doctor noted:

> Much talk is being occasioned by the introduction of flowers as decorations for the church, an innovation greeted with the usual outburst of narrow-minded, conventional hostility on the part of a few. I am told that "the cloven foot of Episcopacy" is suspected in every petal.
>
> (Crockett 1947: 111)

The use of "palm" on Palm Sunday—the last Sunday in Lent—was generally suppressed from about the mid-sixteenth century until well into the nineteenth century (Smith 1994: 243), but, by the century's end, a number of trees were recorded as being known as palm and displayed as such on the day: European silver-fir (*Abies alba*) in Northern Ireland, hazel in Devon and Somerset, larch (*Larix decidua*) in Donegal, and yew (*Taxus baccata*), which was widely used. No doubt flowers were also used to decorate churches at Easter, but few details of these can be found. Presumably, easily available local flowers, such as primroses and daffodils (*Narcissus pseudonarcissus*), were used in country areas, and "flowers of the chaste Trumpet-lily" (now known as altar, or arum, lily, *Zantedeschia aethiopica*), were considered "unsurpassed for the altar" (Burbidge 1874: 202).

Seven weeks after Easter the church celebrates Pentecost or Whitsunday; until the mid- or late nineteenth century there was a custom of decorating churches with birch branches (*Betula pendula*) stuck into holes in the tops of pews (presumably the same holes as held the few sprigs of evergreens at Christmas). This custom, which has never been satisfactorily explained, seems to have had its stronghold in Shropshire (Burne 1883: 350) and still survives at St John the Baptist church in Frome, Somerset (pers. obs. June 2015). It is perhaps germane to note that Lutheran churches in Berlin are decorated with birch at Pentecost, as, presumably they were in the nineteenth century, but no explanation seems to be known for this practice either (pers. obs., June 2017).

The other festival when churches were abundantly decorated was the Harvest Festival or Thanksgiving, which, each year, marked the completion of a successful harvest. The instigator of this service is reputed to have been the Reverend Robert Stephen Hawker (1804–75), who held the first festival in his church at Morwenstow, Cornwall, in 1843. Similar services were held at Elton, near Peterborough, in the 1850s, the festival apparently spreading rapidly throughout the decade (Hutton 1996: 345). By the early 1870s, the Thanksgiving merited much the same attention as Christmas and Easter, and it was recommended that small sheaves of wheat or oats looked "well and very suggestive laid on the altar-table," while tasteful groups of corn, lilies, brightly colored autumn leaves, and fresh green fern fronds looked "well on the pulpit or reading desk." It was said that "ladies, as a rule do these kinds of ornamentation better than anyone else, and seldom fail to make pleasing arrangements." Yellow flowers were to be avoided, their gaudy appearance not being appropriate to the "peaceful harmony" which should prevail in a sacred building (Burbidge 1874: 204–5).

Besides those marked in the Christian year, a number of other festivals using plant materials continued or came into being in the nineteenth century. In 1883 the Primrose League was formed in memory, and in support of the political ideas of Benjamin Disraeli (1804–81), who served as prime minister in 1868 and 1874–80. Primroses were said to be Disraeli's favorite flower, and the League promoted the wearing of them on the anniversary of his death, April 19 (Vickery 1995: 296). In April 1921, a correspondent to the *Westminster Gazette* complained that because of the gathering of primrose flowers, mainly due to the League's activities, seed production was reduced, leading to the gradual disappearance of primroses from the countryside (Vickery 2006: 10).

In June 1912 Queen Alexandra (1844–1925), widow of Edward VII, organized the first Alexandra Rose Day on which artificial roses were sold, raising over £30,000 (almost £3.5 million/US$4.8 million in 2021 money) in support of the Queen's favorite charities (Rensten 1996). Although silk flowers were used in Britain, mainly London, the Danish-born Queen was inspired by the efforts of a parish priest living near Copenhagen, who sold roses from his garden to support the needy children who shared his house (Anon. 1992). Alexandra Rose Day is believed to be the first "flag day" to be held in the British Isles, but since 1912 such days have proliferated, the most prominent being the Royal British Legion's annual Poppy Appeal in support of ex-service personnel and their families.

Since early in the nineteenth century, if not earlier, the red or corn poppy (*Papaver rhoeas*) has been associated with soldiers killed in battle. Red poppies which followed the ploughing of the field of Waterloo after the Duke of Wellington's victory were said to have sprung from the blood of the fallen soldiers, an idea reinforced by the fact that at that time British soldiers wore red tunics (Thiselton Dyer 1889: 15). Later, in the First World War, during the second battle of Ypres in May 1915, a Canadian doctor, Colonel John McCrae (1872–1918), wrote his now famous poem, "We Shall not Sleep," later called "In Flanders Fields." Published in England in December 1915, the poem caught the eye of Moina Michael (1869–1944), an American teacher; she too wrote a poem, "We Shall Keep the Faith," and started to wear an artificial poppy to "keep faith in those who died" and became "the Poppy Lady." In November 1918 a French woman, Anna Guérin (1878–1961), decided to get such poppies manufactured in France and sell them with profits being used to help people returning to war-devastated areas (see for example Leonard 2015 and Saunders 2014).

Other things were happening elsewhere. In June 1918, when members of Papton Adult School visited the Derbyshire home of the socialist philosopher and activist, Edward Carpenter (1844–1929), he spoke on "War and Reconstruction":

> … literature was distributed along with red poppies. The Millthorpe [where Carpenter lived] poppies were symbols of a grassroots patriotism, for the mighty wartime state had overlooked the wounded men discharged from the forces, and two left-wing organisations had taken on their welfare: the National Association of Discharged Soldiers and Sailors (N.A.D.S.S.) and the National Federation of Discharged and Demobilised Soldiers and Sailors (N.F.D.D.S.).
>
> (Rowbottom 2008: 395)

The British Legion, a charity devoted to the welfare of military veterans, organized its first Poppy Day in 1921, using poppies imported from France. The success of this event, which raised £106,000 (well over £5 million/US$6.9 million in 2021 money), led the Legion to establish its own poppy factory, which opened in June 1922.

Last examples considered here concern plant material used to decorate homes and businesses for Christmas. Mistletoe (*Viscum album*) was in such demand that in the 1870s the owner of Grimsthorpe Park, Lincolnshire, employed an additional fourteen "watchers" to protect her property, prevent damage to mistletoe-bearing trees and disturbance of deer (*Notes and Queries*, ser. 5, 5: 126, 1876). Holly, ivy, and other evergreen shrubs were gathered for decorating homes, but for the upper and middle classes the Christmas tree became the main ornament. The first of such trees in England were those introduced by Hessian soldiers serving in the army of King George III (1738–1820) and by German merchants who had settled in British cities. In 1800 Queen Charlotte (1744–1818) celebrated "with a German fashion," having a tree with lighted tapers, decorations, and presents for children attached to it. Then, in 1841, a tree set up by Queen Victoria and

FIGURE 6.7 "Hearty Christmas Greetings." Postcard, printer unknown, sent to Master Franklyn North, Havant, Hampshire, England, December 23, 1906. Photo by Roy Vickery.

Prince Albert (1819–61) at Windsor was widely illustrated in the press, establishing an enduring tradition (Vickery 2010: 173).

Inevitably it is difficult to ascertain when such practices and beliefs started, or when they died out, and therefore work out what can be considered to be characteristic of the period. Many of the usages mentioned here were established well before 1800, and some have continued, or have been revived in recent years, but, in most cases, it is during the nineteenth century that they were most fully and unselfconsciously observed.

CHAPTER SEVEN

Plants as Natural Ornaments

CLEMENS ALEXANDER WIMMER

In the nineteenth century, the new bourgeois class joined the nobility in influencing the world of plants and gardens. The rise of horticultural societies, great nurseries, horticultural exhibitions, and plant-lovers' associations was in large part associated with the rise of the middle classes. Whereas the aristocrats of the eighteenth century could be said to have promoted gardens largely as works of art, the newly rich were interested in plants whose culture required modern science and technology such as heated glass houses and chemical fertilizers. In the gardens themselves, idealized nature was no longer to be celebrated, as zeal for manifestly interventionist horticulture, exotic botany, and obsession with novel introductions by plant collectors came to dominate. The new collection-based horticulture led to recognizably distinct elements: flower garden, arboretum, pinetum, fernery, heathery, palm house, and so on. The movement was due not only to the wealthy class of industrialists and bankers but also to the lower middle class, while even working men expressed enthusiasm for plants in their little gardens or in window-sill gardening. Never had the relationship between humans and ornamental plants been more intense and widespread.

HORTICULTURAL SOCIETIES AND JOURNALS

The creation of a British horticultural society was first suggested in 1801 by the banker John Wedgwood. In 1804, Sir Joseph Banks (1743–1820) called together six men, among them the Kew gardener William Townsend Aiton (1766–1849) and the botanist Richard Anthony Salisbury (1761–1829), to establish the society. From 1827 the society held fêtes (breakfast parties) at its garden in Chiswick, which from 1833 were associated with modest shows of flowers and vegetables. These soon grew soon into large events patronized by up to fifteen thousand people. In 1858, Albert, Prince Consort, became president, and a new Charter in 1860 altered the Society's name to Royal Horticultural Society (Elliott 2004). The Société d'agriculture et de botanique de Gand (also Koninklijke Maatschappij voor Landbouw en Plantkunde te Gent) was established in 1808 (Van Damme-Sellier 1861: 8) by a circle of gardeners from Ghent, then part of France. Despite its name, which perhaps reflected the French government's dislike of gardens as luxuries, a result of the revolution, it was essentially a horticultural society, as witnessed by its membership, the plants exhibited and the business transacted. Following the British example, the Verein zur Beförderung des Gartenbaues in den königlich preußischen Staaten was established in 1822, from 1910 known as the German Horticultural Society (Deutsche Gartenbaugesellschaft). The founder was no gardener but Prussia's minister for clerical, educational, and medical affairs, Carl Freiherr vom Stein zum Altenstein (1770–1840) who was personally interested in gardens and plants, whereas horticulture was also regarded as an important industry contributing to the economic wealth of the state. From the 1820s onward, a large number of national and regional horticultural societies were established in Europe and elsewhere. Commercial gardeners often formed their own associations to distinguish themselves from the amateur societies.

The publication of horticultural magazines promoted gardening and new plants. Some influential periodicals were issued by large nurseries such as Loddiges' *Botanical Cabinet* (1817) and van Houtte's *Flore des Serres* (1845). The interest was so great that some journals were published weekly, for example, *Allgemeine deutsche Garten-Zeitung* (1823) and *Gardeners' Chronicle* (1841). Trials to test the qualities of plants were one of the main objectives of agricultural and later horticultural societies. So experimental gardens became necessary as in Edinburgh (1816) and London (1818). In 1859, the Horticultural Society of London established a floral committee to examine and decide the merits of new plants introduced from foreign countries or bred by domestic horticulturists. First and Second Class Certificates were created (the latter replaced by the Award of Merit in 1888). Collectors, breeders, and commercial gardeners created their own private trial grounds, too. This is the origin of today's Wisley Gardens, founded by the industrialist and plant lover George Fergusson Wilson (1822–1902) in 1878.

STYLES OF GARDENING

As in the arts, in the nineteenth century there were several styles of planting and gardening. The traditional landscape garden and the separate flower garden continued to co-exist, though refined according to new principles of taste and scientific observation. Due to the immense increase in available plants by introductions and breeding, a stronger differentiation in planting design was necessary to display them. In the collection-based gardens such as the arboretum, rose garden, and rock garden, plants were not chosen to express design ideas, but designs were chosen to show off the plants. John Claudius Loudon (1783–1843), the greatest gardening writer of his time, very clearly characterized the new approach in 1832: "Mere picturesque improvement is not enough in these enlightened times" (Loudon 1832). Nature as the model became obsolete, as science and art came to reign in gardens, with great emphasis on the individual plant. Likewise, the painter Ludwig Richter (1803–84) recalled a paradigm of German romantic circle in Rome in the 1820s as "to go into the characteristic details of the vegetable kingdom"—"We lost our hearts to each blade of grass, to each humble branch …" (Wimmer 2018: 223).

Garden design was no longer a matter of uniting elements of nature in a harmonious *Gesamtkunstwerk* but bringing a vast number of plants and special gardens together to constitute an impressive collection. Contrasts between the elements, in form, color, and structure were no longer abhorred but indeed espoused. This is what, in 1832, Loudon termed the *gardenesque* style of gardening, which was to supplant the traditional picturesque. A garden, according to Loudon, must not imitate a landscape, but demonstrate botanical and horticultural advances. Trees, shrubs, and herbs should therefore be separated to display their individual characters. To avoid giving a false impression of nature, native plants should be excluded from the garden.

FIGURE 7.1 Grouping of fir and birch. From Izabella Czartoryska, *Myśli różne o sposobie zakładania ogrodów*, 1805.

First, Humphry Repton (1752–1818), the most successful landscape gardener of the Georgian period, had promoted plain flower beds in his pleasure grounds, and specialized gardens including rose gardens ("rosaries"), in his park designs like those (1813) for the Neo-Gothic Ashridge House, Hertfordshire, England. On the Continent, landscape architects like Peter Joseph Lenné (1789–1866) and Prince Hermann von Pückler-Muskau (1785–1871) in Germany, and Jean-Pierre Barrillet-Deschamps (1824–73) in France were essentially Loudonian practitioners, as was Andrew Jackson Downing (1815–52) in the United States. Some promoted flower gardens in the landscape style, often called pleasure-grounds (Bosse 1829; Gilpin 1832; Jackson 1816; Pückler 1834). The establishment of special gardens for particular kinds of plants is closely related to the application of different styles within a single site. Repton suggested that his design for Ashridge, with gardens in different styles brought together, was as one might collect paintings by various artists or books by a range of authors. These notions chimed well with the acquisitive urges of the newly rich. Depending on the purpose and situation of the gardens, all styles were permitted (Loudon 1822; Reider 1832b). From today's perspective, the term "eclectic garden" (*eklégein* means "to select" in ancient Greek) perhaps best embraces these phenomena, as was first suggested by a reviewer in *La Revue Française* in 1828: "Today eclecticism is so fashionable, and it is everywhere so well-received, why not introduce it into gardens?" (Wimmer 2018: 222f.).

About the middle of the century, in trying to achieve more naturalness, consideration began to be given to the natural sciences, especially to plant geography. The scientific writings of Goethe (1749–1832) and Humboldt (1769–1859) were influential. In mid-century, German landscape gardeners tried to group plants according to their geographical origins. In the garden, a "Reunion of Science and Art" seemed to be possible. Humboldt (1847: 98) himself referred to garden design: "The physiognomic formation of the plants and their contrasting composition is, however, not merely an object of the study of nature, or an impulse for this; the attention given to the physiognomic beauty is also of great importance for landscape gardening."

Urbanization, alienation, and destruction of the natural environment made compensation for unhealthy living conditions a main task for the garden. Romanticist movements directed against the consequences of industrialization yearned for an idealized, unspoilt nature. John Ruskin (1819–1900) did not value his mother's exotics, carpet bedding, and hot-house plants. He loved wildflowers, old-fashioned flowers, and wild roses. As a nineteen year old, he wrote: "And the flower-garden is as ugly in effect as it is unnatural in feeling; [...] But, in laying out the garden which is to assist the effect of the building, we must observe, and exclusively use, the natural combinations of flowers …. All dahlias, tulips, ranunculi, and, in general, what are called florist's flowers, should be avoided like garlic" (1838: 493f.).

The so-called Pricean Revival or New Picturesque in England and Scotland was epitomized by the reissue of Sir Uvedale Price's *On the Picturesque* in 1842 and in new garden treatises (Kemp 1850; Major 1852; Macintosh 1853; Smith 1852). In Germany Hermann Jäger (1815–1890), Gustav Meyer (1816–77), and Theodor Rümpler (1817–91) were similar advocates for the return to picturesque landscape, while in France, this arose later in the writings of Edouard André (1840–1911).

In 1861 the Arts and Crafts movement was initiated in England, aiming to reassert the importance of craftsmanship and naturalness above Victorian artificiality and factory goods. Forbes Watson (1840–69), a disciple of Ruskin, and William Robinson (1838–1935) called for plants to be used individually, for their own sake, instead of filling

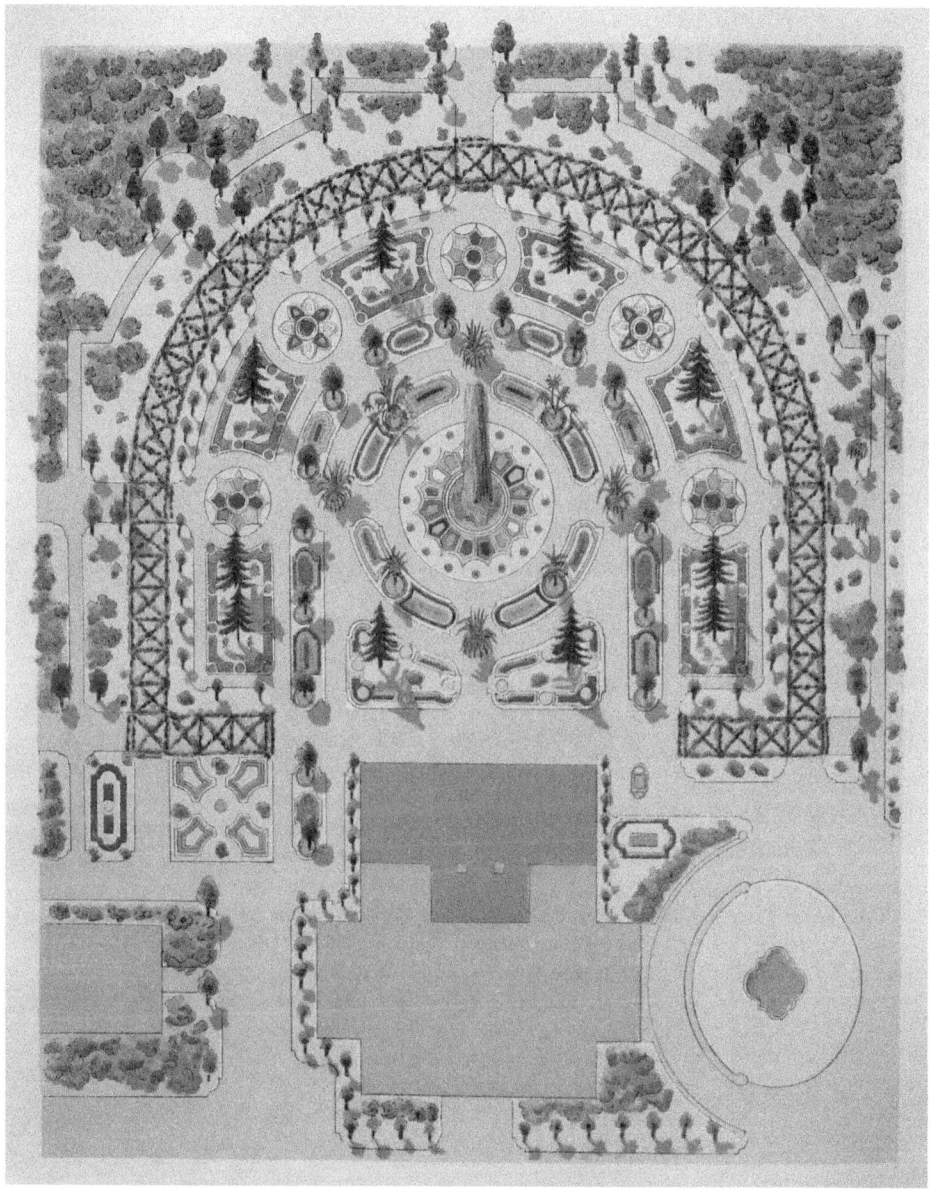

FIGURE 7.2 Arnold von Regel, Design for Hungerberg, 1881. From Регель, *Арнольд: Изящное садоводство и художественные сады: историко-дидактический очеркъ*, 1896.

bedding patterns (Elliott 1986: 151f.). From 1870 onward, William Robinson made "wild gardening" popular. He criticized the geometrical gardening, formal bedding, and bedding plants, but did not condemn exotics as such, only their isolated use, while he pleaded for a combination with native plants appropriate to their habitat.

The 1870s brought a new appreciation of old-fashioned and native flowers in England. Forbes Watson wrote that a wild rose was a work of God, and to prefer a cultivated variety was an impiety (Watson 1872). William Morris (1834–96) also defended wild roses against garden roses (1882: 88). There were similar trends in Germany (Bratránek 1853; Hallier 1891). The German Heimatschutz movement, which began in 1880, advocated the values of native landscape and its characteristic plants. In this context, there was a special appreciation of traditional garden plants such as country-garden flowers, linden, horse-chestnut, yew, lilac, and wild roses. In 1913, Fritz Graf Schwerin (1856–1934) called the love of colorful cultivars appropriate only to beginners and peoples "at a low cultural level." Experienced and educated people would recognize these as being in bad taste and never beautiful (Silva Tarouca 1913: 97–9). Willy Lange (1864–1941) coined the term "Naturgarten," meaning to make a "re-creation of the plant communities based on the model of domestic nature ... but artistically increased" (1907). Lange, in general, rejected zoning and plant-geographical restrictions.

The planting design of the neo-romantic period was characterized by the reunion of the different habitats of the plants in a single whole, the range of specialized gardens becoming restricted. The naturalism in nineteenth-century gardening was not the same as in the eighteenth century. It was a way to the domestication of nature in a time characterized by comfort and industrialization when Nature as "the other" was more remote than ever. As compensation, a convincing, perfect representation of the perceived rough original was aimed at. Planting should look as natural as possible even if gathered in

FIGURE 7.3 Beatrice Parsons, Group of flowering shrubs at South Lodge, Horsham. From J. G. Millais, *Rhododendrons in which is set forth an account of all species of the genus Rhododendron and the various hybrids*, 1917.

a most artificial way, as, for example, in a greenhouse. On the other hand, the "architect's garden" appeared in England as a counter-movement to "wild gardening." In doing so, a garden culture of the "good old days," conceived as homogeneous and original, was to be revived. Architects were leading exponents (Blomfield 1892; Sedding 1891; Triggs 1902). Art in the garden was not considered less acceptable than art in the house (Sedding 1891: 156). *Architekten-Gärten* or *Architekturgärten* were promoted in Germany, too.

Eclectic Planting

The mixed planting scheme for shrubbery and flower beds was abandoned bit by bit after 1800. In 1822, the so-called "select" or "grouped" manner was widely accepted. In the mixed plantation "the object is so to mix the plants, as that every part of the garden may present a gay assemblage of flowers of different colours during the whole season." In the select manner, however, "the object is limited to the cultivation of particular kinds of plants, as, florist's flowers, American plants, annual, bulbs, &c." avoiding mixture (Loudon 1822: 904), but at the same time, contrasts were intensified. Now specimen trees attained central significance in the gardenesque style. The diversity aimed at came from the numerous new species and garden cultivars, led by the purple beech, the cedar, and *Sequoiadendron giganteum*. Such picturesque grouping also included the close coordination of native trees, to lead to oblique stems whether of the same species, or another. Sometimes, several trees were placed in one planting hole.

Where flower gardens were created, a new feature was the placing of showy perennials or flowers such as delphiniums, dahlias, sunflowers, and peonies either as specimens or in small groups of one species in the lawn, in the same way as trees and shrubs had been placed in traditional parks. Flower beds were filled with one species of florist's flower, sometimes bordered by one or two other contrasting species. The outline was usually oval, sometimes circular, wave-, or tongue-shaped (Jackson 1816; Reider 1832b; Thomson 1868). Around 1820, the outline became more sophisticated, taking the shape of rosettes, fans, palm leaves, clover leaves, stars, and heraldic devices. These beds were particularly suitable for the use of historical styles, to accord with the architecture of the house. Some designers also used broderie models from the Baroque period, filling the leafwork forms with flowers instead of gravel.

Around 1850, bedding plants grown in the greenhouse become predominant, whereas perennials and hardy annuals were pushed into the background. For winter effect, potted evergreens such as species of *Aucuba, Ilex, Cotoneaster, Skimmia, Mahonia,* and *Gaultheria* were placed in the beds. In its most elaborate form, the circular bed was divided into several colored sections of bedding plants framed by edgings of other plants without any spaces of lawn or gravel between: the carpet bed. It is not certain where the carpet bed first appeared. Pückler's cornucopia bed, executed in Muskau, Saxony in 1826, and figured idealizedly in his book in 1834, was a precursor. One of the earliest was that in Linton Park, Kent, in 1861, disscussed in *La Belgique Horticole* (1865). In the same volume, carpet beds from Cologne were pictured, but the term "carpet bedding" did not appear in print until 1868 (*Gardeners' Chronicle* 1868: 487). Among the carpet plants were new introductions from tropical America such as *Alternanthera* spp., *Iresine herbstii, Cuphea ignea*, echeverias, begonias, and petunias, together with earlier-introduced plants from elsewhere like *Tanacetum parthenium* 'Aureum,' *Helichrysum petiolare, Lobelia erinus, Lobularia maritima, Perilla frutescens, Coleus scutellarioides, Glandularia* spp., *Ageratum houstonianum, Cerastium tomentosum, Jacobaea maritima, Antennaria dioica*, and variegated pelargoniums. The center was often accentuated by large-leaved "subtropical" plants such as *Canna, Colocasia, Fatsia, Yucca, Dracaena, Arundo donax,* or *Ricinus*. In the 1880s, carpet beds became more and more complex and grew into a three-dimensional form (Elliott 1981; Wimmer 1991).

Romantic and Neo-romantic Planting

Certain romantic features persisted throughout the century. Landscape gardeners, turning away from eclectic and plant-geography principles, sought a subjectively produced naturalness in the garden. Instead of geographical origin, the focus was on the natural grouping of plants. The aesthetic principles of the picturesque developed in the eighteenth century were readapted, even if the range of plants had become much larger, permitting more striking effects (Wimmer 2018). When the formation of picturesque groups, originally developed for woody plants, was adapted to perennials, so-called foliage plants, including ornamental grasses, came into use. For the first time, perennials were regarded in terms of their form (instead of flower color), height, and flowering time. This fashion had its origin in the gardens of Berlin and Potsdam. As early as 1817, royal gardener Ferdinand Fintelmann (1774–1863) planted *Ricinus communis* and *Heracleum* spp. next to *Zea mays* and *Arundo donax* (Fintelmann 1834). From 1828, his colleague, Hermann Sello (1800–76) collected and improved foliage plants including variegated cultivars. He received new introductions like *Monstera deliciosa* directly from the tropics. In the 1850s and 1860s, the "Phyllomania" spread from Germany to France and to England, where beds of foliage plants were called German beds (Barr and Sugden 1862: 93). Grasses were also important in Robinson's campaigns. In his book *Gleanings on French Gardens* (1869), he praised the "subtropical" gardening he saw in Paris. With other foliage plants, grasses served as an easy-care outdoor substitute for greenhouse plants. He used grasses in solitary clumps as well as grouped, with other grasses or with shrubs, perennials, or sunflowers.

Woodland embellishment by underplanting trees with exotic shrubs, especially rhododendrons, was documented in England as early as 1830 (Elliott 1986: 93). According to Jäger (1845), lilies-of-the-valley, snowdrops, Turk's cap lilies, cowslips, anemones, and hepaticas were to be planted as ground cover. In 1858, he also recommended grasses, ferns, woodruff, heathers, blueberries, larkspur, periwinkle, and others in large quantities. Such attempts, however, were rather unusual until Rümpler and Robinson made them popular. Perennials were to be planted at the edge of the shrubbery, so that it had the appearance of being in the wild state (Jäger 1845: 72). Jäger mentioned single plants and groups of three to six specimens for the lawn, including lilies, delphiniums, *Rudbeckia, Phlox,* ferns, ornamental grasses, and foliage plants (1858: 418, 420). Robinson compiled lists of plants suitable for naturalization, and, after visiting the United States in 1870, considered the habitats of perennial plants as models for garden design (1870: 128–31; 1883: 156).

The Danish landscape architect Jens Jensen (1860–1951), who emigrated to Chicago in 1885, was inspired by the American prairies to use perennials on a large scale. This type of perennial planting, known as the Prairie Style, was more suited to large US estates than to European gardens. The term originated in the Midwest of the United States as a designation of a style of architecture (Prairie School). With regard to landscape architecture, it was not initially associated with perennials, but advocated the use of indigenous plants, especially shrubs. In 1901 the landscape architect Chalmers in Wisconsin laid out for the first time a large-scale perennial plantation of *Phlox paniculata, Iris versicolor, and Hibiscus grandiflorus*. According to Wilhelm Miller (1869–1938), the Prairie Style was "based on a geographical, climatic and scenic unity, using three recognized design principles, preserving the natural scenery, restoring the local vegetation, and repeatedly applying a predominant line," which means the horizontal of wide landscape (1915: 5).

Natural-shore planting was promoted early by Jäger (1858: 434–40) who recommended "largest plenty of flowering" not only for ponds and basins, but also for water margins in pleasure-gardens and even in parks. The abundant natural shore vegetation evoked the tropics and "delivered the sweetest pleasure of landscape." Rümpler, Robinson, André, and Jekyll discussed the subject extensively. Wall gardening became a popular genre in England from the 1890s. In her book *Wall and Water Gardens* (1900), Jekyll advocated for the planting of dry stone walls and stairs. Soon afterward, planting in pavement joints became fashionable.

The composition of ornamental lawns was much discussed after Francis Russell, 5th Duke of Bedford (1765–1802) of Woburn Abbey experimented with grasses (Sinclair 1816). By the 1830s it was realized that only a mixture of grass species gave the best results. After a visit to Britain, Pückler (1834) recommended the inclusion of herbaceous plants and even mosses. The young Ruskin spoke of his vision of an alternative flower garden of "the wild violet and pansy sown by chance" on the lawn, "relieved by a few primroses" (1838: 494). The planting of bulbs and wild plants in the lawn, a Scottish practice at the beginning of the century, was recommended by Sir Thomas Dick Lauder (1784–1848), and, around 1850, some English gardeners were experimenting with it. The planting of primroses and violets by John Fleming (*c.* 1822/3–83) in a valley at Cliveden near London, in the 1860s was influential (Elliott 1986: 94). Dutch bulb wholesalers, starting with Jacob Heinrich Krelage (1824–1901) at the Paris World Exhibition of 1889, popularized the mass planting of tulips (Pavord 1999: 258f.).

Reformers rejected the labor-intensive carpet bedding and the uniform mass planting. "The beautiful flower forms are degraded to mere color, which forms a design, without regard for the natural form and beauty of the plant" (Boulger 1878: 4). Besides several attempts to make flower beds less artificial, the old mixed or herbaceous border which had survived in gardens where growing bedding plants was not feasible, was revitalized. Robert Errington (1790–1860), a gardener at Oulton Park, England, worked for the renaissance of the border system from 1845 onward (Elliott 1986: 159).

The Baroque principle of changing color, height, and flowering time was taken up but modified according to a new aesthetic. Instead of single specimens, so-called "troops" or "drifts" of one species or cultivar were planted. According to the first method, the borders comprised short, repetitive modules. Several specimens of a species were planted in geometrical clusters, the length of which was approximately equal to their depth, or at most triple that. Such a plan came from Robinson in 1879, the perennials designated by numbers being grouped in groups of four (*Gardening Illustrated* 1: 362). Later the groups were represented in the planting plans as circles, squares, rectangles, triangles, rhombuses, trapezoids, or octagons. Single specimens could be added to the clusters. This method became common in Germany under the name "rhythmic planting" (Schneider in Silva Tarouca 1910). In 1882, Gertrude Jekyll (1843–1932) devised another method for planting borders. Instead of endless repetition of one and the same planting scheme, she proposed informal, oblong, streamlined areas, the so-called color drifts. Others applied quite amorphous, or rather rounded, planting areas within the border (Foerster 1911; Goos in Silva Tarouca 1910).

The Use of Color

Garden writers of the nineteenth century used established theories of color. Their focus was the thesis that harmony is achieved by contrasting colors. This corresponds to the color doctrines of Isaac Milner (1750–1820; 1803), Goethe (1810), and Michel-Eugène Chevreul (1786–1889; 1839), according to which each color calls for a contrast, containing the other two basic colors (yellow/violet, green/red, blue/orange). These combinations were called "harmonious." Neighboring colors in the color circle (yellow/orange, yellow/green, green/blue, etc.) reveal "characterless" combinations, the other ones (yellow/blue, yellow/red, green/orange, green/violet, etc.) "characteristic" or "expressive" combinations. Non-harmonizing colors could be used by interposing white. Contrasting, especially by flower and foliage colors, was an essential element of nineteenth-century planting design. Color was increasingly considered as more important than form. From 1840, attention was also paid to fall (autumn) colors. The trend toward violent contrasts reached a climax around 1850.

Eduard Petzold (1815–91) wrote the first books devoted exclusively to color in the garden (Petzold 1853). Referring to Goethe, he dealt mainly with flowers, but also recommended contrasting combinations of trees such as gold and silver willows with dark conifers. The purple beech was the most popular tree with colored foliage, being particularly beautiful "standing alone before groups of firs" (Regel 1855: 94). Japanese maples and the purple-leaved plum, *Prunus cerasifera* 'Pissardii', were newly introduced red-foliaged trees. To increase the contrast, white-variegated cultivars, especially *Acer negundo* 'Variegatum', were used. The acme of such were plants with tricolored foliage such as *Fagus sylvatica* 'Purpurea Tricolor' (1873). Further embellishment was furnished by woody plants with bright red autumn colors, especially from North America, e.g. *Quercus rubra* and *Q. coccinea*. For flower beds, red colors were to be used at the margins to contrast with the surrounding lawn.

Later in the century, color theory was much refined (Jäger 1858; Thomson 1868; Wilkinson 1858). In England there was vigorous discussion about its correct application in the garden. Comprehensive tables were compiled listing suitable plants for several shades of color according to flowering times. The Neo-Romanticists redefined the role of color in the garden, no doubt influenced by garden painting (Elgood 1904; Lange 1907). The hitherto customary strong and large-scale contrasts were rejected as unnatural. According to Camillo Schneider (1876–1951), "we want to bring out the plants' shape and color to their best advantage and work with that" (1904: 148). From observations of nature and of cottage gardens (Lange 1907; Lichtwark 1909) the strict application of color theories was questioned (Jekyll 1882, 1908) or even rejected (Foerster 1911; Meyer-Ries 1911; Sedding 1891). Instead of complementary colors, adjacent ones were put together again, bright colors and contrasts being avoided, pastel tones preferred. Gertrude Jekyll combined quite undogmatically, but with great sensitivity, color nuances and had a loathing of indiscriminately colourful mixes (Elliott 1986: 148–51).

TECHNICAL ADVANCES

The relationship of humans to the garden changed fundamentally in the nineteenth century. Labor became more and more expensive, and garden owners had to reduce staff and use their own hands (Wimmer 20121). Loudon wrote that some tasks with conventional equipment would be "insufferably tedious, and others inconveniently cumbersome," which is why machines were invented (1822: 322). First in agriculture, later also in horticulture, tools, and machines that were suitable for less experienced laymen were developed so as to allow them to work more efficiently. The aim was to "work with reduced effort and time, but still with increased performance" (Neubert 1879: 7).

Tools and their Improvement

Tools should be easy to carry and also usable by women and children. Loudon (1824), in his encyclopedic zeal, described and illustrated more than 220 gardening implements, while Boitard (1833) illustrated some nine hundred. Invented by Antoine François Marquis Bertrand-Molleville (1744–1818) in 1807 (Boitard 1833; Pellerin 1996: 99), modern secateurs (*sécateur*) replaced the pruning knife. Most garden tools were made by local craftsmen, often to order, but, gradually, production became mechanized. By the end of the century, the traditional range of gardening tools was extended to technical equipment and machines, for example, rubber hoses, hose-carts, pumps, sprinklers, wheel hoes, seed drills, lawnmowers, atomizers, and poison syringes. The transport of living plants from their native places to European gardens was a serious problem (Desmond 1979). A great advance, however, was the invention of Nathaniel Bagshaw Ward (1791–1868) in 1833, the famous glassed Wardian case.

Watering with watering cans is time consuming, so attempts were made to improve this situation and save personnel costs through the use of rolling water vessels or hoses. In 1837, a "hydraulic engine" (or "water pump") for the garden was introduced. For the first time, this ensured a continuous water jet uninterrupted by the suction phase (Neubert 1857). Garden pumps, however, did not produce the necessary pressure for large-scale irrigation systems. In the 1840s the first public water pipes were built in England and Hamburg; Paris and Berlin followed in the 1850s. These were used for irrigation of public gardens, too. At the end of the century most municipal gardens had a water pipe.

The first cast-iron spindle mower was patented in England in 1830 and produced by Ransomes, Ipswich, from 1832. The intention was that the landed gentry should use it personally, thus accomplishing "an amusing, useful and healthy job." Ransomes sold more than a thousand lawnmowers by 1840, and by 1858 more than seven thousand. In 1868 the American Amariah Millar Hills patented a lawnmower, marketed as the Archimedean Lawn Mower (because of the spiral blades). Due to its small size, it was considerably lighter, easier to handle, and cheaper.

Fertilization and Plant Protection

Agriculturists analyzed the components of soil and the effects of minerals on plant nutrition. According to the doctrine of the chemist Justus von Liebig (1803–73), the "father of the fertilizer industry," nitrogen (N), phosphorus (P), potassium (K), and trace minerals were the essential plant nutrients. Liebig (1840) argued against the prevailing theory of the essential role of humus. Fertilizers were believed to act by breaking down humus, making it easier for plants to absorb. Liebig promoted the idea that chemistry could revolutionize agricultural practise, increasing yields and lowering costs. This was welcomed by gardeners, chemical fertilization becoming a topic in garden literature. Potash from Canada, seabird guano and saltpeter (potassium nitrate) from Peru, including the Atacama Desert extending into Chile, were exported to Europe in large quantities.

FIGURE 7.4 Archimedian Mower by Williams & Cie. From *Annuaire du Commerce*, 1878.

Insects and snails, according to Loudon (1822), were mainly killed by handpicking. With regard to red spider, he recommended fumes of sulfur and ammonia gas. Powdered sulphur was applied to mildew, or a mixture of sulfur and water was applied with a syringe (Nicholson 1884: IV: 366). As downy mildew and other diseases attacked the vineyards of Europe, Pierre-Marie-Alexis Millardet (1838–1902) (1885), a botanist from

Bordeaux, recommended the so-called Bordeaux mixture, which contains copper sulfate. It was soon introduced into horticulture (Thompson 1906: 129f.). In 1897, the chemist Georges Truffaut (1872–1948), from a family of gardeners, founded the Établissements Georges Truffaut in Versailles, which became famous for a large range of articial fertilizers, insecticides, herbicides, mole poisons, and vaporizers. Chemical poisons with names like Thanatophore or Nécrol replaced the traditional organic remedies.

The Glasshouse

Glass played an important role in nineteenth-century gardening. No longer restricted to conservatory windows and hotbeds, it began to cover complete buildings and an increasing part of the surface of gardens. Traditional glasshouses were lean-to houses with the back wall of masonry, but the new glasshouse could be glazed on all sides. The resulting free-standing structure allowed for positioning in the center of a garden instead of at the margin. The use of iron frames, as proposed by Loudon in 1817, admitted more light and enabled more elegant building structures. The front, or all sides, could be in the form of an arch, or the central part of the building could have a cupola or dome. The glazed ridge and furrow roof was another of his inventions. Greenhouses were generally heated by hot-air flues. Moll first described a hot-water heating system built in Holland (1829). Heating by vapor and steam was also developed but was not very successful. Lighting by gas or electricity started much later (Muijzenberg 1980).

Hot-houses were mainly used for forcing fruits and vegetables, for grapes and pineapples, but increasingly for ornamentals, too. The collecting of pelargoniums in the first decades of the nineteenth century made necessary houses providing much light during winter. Later in the century, the cultivation of tropical plants like palms, orchids, and *Victoria amazonica* led to glasshouse types never seen before. Finally, the chrysanthemum mania required special houses.

The most admired of all greenhouses was the Crystal Palace, 360 m (393 yards) long and covering 7.28 ha (about 18 acres), built by Joseph Paxton (1803–65) for the Great Exhibition of 1851. He intended it as a winter park for Londoners. It inspired palmhouse building throughout Europe, notably in Paris, Berlin, Vienna, Frankfurt, Madrid, and Brussels. A wintergarden with some tropical plants became a common part of the bourgeois estate. In order to maximize usage all year round, plants were planted in the ground and arranged in landscape style with picturesque groups amid wandering paths, giving an impression of a natural "jungle." One of the most exotic examples was the wintergarden of the Bavarian king Ludwig II on the roof of the Munich Residenz (1870)—with palms, roses, orange trees, a heated sea, swans, parrots, a grotto, and plants of sheet metal, the whole artificially lit in color (Stephan 2010).

NURSERY FIRMS AND THE INTERNATIONAL TRADE IN ORNAMENTAL PLANTS

The period is characterized by diversification and specialization of trade. Some leading international merchant gardeners like Loddiges, Veitch, van Houtte, Haage & Schmidt, and Lemoine continued to offer the whole range of plant material from seeds and fruit trees to other woody plants, perennials, and greenhouse plants. Increasingly, however, firms specialized with groups such as bog plants, conifers, roses, glasshouse plants, or perennials.

Great Britain

In England, the German Conrad Loddiges (1738–1826) was famous for his Hackney nursery, which received plants directly from America and Siberia. Under his sons William (1776–1849) and George Loddiges (1786–1846) the collection of the company exceeded that of Kew Gardens. Loddiges received from the Horticultural Society the introductions of Wallich (India) and Douglas (North America) to propagate. In the middle of the century James Veitch (1792–1863), Exeter, ran the most important English nursery. "What ever other nurseries had done before them, the family Veitch did more and made it better, whether the spread of new varieties [cultivars], the sending of collectors, the publication of catalogues, the winning of prizes, the training of the gardeners, the volume of collections or the variety of the cultivated terrain, the Veitchs were overly successful in every field" (Willson 1982: 50). From 1840 onward, James Veitch sent out a total of twenty-two of his own plant hunters. Lindley wrote in 1851 that in Exeter "will be found more new and valuable plants than in any place in Europe" except Kew (1851: 579). John Gould Veitch (1839–70) was a collector himself, in 1860 introducing plants from China and Japan. His brother Sir Harry James Veitch (1840–1924) commissioned Wilson to collect in China (Veitch 1906).

The Waterer company in Knap Hill, Surrey, became famous after 1810 for its rhododendrons. Joseph Dalton Hooker of Kew entrusted the company of John Standish (1814–75) and Charles Noble (b. 1817) with his rhododendrons from the Himalaya. From 1838 Standish was crossing the new species with older ones. He also dealt with plants that Fortune brought from China and Japan (Willson 1989). Thomas S. Ware (1824–1901) was the principal specialist for perennials. His firm, founded in 1860 in Tottenham, was surpassed later by "Kelways of Langport," which opened in 1851 by James Kelway (1815–99) and was continued by his son William (1839–1933). Their rich range of perennials, including their own cultivars of delphiniums, gladioli, pyrethrums, and peonies, is documented in Kelway's *Manual of Horticulture*. William Bull (1828–1902) of Chelsea competed with his neighbor Veitch. First he collaborated with Verschaffelt (Belgium), whose introductions he distributed exclusively in Britain, for example introducing "Coleus verschaffeltii" (a form of *C. scutellarioides*) from Java in 1861. Indeed, he specialized in tropical plants, mainly orchids, and opened a large wintergarden for the public (Taylor 2018).

Germany

James Booth (1772–1813) from a Scottish nursery came to Klein Flottbek near Hamburg in 1795 at the invitation of the landowner Caspar von Voght (1752–1839). In 1831 he imported original Douglas (see below) seeds of *Pseudotsuga menziesii* (Douglas fir). Loudon reprinted Booth's catalog in 1838, considering it one of the best in the world: it included 1,100 roses.

Indeed, the period 1870 to 1914 is regarded as the Golden Age of ornamental plant breeding in Germany. In Erfurt, many important nurseries were established, notably F.A. Haage, Ernst Benary, F.C. Heinemann, and Haage & Schmidt, such that the city was the "metropolis of the German seed trade" (Regel 1856). Some produced shrubs, greenhouse plants, and perennials on a huge scale. From 1863 to 1900 Franz Späth (1838–1913) developed his small family-run nursery in Berlin into the world's largest tree nursery. He concentrated on the breeding and introduction of trees (Späth 1930: 1:42).

From 1870, famous nurseries such as Frahm & Timm (1870), Sievers (1883), von Ehren (1885), and the rose nurseries Kordes (1887) and Tantau (1895) dealing exclusively with ornamental trees and shrubs were established in Schleswig-Holstein. Toward the end of the century, companies such as Goos & Koenemann in Niederwalluf (1885), Arends & Pfeifer in Ronsdorf (1888), Otto Mann in Leipzig (1890), Nonne & Hoepker in Ahrensburg (1891), Heinrich Junge in Hameln (1896), Karl Foerster near Berlin (1907), and Kayser & Seibert in Roßdorf near Darmstadt (1909) devoted themselves completely to perennials.

France

The Baumann nursery in Bollweiler (Alsace) was of international repute (Christnacher et al. 2015). Founded as a fruit-tree nursery in 1740 by Johann Baumann (1708–59) and continued by his son Franz Josef (1751–1837), the nursery covered 30 ha (about 74 acres) by 1822. Loudon (1838) ranked Baumann and Booth the most important nurseries in the German-speaking world. Napoleon (1769–1821) promoted fruit-growing and silviculture, requiring in 1802 that at least one tree nursery be in each prefecture, the occupied areas of Germany too having the French-founded *pepinières départementals*. After Napoleon's fall, the French foundations were taken over by the Prussian government, and state nurseries were established in other places, for example in Potsdam in 1823.

The Vilmorin-Andrieux company of Paris, operating since 1780, originated as a seed shop in 1743. During the nineteenth century it was probably the most famous seed company in France, but plants, trees, and shrubs were also sold. André de Vilmorin (1776–1862), owner from 1804, was interested in dendrology and created arboreta in Verrières-le-Buisson and Les Barres near Mogent-sur-Vernisson (Loiret). Early in the 1830s, the Simon family founded a tree nursery in Plantières-lès-Metz, which became Simon-Louis Frèrcs by 1841. In 1872, under Léon Simon (1834–1913), it occupied 40 ha (about 99 acres). The firm surpassed even Booth in its plant range. In 1850 Victor Lemoine (1823–1911) founded a remarkable establishment in Nancy, becoming particularly renowned for breeding *Chaenomeles*, *Clematis*, *Deutzia*, *Hydrangea*, *Philadelphus*, *Spiraea*, *Syringa*, and *Weigela*. Lemoine achieved the greatest success with double-flowering lilacs like *Syringa* 'Madame Lemoine' (1890). In 1856 Carl Huber (1819–1907) from Baden founded Charles Huber frères et Compagnie in Hyères (Côte d'Azur) whose main business was the introduction of palms, eucalyptuses, and other tropical plants (Mabberley 1985b). The demand in this region was great because of the number of villa estates owned by the wealthy, but the spreading of Australian eucalyptus caused ecological problems still evident today.

Netherlands and Belgium

After the French Revolution, bulb cultivation occupied many in the Netherlands with Haarlem being the global center for bulb production. The Golden Age was between 1860 and 1914, when bulb production increased from 1 to 25 million kg (55 million lbs). In 1811 Ernst Heinrich Krelage (1786–1855) from Germany opened a nursery in Haarlem, in 1837 acquiring much of the business of his neighbor George Voorhelm Schneevogt (1775–1850), and becoming the largest florist and bulb breeder there. Through the marketing of the so-called Darwin tulips in 1885 his son Jacob Krelage became the "tulip king" (Dwarswaard and Timmer 2010; Timmer 2009).

Lambert Jacob (1790–1873), the founder of Belgian horticulture, married the daughter of the florist Makoy of Liège in 1810. From then on he styled himself Jacob-Makoy and built up a nursery along English lines. Louis van Houtte (1810–76) made Ghent a continental center for plant trade, founding his company in 1839 after a Brazilian journey of 1834–6, during which he collected orchids. By 1865, it was the largest private plant company in Europe to deal in all kinds of plants.

Jean Linden (1817–98) from Luxembourg discovered his love for tropical plants during three expeditions to South America in 1835–45. On his return, he founded an enterprise specifically for the introduction of new plants, moving from Luxembourg to Brussels in 1855. He introduced thousands of plants, especially orchids and palms, but also in 1858 the colorfully leaved *Begonia rex* from Assam. In 1850 Ambroise Verschaffelt (1825–86) inherited a Ghent nursery from his father: palms and tropical foliage plants, camellias, and rhododendrons were his specialty.

The United States of America

The famous old nurseries of the families Bartram in Philadelphia and Prince in Long Island, New York, continued business to 1823 and 1865 respectively, but Ellwanger & Barry in Rochester, New York, was the pre-eminent US nursery of the century. George Ellwanger (1816–1906) from Württemberg, Germany, emigrated in 1835 and founded the Mount Hope Nursery, Rochester, in 1839 (Taylor 2014: 106–9). In 1838 in Philadelphia, in partnership with Henry Hirst, Henry Augustus Dreer (1818–73), son of a German immigrant, founded a seed business, later extended to bulbs and plants, to be continued by his son William F. Dreer (1849–1918). In 1878 another influential company was established in Little Silver, NJ, by John Thompson Lovett (1852–1922).

PLANT INTRODUCTIONS

After examining botanic garden catalogs, Gregor Kraus (1841–1915; 1894) deemed the period 1772 to 1805 as the era mainly of Cape and the first Australian plant introductions. The next period was dominated by plants largely from temperate East Asia, though other regions, especially western North America for trees and shrubs, tropical America for bedding plants and orchids, and the Himalaya for rhododendrons were also important sources of new plants. Though the motive was often strategic positioning, the focus on plant collecting was new in the expeditions from 1770 to 1830. Some participating botanists or gardeners became famous as plant hunters; many paid for this with their lives. Thousands of new species were collected and named, while their introduction and rapid distribution were facilitated by improved infrastructure.

The quest for new plants became so great that royal families, government organizations, botanical gardens, horticultural societies, commercial firms, and private amateurs competed in sending out collectors. The main interest of garden enthusiasts in the early nineteenth century was glasshouse plants, among them Chinese chrysanthemums, peonies, tender roses, camellias, and azaleas, but later tropical plants became fashionable.

Commercial Firms and Private Patrons

Introductions by European nurseries traditionally came from personal connections with foreign amateurs, botanists, and traders. Loddiges, for example, received Russian plants

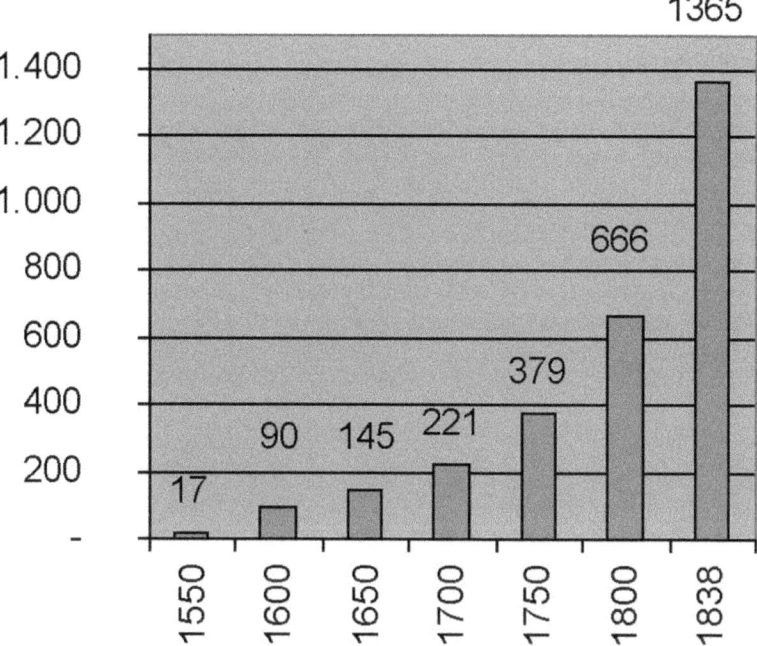

FIGURE 7.5 Numbers of woody plant species introduced to Britain according to Loudon (1838).

from John Busch (c. 1730–95), American ones from William Bartram (1739–1823) and François André Michaux (1770–1855). The nursery of Lee & Kennedy in Hammersmith, London, is said to be the first who paid, at least in a syndicate with other sponsors, its own collectors, though often the collectors themselves were not given due credit (Coats 1969).

The physician Philipp Franz von Siebold (1796–1866) from Würzburg lived in Japan from 1823 until 1830. Up until then, for political reasons, only a very few plants, such as *Magnolia liliiflora* (1790) and *Lonicera japonica* (1805), had been introduced from Japan. In Leiden (Netherlands) in 1842 von Siebold founded a commercial nursery for Japanese plants that he had collected in Japan or ordered thence subsequently. After 1854, Japan allowed the British unfettered access while von Siebold stayed there 1859–62. He is said to have introduced over 1,100 species, important being *Acer palmatum* 'Atropurpureum,' *Berberis thunbergii*, *Cercidiphyllum japonicum*, *Euonymus fortunei* var. *radicans*, *Deutzia scabra*, *Hamamelis japonica*, *Hydrangea paniculata*, *Malus floribunda*, *Paulownia tomentosa*, *Spiraea thunbergii*, *Thujopsis dolabrata*, and *Wisteria floribunda*.

The exploration of the inner parts of Central America led to a hitherto unknown orchid mania (Reinikka 1995; Vadon 2014; Nesbitt this volume). Conrad Loddiges (1738–1826) was the first nurseryman to realize the economic value of orchids, publishing an orchid catalog as early as 1812 and by 1842 offering more than 1,600 species. Orchids as fashionable luxury items made for good business, with van Houtte of Ghent, Veitch of Exeter, and Jean Linden of Brussels becoming the greatest competitors.

In 1849, Veitch commissioned William Lobb (1809–64) to collect seeds in California. In 1853 Lobb learned of a giant tree just discovered in the woods and immediately sought it out, not resting until he could send two seedlings besides seeds to Veitch.

Lindley promptly named the tree, now called *Sequoiadendron giganteum*. In 1849–53 the nurseryman Thomas Lobb (1811–94), William's brother, collected in India and the Himalaya for Veitch, but Veitch's most successful collector was Ernest Henry Wilson (1876–1930), who was first commissioned in 1899 and introduced *Davidia involucrata* (1901) and *Lilium regale* (1903). William Bull (1828–1902) sent out his own collectors to Colombia and Liberia.

Competing collectors often brought back the same species at the same time, so it is difficult to discern the true introducer as with *Parthenocissus tricuspidata* 'Veitchii', which was gathered by both von Siebold and John Gould Veitch (1839–70) in 1860/1. In 1862 *Magnolia stellata* was sent almost simultaneously by Carl Maximowicz (1827–91), George Rogers Hall (1826–99), Richard Oldham (1837–64), and Veitch to Russia, the United States, and England, respectively.

Institutions as Patrons

During the eighteenth century, knowledge and exploitation of plants became "a fundamental instrument of imperial expansion and government control" (Batsaki et al. 2016: 4). For the major colonial powers, Britain and France, gathering new economic plants was as important as the slave trade. "Between 1770 and 1820 Britain alone had 126 official collectors in the field and a network of informal suppliers and transporters" (Batsaki et al. 2016: 4). Botanical gardens represented foreign plants as an expression of imperial power. If situated in the colonies, they served as laboratories for acclimatization and as temporary storage depot in the transport process (Nesbitt this volume).

Joseph Banks, unofficial director of the Royal Botanic Gardens Kew, was instrumental in the transferring and acclimatizing of plants and in establishing key botanical gardens in the British colonies. The appointment of [Sir] William Hooker (1785–1865) as the first real Director of Kew in 1841 saw a steady rise in both the concept of Empire and the imperial role of botanical gardens. Britain was not alone. When the Anglo-Dutch Treaty had come into effect, the Netherlands sent officials to resume control of what became Dutch East Indies (modern Indonesia) in 1816. A year later, Buitenzorg Plantentuin, now Kebun Raya Bogor, Java, was established. The French also founded botanical gardens in their colonies, as did Belgium and Germany, if on a smaller scale. From 1855 to 1892, St. Petersburg Botanical Garden (now officially the Russian Academy of Sciences Vladimir Komarov Botanical Institute's Botanical Garden of Peter the Great), under the direction of Eduard von Regel (1815–92), was an important center of introduction of Eastern plants. Its work, too, had strong imperial connotations.

The Horticultural Society of London concentrated on introductions from China. Banks asked merchants to acquire plants in Guangzhou (Canton) nurseries. So *Paeonia suffruticosa* 'Papaveracea' (1802), many chrysanthemum cultivars and the first tea rose, 'Hume's Blush Tea-Scented China' (1808) reached England. In 1804 Banks sent William Kerr (d. 1814), who introduced *Kerria japonica* 'Pleniflora' (1805), as a collector. Horticultural societies, in general, were not financially able to promote major expeditions. From 1820 to the 1840s, the London Society awarded medals to sea captains who had successfully shipped to England foreign plants such as chrysanthemums, *Wisteria sinensis* (1816) and *Primula sinensis* (1820) (Elliott 2004: 197). The Scottish gardener David Douglas (1799–1834) was sent by the Society to the east coast of the United States in 1823/4. In 1824–7 he traveled under the auspices of the Hudson Bay Company, sailing around Cape Horn. He introduced more plants than anyone before him (Lemmon 1968;

Morwood 1973), the European garden landscape being markedly changed in color and shape by his introductions. He sent 360 kinds of plants to the Horticultural Society, among them *Ribes sanguineum, Lupinus polyphyllus*, and *Pseudosuga menziesii*. In 1836, Carl Theodor Hartweg (1812–71) traveled to Mexico to collect orchids and conifers under the Society's auspices.

The Nankin Treaty of 1842, which ended the Opium War, resulted in greater freedom for British travel in China. The Horticultural Society sent the Scottish gardener Robert Fortune (1812–80) to Hong Kong in 1843–6. He was to seek out tea plants, lotuses, blue peonies, and other mainly hardy plants, but he exceeded all expectations (Watt 2017; Nesbitt this volume). Other expeditions followed in 1848, 1853–65, and 1860–2. He earned the confidence of the gardeners of Shanghai and succeeded in traveling to the interior. Among his shipments to London were 120 new plants including *Anemone hupehensis* var. *japonica, Dicentra spectabilis, Forsythia viridissima*, and *Jasminum nudiflorum*. Due to financial difficulties in the second half of the century, the Society could promote no important exploration, but in 1914 it sponsored the Tibetan journey of Reginald Farrer (1880–1920) and William Purdom (1880–1921), then George Forrest (1873–1932) in 1916 (Elliott 2004: 198–207).

The increasing influence of the British in India was accompanied by further exploration and introduction of the local flora (Desmond 1992). From 1793 to 1842, William Roxburgh (1751–1815) and his successor Nathaniel Wallich (1786–1854), as directors of the botanical garden in Calcutta, sent 190,000 plants to England. A special interest was Himalayan plants which promised to be suitable for the European climate.

From 1860 onward, French missionaries, notably Armand David (1826–1900), Jean Marie Delavay (1834–95), Jean André Soulié (1858–1905), and Paul Guillaume Farges (1844–1912), were important introducers of some well-known trees from distant parts of China, Mongolia, and Tibet. In 1881, Augustine Henry (1857–1930) from Ireland went to China for nineteen years as a physician of the Chinese Imperial Maritime Customs (Pim 1984). Enticed away from Veitch, Wilson undertook four expeditions to Central China, financed by the Arnold Arboretum, Boston, USA, in 1906–11. Wilson, who discovered 4 genera, 521 species, and 356 infraspecific taxa, is called the greatest plant-hunter of all (Sargent 1913–17). We owe to him such ubiquitous shrubs as *Kolkwitzia amabilis* (1901), *Viburnum rhytidophyllum* (1900), *Cotoneaster salicifolius* 'Floccosus' (1908), *Berberis julianae* (1900), *Lonicera ligustrina* var. *yunnanensis* (1908), and *L. pileata* (1900).

BREEDING PROGRAMS

With introductions from China and Japan, countries with an ancient garden tradition, Far Eastern garden cultivars came to Europe. Because of this genetic material, which had been fostered for centuries, the breeding of cultivars in Europe developed more rapidly than ever before. The centers of European breeding were in England (Low & Standish, Barr & Sugden), Belgium (Mackoy, van Houtte, Verschaffelt) and France (Cels, Leroy, Lemoine). Germany, with a few exceptions (dahlias, pelargoniums) became significant only in the late nineteenth century.

Roses and other Shrubs

Among the hardy trees and shrubs, rhododendron, lilac, and conifers showed the strongest increases in cultivars. Rose breeding developed rapidly. In 1810 and 1812 in France the merchant gardeners Jacques-Louis Descemet (1761–1839) and Jean-Pierre Vibert (1777–1866) began to sow thousands of seeds to raise new cultivars (Joyaux 1998; Vibert 1824). In 1815, Vibert opened his own rose nursery, which is said to have produced six hundred new cultivars. Artificial crossings, however, were only gradually developed. With the introduction of *Rosa chinensis* and *R. gigantea* or their hybrids (*R.* × *odorata*) around 1800, significant advances in rose breeding became possible, the combining of these diploid "tea" roses, which have an extended flowering season, with the tetraploid "cabbage" roses of the West yielding the modern roses. The empress Joséphine promoted rose breeding in France and on her death in 1814 there were about 250 rose cultivars in her garden at Malmaison near Paris. Luigi Villoresi (1779–1823), garden director of the Villa Reale di Monza, who worked for Eugène Beauharnais (1781–1824), a son of Joséphine, was the first Italian rose breeder (Dickerson 1992: 35; Pisoni 2005).

The Remontant Group (1837) of roses became the most important group after the Noisette Group (1817) and Bourbon Group. About four thousand remontant cultivars were produced before 1900, most of them in France. Crossing of remontant roses with tea roses gave "Hybrid Teas," which combined winter hardiness with multiple flowering, becoming the most popular rose group, beginning with 'La France' (Guillot in 1867). For the purposes of romantic, "natural-looking" gardens, wild roses and so-called shrub or park roses were favored, used alone or grouped with herbaceous perennials. Peter Lambert (1859–1939) from Trier became famous for his multiple flowering shrub-roses, beginning with 'Trier' (1904). The park roses were derived from wild species roses and included 'Conrad Ferdinand Meyer' (1899), 'Hansa' (1905), 'Roseraie de l'Haÿ' (1910), and 'Max Graf' (1919). Climbing roses including favorites like 'American Pillar' (1902), 'Dorothy Perkins' (1901), and 'Paul's Scarlet Climber' (1916) became popular, being grown against buildings or up into trees.

The first new cultivars of lilac (*Syringa* spp.) were 'Alba Plena' (1823), 'Charles X' (1830), 'Prince Notger' (1840), and 'Colmariensis' (1846). Around 1850 intensive breeding began such that the number of cultivars in commerce rose from five in 1858 to 450. The most important breeder was Victor Lemoine in Nancy, who from 1876 to 1927, with his son Emile, brought 153 cultivars on to the market (Fiala 1988; McKelvey 1928).

The first new hardy rhododendron cultivars were *Rhododendron* 'Azaleoides' (1801), *R. catawbiense* 'Grandiflorum' (about 1820), *R. ponticum* 'Album' (before 1825), *R.* 'Cunninghamii' (before 1830), and *R.* 'Cunningham's White' (before 1838). In Germany the breeding of hardy rhododendron cultivars began in 1877 with Traugott Jakob Hermann Seidel (1833–96) of Dresden.

Trees

Some new cultivars of trees, marked by their habitat or color, were already known in maple, beech, oak, ash, willow, robinia, magnolia, and broom. Several cultivars of beech (*Fagus sylvatica*), especially Cuprea Group (copper beech, 1680) and later Atropunicea Group (purple beech), 'Aspleniifolia' (1804) and 'Pendula' (1811) were often planted in the gardens. New cultivars included *Fraxinus excelsior* 'Aurea' (before 1800) and *Corylus maxima* 'Atropurpurea' (1825). After 1850 the number of cultivars increased significantly and the peak of the use of red- and yellow-leaved shrubs was in the second half of the nineteenth century. Among hardy plants *Acer negundo* 'Variegatum' (Bonamy in 1852),

FIGURE 7.6 Maples by Fritz Graf von Schwerin. From *Mitteilungen der deutschen dendrologischen Gesellschaft*, vol. 5, 1896.

A. palmatum 'Atropurpureum' (von Siebold in 1860), *A. pseudoplatanus* 'Purpureum' (*c.* 1828), 'Leopoldii' (1864) and 'Atropurpureum' (1883), and *Corylus avellana* 'Fuscorubra' (1887) were particularly popular.

In the late nineteenth century bizarre and colorful cultivars of conifers gained considerable importance. Between 1843 and 1862 a number, including 'Argentea' and 'Aurea,' of *Chamaecyparis pisifera* were introduced from Far East gardens. Selected in European cultivation from the new conifers of the American west coast were *Xanthocyparis nootkatensis* 'Glauca' (before 1858), *Chamaecyparis lawsoniana* 'Argentea' (before 1862), and *Picea pungens* 'Glauca' (1883).

Tender Perennials

The greatest increase in cultivars of perennials was of the tender dahlias, chrysanthemums, and gladioli. The development started with dahlias and chrysanthemums, whose genetic variability was already considerable at the time of their introduction from Mexico and China, respectively. From 1805, systematic sowings produced different forms of *Dahlia coccinea* and *D. pinnata*. In 1808, a fully "double" cultivar (ball dahlia) arose in Karlsruhe, Germany. Hibberd (1857) speaks of a dahlia mania beginning in 1815, comparable to the tulip mania of the seventeenth century. Christian August Breiter (1776–1840) in Leipzig listed 55 cultivars in 1807, Lee & Kennedy had 200 by 1822, and an exhibition in Dessau, Germany in 1842 brought together 1,596. Dutch, Belgian, German, and French gardeners played the leading roles in breeding, the work of Christian Deegen (1798–1888) and Johann Sieckmann (1804–89) in Köstritz being significant in raising the Lilliput or German Dahlias around 1850. The latter, with their perfect round heads, corresponded best with the then prevailing ideal of the flower. The change in the dahlia at the end of the nineteenth century clearly reflects the change of garden taste. A Mexican cultivar grown as *Dahlia juarezii*, flowering in Europe for the first time in 1864, was the starting point for the development of "Cactus" dahlias, which have an irregular, loose appearance, as opposed to the previous ideal ball dahlia. The new form was exhibited in England before 1880 and has been further improved since then. Likewise, from 1897 onward, the semi-double "Decorative" dahlias and the almost disorderly, half-filled "Collarette" dahlias, which had originated in France in 1900, chimed with the new ideal (Dean 1903; Foerster and Schneider 1927).

The chrysanthemum of the nineteenth century was only a greenhouse plant in northern climes. From 1789 different garden forms of *Chrysanthemum indicum* were introduced from China. In 1821, Sabine already knew of twenty-three. England and France competed in breeding. In 1830 seeds were first gathered in the south of France and shortly afterward also in England, the beginning of European breeding. Wredow (1837) listed thirty-seven English cultivars; the first chrysanthemum exhibition was held in England in 1846, and the first Chrysanthemum Society was founded. Fortune's imports of small-headed (1846) and striped cultivars (1862) from China diversified the range further. As in the case of the dahlia, the irregular, disfigured flower-heads of the imported cultivars became perfectly spherical, regular structures through selection. John Salter's (1798–1874) collection in Hammersmith included 300–400 cultivars in 1846 and nearly 750 in 1860 (Salter 1865) while Burbidge (1885) listed almost 2,000.

The first hybrid "geraniums" (*Pelargonium* spp.) were noted by Dumont de Courset (1746–1824) in 1802. From 1815 pelargoniums in England were deliberately bred, notably by Robert Sweet (1783–1835) who stimulated German, French, Belgian, and Dutch gardeners to compete (e.g. Trattinnick 1825–43). *P.* × *domesticum* or "regal pelargoniums" were derived largely from *P. cucullatum*. By 1830, there were some one thousand cultivars and by 1850, the popularity of pelargoniums surpassed almost all other garden plants, to rival the rose and the carnation (Martinsson 2001). When the mass planting of "geraniums" arose around 1840, pelargoniums became primarily bedding plants. Shirley Hibberd (1825–90) spoke of a "scarlet fever" which led to a sort of mania in 1840–55. The scarlet pelargonium was the "Queen of the Bedding Plants." Breeding now focused on bright color, compact habit, and copious flowering, and the main interest shifted to *P. zonale* hybrids. The red dwarf cultivars "General Tom Thumb" (1842) and "Paul Crampel" (1889) were particularly famous bedding plants. In 1869 there were already sixteen "double" cultivars and, after 1850, fancy-leaved cultivars became popular. A tricolored cultivar appeared in England in 1851; "Mrs. Parker" (1880) and later cultivars combined the desired properties of double flowers with variegated foliage. *Pelargonium peltatum* became important as a basket and balcony plant in the late nineteenth century; in 1872 the Saxonian gardener Oscar Ebert selected a double-flowered form (Wimmer 2002).

Hardy Perennials

Only a few cultivars of hardy perennials were deliberately bred in the first half of nineteenth century. *Paeonia lactiflora* was introduced about 1805 as Chinese cultivars, but only in 1822 were seeds first obtained—in France, where Nicolas Lémon (1787–1836) and his son Jean-Nicolas (1817–95) introduced his cultivars from 1836. In 1844, thirty cultivars were available in Paris (*Le Bon Jardinier* 1844: 862). Belgian gardeners also bred new cultivars such that in 1868 Haage & Schmidt could offer 106 French and Belgian cultivars for sale.

A white form of *Phlox paniculata* arose in England in 1813. In 1824 the commercial gardener George Wheeler (*c.* 1791–1878) selected seedlings "Wheeleri" (pink), "Shepherdii" (purple), and "Ingramiana" (purplish-blue). In the 1820s, French gardeners such as Nicolas Lémon then began phlox hybridization and new hybrids were also raised in England, Belgium, and the United States, including those with *P. maculata* as a parent. German phlox breeding began in the 1860s, when in 1861, Laurentius in Leipzig could list 179 French phlox cultivars.

Only a few unnamed color forms of *Iris* × *germanica* were known before 1800 and Bosse (1829) could record only seven. In 1840, however, Lémon had more than a hundred named cultivars, raised from *I. pallida* and *I. variegata*. The first cultivars of perennial delphiniums were introduced by Lemoine in 1856, and by Deegen in 1862. Bosse (1859) listed eleven cultivars of *Delphinium elatum* and twelve of *D. grandiflorum*. About 1886 James Kelway in England produced the first white one.

Bulbs

Many daffodils and fritillaries were mentioned in the *Bon Jardinier* for 1821, but few were named. Loudon (1824) listed eight *Narcissus* cultivars, while the catalog of the London nursery of James Carter & Co. (1842) had twelve with French and Dutch names. Loudon (1822) and Glenny (1847) ranked daffodils and fritillaries, next to hyacinths and tulips, as florists' flowers.

Bulbs, largely tulips and hyacinths, were cultivated mainly in the Netherlands, and by 1900 were exported all over the world. As tulips and daffodils increased so hyacinths declined. In England a second breeding center emerged with Cottage Tulips, which resembled the striped cultivars of Dutch still-life paintings, becoming commercially valuable from nurseries, while new forms were selected: Darwin Tulips (Krelage 1886), Rembrandt Tulips (1892), Mendel Tulips (1915), and Triumph Tulips (1918) (Pavord 1999; Timmer 2009).

Annuals

Because of the increasing demand for bedding plants, annuals became florists' flowers to be improved by breeders. In 1812 Lady Mary Elizabeth Bennet (1785–1861) first presented her pansy hybrids (*Viola* × *wittrockiana*), introduced into the trade next year by the merchant gardener Lee. The florists' flowers listed by Main (1835) were still the same as those of Maddock (1792), while Loudon (1822), Reider (1831), Schmidlin (1843), and Glenny (1847) extended the term florists' flower to several other genera.

In England four hundred pansy cultivars were reported as early as 1833. Between 1832 and 1837 two hundred cultivars of calceolaria were grown, and twenty petunia cultivars were offered in Hamburg in 1834. Hybrids from the various newly introduced *Glandularia* (*Verbena*) species were raised after about 1840. The middle of the century also saw the beginning of breeding (Heynhold 1840) of *Erythranthe* (*Mimulus*) (with five cultivars), *Zinnia elegans* (seven) and *Pericallis* × *hybrida* (*Cineraria*) (fifty-seven).

In the second half of the century more species of ornamental annuals were included in breeding programs, notably, as offered in the Haage & Schmidt catalog of 1869, *Antirrhinum majus* (57 cultivars), *Callistephus sinensis* (aster, 685), *Clarkia pulchella* (23), *Dianthus chinensis* (pink, 46), *Lobelia erinus* (26), *Matthiola incana* (stock, 378), *Phlox drummondii* (23), *Salpiglossis sinuata* (32), *Tagetes erecta* including *T. patula* (20), and *Tropaeolum* (nasturtium, 47). The sweet pea, *Lathyrus odoratus,* became very popular in England and North America with the number of cultivars greatly increasing after 1860; it was the button-hole favorite of the Edwardians. The first influential breeder was James Carter & Co. of Reading, England; offering two hundred cultivars in special catalogs was not unusual. Another new field for breeding was opened after 1860 with a fashion for tropical hot-house foliage plants, especially species of *Alternanthera, Amaranthus, Begonia, Celosia, Coleus,* and *Ricinus*, especially with their red-leaved forms suitable for carpet bedding.

CONCLUSION

The nineteenth century surpassed all preceding times in popularizing the world of plants and gardens for most European people, whereas in the eigthteenth century it was more restricted to a small wealthy class. The First World War ended this, the most ambitious period of plant amateurism ever seen in Europe. Great gardens both public and private needed to be reduced as the costs of labor increased. Extensive greenhouse culture was stopped, and labor-intensive bedding schemes were curtailed. In the era after 1918 the garden community was divided into two camps. One continued to dream neo-romantic thoughts of a rich-flowering garden as a counter-world, whereas the other insisted on a merely functional green space which met the most urgent horticultural needs of modern life.

CHAPTER EIGHT

The Representation of Plants

H. WALTER LACK

INTRODUCTION

The nineteenth century is often called the "long" century, considered by many to have actually begun in 1789 with the outbreak of the French Revolution and to have ended with the First World War. The first event established the concept of egalitarian and universal citizenship in Europe and profoundly altered the course of history, the second marked—in the words of the diplomat and historian George F. Kennan—"the great seminal catastrophe" of the twentieth century leading to the loss of the long-established balance of power on a global scale (Kennan 1979). All over the world the nineteenth century was the beginning of the modern era resulting in the abolition of slavery, unprecedented urbanization, better education for many, and a growing importance of science and technology. All these developments influenced the way in which plants were represented in the fine arts, the applied arts, and in the sciences.

THE REPRESENTATION OF PLANTS IN THE FINE ARTS

In marked contrast to botanical illustration, which falls within the realm of the sciences (see below), the representation of plants in the fine arts is generally set in an artistic or "creative" context. Plants form non-repetitive elements of a composition, often selected to create or intensify an emotion or feeling and may or may not be represented true to nature—with recognizability not being a key criterion. A well-known work of the Biedermeier period, the painting *Vorfrühling im Wienerwald*, also known as *Veilchenpflücker im Wienerwald* (1858) by Ferdinand Waldmüller (1793–1865) in the Germanisches Nationalmuseum, Nürnberg, is a good example. It shows children in early spring in the Vienna Woods under beech trees not yet in leaf, with a limekiln in the distance. Two girls hold bunches of violets, while a few flowering primroses and violets are shown in the foreground. The violets and primroses are not the focus of the composition but are merely elements evoking early spring. Moreover, the violets are not identifiable as any particular *Viola* species, of which there are upwards of five hundred. Similar examples can also be found in the fields of sculpture, literature, or music as well as in the applied arts. A more extreme example, *Das Liebespaar* or *Der Kuss* (1908) by Gustav Klimt (1862–1918), an iconic

FIGURE 8.1 Gustav Klimt, *Der Kuss*, 1908. Österreichische Galerie Belvedere, Vienna. Photo by DeAgostini/Getty Images.

work of the Jugendstil movement (see below) now housed in the Belvedere, Vienna, depicts two lovers in passionate embrace. The hair and dress of the kneeling woman is shown with representations of flowers, but their identity cannot be determined—clearly the artist had intended them as decorative elements and was probably not depicting any particular plant species.

The Representation of Plants in Paintings

Here paintings are discussed according to the orthodox hierarchy of genres—i.e. historical painting, portrait, genre painting, landscape, animal painting, and still life, of which historical painting and animal painting are not important in this book—and also according to style, the various movements of the nineteenth century, though these approaches are obviously interwoven.

Genres

Portraits are artistic representations of people displaying their likenesses and personalities, often reinforced by carefully chosen ancillary images. When plants are depicted, they sometimes make a special reference to the sitter (see also below). For example in the Suermondt-Ludwig-Museum in Aachen, an 1805 canvas by Robert Lefèvre (1755–1830) shows Joséphine, Empress of the French, wearing a diamond tiara and a long dress, typical of the Empire style, and, standing next to her, a vase of flowers on a table (Lack 2004). While all this—including her right hand pointing to a large book open and lying flat on the table—is rather conventional, the fact that there is a representation of a live sprig on top of the book is not. Although at first glance unexceptionable, it seems very likely that the portrait shows the Empress with the new book that she had sponsored—Etienne-Pierre Ventenat's *Jardin de la Malmaison* (1803–5). In this work, which documents the rarities cultivated in her garden at Malmaison near Paris, the incense plant (*Calomeria amaranthoides*), a spectacular Australian species, was described as new to science and illustrated for the first time. This is the very plant with its tiny, few-flowered pendent inflorescences that Lefèvre represented. The generic name of the plant is derived from the ancient Greek *kalos* (beautiful) and *meris* (share) and is the literal translation of Joséphine's married name Buonaparte in Greek (Heine 1967). Similarly, in the Herzogin-Anna-Amalia-Bibliothek in Weimar, is a well-known portrait of Johann Wolfgang von Goethe with his amanuensis (assistant) John (1831) painted by Johann Joseph Schmeller (1796–1841). Goethe, standing in a long coat, is shown in his study in Weimar, John sitting at a writing table, but the two potted plants on the window sill are more interesting. The right-hand specimen fastened to a stick with a red band is one of Goethe's favourite plants, the Madagascan *Kalanchoe pinnata*. This succulent plant had been introduced from the Royal Garden at Kew, England, to Hanover, then to Weimar, and Goethe had intended to publish a paper on it (Habrich 2013). Many allusions are less clear, a good example being the portrait of the French actress Sarah Bernhardt as Princess Lontaine by Alphonse Mucha (1860–1939), which appeared in print in the art magazine *La Plume* in Paris in 1897. It shows Bernhardt's head crowned with white lilies, but it is not known if these refer to the role she played or to her personality, or indeed if she had a special liking for these flowers.

Genre painting is the pictorial representation of scenes and events in everyday life, either realistic, invented, or romanticized by the artist. Since plants are omnipresent in everyday life, numerous paintings belong to this genre, with Waldmüller's *Vorfrühling im Wienerwald* (see above) as a good example of a realistic approach. The same idea had been used by Erasmus Ritter von Engert (1796–1871) whose *Wiener Hausgarten* (c. 1828–30), now in the Nationalgalerie in Berlin, shows a young woman sitting, while knitting and reading, in the peaceful atmosphere of a back garden, where mallows (*Malva sylvestris*) and sunflowers (*Helianthus annuus*), as well as grapevine shoots and other less clearly recognizable plants are depicted (Friedrich-Sander 2006). There is less realism in a painting by John Fredrick Lewis (1805–76) entitled *In the bey's garden* (1865) in the Harris Museum and Art Gallery in Preston, England. A young woman in a long Victorian dress with references to orientalism, that was fashionable at the time, is shown standing in full light in a partly imaginary flower garden planted with a profusion of poppies (*Papaver rhoeas*), Madonna lilies (*Lilium candidum*), tiger lilies (*L. bulbiferum*), and delphiniums as she is putting a Madonna lily in a tall ceramic vase. A more realistic scene is captured by Childe Hassam (1859–1935) in her painting

FIGURE 8.2 Alphonse Mucha, Sarah Bernhardt, *La Plume*, 1897. Alphonse Mucha Museum, Prague. Photo by Fine Art Images/Getty Images.

Summer evening (1886) in the Florence Griswold Museum, Old Lyme, Connecticut (Hennig 2006). A young woman in a long white dress is shown sitting and tending the pinkish flowers of a potted geranium (*Pelargonium* sp.) on the sill of a window opening on to a plain.

In landscape painting—the depiction of natural scenery, especially where the main subject is a wide view and the sky—the representation of plants has a subordinate role. A few examples are presented here, all views of natural scenery near Vienna. A monumental specimen, about 450 years old, of the Austrian pine (*Pinus nigra*) near Mödling is the main subject of *Die breite Föhre* (1838) in the Fürstliche Sammlungen, Vienna, by Ludwig Schnorr von Karolsfeld (1788–1853). The painting shows the view towards the Vienna plain, which is dwarfed by the expansive sky; it is of historic and scientific interest, since the tree was felled in 1997, its trunk now being exhibited in the Niederösterreichisches Landesmuseum in St. Pölten. Since such large specimens have become extremely rare, Schnorr's painting can be seen as a parallel to the study of English elm trees by John Constable (1776–1837), in the Art Institute Chicago, as the trees have almost disappeared from the British landscape. Klimt documented a thicket of birch trees on a square canvas entitled *Birkenwald 1* (1903) in the Staaliche Galerie Neuer Meister in Dresden, while the expressive painting *Vier Bäume* (1917) in the Belvedere by his colleague Egon Schiele (1890–1918) focuses the eye on a stunning red sun setting behind a chain of mountains in the background. In the foreground of this much-reproduced work are four horse-chestnut trees (*Aesculus hippocastanum*) with colored autumn foliage, one of them having shed most of its leaves. Opinions are divided as to whether the bare tree might symbolize Schiele's position on the outer fringe of the circle of the established artists in Vienna. It may be hypothesized that he was influenced by the work of Lawrence Alma-Tadema (1836–1912), one of the most successful, financially, of all Victorian painters but renowned for the accuracy of his depiction of flowers, reminiscent of the meticulous representations of the Dutch Masters.

A still life depicts "lifeless" (or motionless) subject matter, either natural, man-made, or a combination of both. Often this distinct genre gives the artist considerable freedom in the arrangement of the different elements within a composition, quite different from landscapes and portraits. Furthermore, still life tends to have an affinity to botanical illustration (see below), as is clear in what is perhaps the most refined still life of the nineteenth century. The huge *Huldigung an Jacquin* or *Jacquins Denkmal* (1822), a masterpiece by Johann Knapp (1778–1833), in the Belvedere, Vienna, has many, very accurate representations of different plant species which are still the subject of research. Its complex, and not yet fully elucidated, iconographical program reflects both plant and, to a minor extent, animal diversity on all five continents, with representations of all twenty-four classes of plants as recognized by the eighteenth-century Swedish naturalist Carl Linnaeus, as well as some minerals. Many colleagues, correspondents, and associates of Nikolaus Joseph Freiherr von Jacquin (1727–1817), the second professor of botany and chemistry at Vienna University, are referred to indirectly in the painting—either by plants described by them as new to science or by plants with names dedicated to them (Riedl 1976). Similarly, an unnamed fruit-piece (1834) of Pierre-Joseph Redouté (1759–1840) in Compiègne Castle, France, depicts a profusion of different kinds of fruit, very true to nature and easy to identify, while the famous *Zonnebloemen* by Vincent van Gogh (1853–90) show a single species as a bunch of flowers. There are several variants of this work, showing three to fifteen flower heads at different stages of maturity and/or belonging to different cultivars, in different collections, that in the Van Gogh Museum in Amsterdam being dated 1889. A much more modern approach can be seen in the still life *Le vase blue* by Henri Matisse (1869–1954), annotated "Tanger 1913" and now in the Pushkin Museum in Moscow, showing calla lily (*Zantedeschia aethiopica*), an iris, and a flowering branch of acacia.

FIGURE 8.3 Vincent van Gogh, *Zonnebloemen*, 1889. Van Gogh Museum, Amsterdam. Photo by DeAgostini/Getty Images.

Styles

While neoclassicism has already been exemplified in Lefèvre's portrait, an iconic work of romanticism is *Der Tetschener Altar*, an altarpiece of 1808–9 in the Galerie Neuer Meister in Dresden; it comprises a gilt frame and an oil-painting, both by Caspar David Friedrich (1774–1840). Friedrich shows a cross bearing Christ's body, at the top of a mountain covered with conifers, all against the setting sun. The pen and ink sketches for the trees, mainly Norway spruce (*Picea abies*) and Scots pine (*Pinus sylvestris*), are mostly in the Nasjonal Museet for Kunst in Oslo and reveal the artist's method: putting the painting together like a mosaic, very much in the same way Knapp did later with the plant watercolors now in the Museum für Angewandte Kunst in Hamburg and used to create *Huldigung an Jacquin* (Cuveland 2006; see above). As regards realism, a movement attempting to depict subjects accurately, *Ophelia* (1852) by John Everett Millais (1829–96) in the Tate Britain, London, has a similar significance. This figure from Shakespeare's *Hamlet* is shown floating in a river, before she drowns—together with the plants mentioned in his lines "there with fantastic garlands did she come of crowflowers nettles daisies and long purples" (Act 4 scene 7). Of these four flowers, three are interpreted by the artist as water crowfoot (*Ranunculus aquatilis*), daisy (*Bellis perennis*), and purple loosestrife (*Lythrum salicaria*). Several more are included in this painting, possibly referring to the Victorian language of flowers (see Chapter 6 of this volume). With failing eyesight Claude Monet (1840–1926) produced a long series of large canvases titled *Les nymphéas*, of which no fewer than forty-eight from the years 1903–8 were shown at a single exhibition in Paris on May 6, 1909. They are now dispersed all over the world, one of them being in the Museum of Fine Arts in Boston. Painted almost exclusively in his own water garden at Giverny near Paris under different light conditions, defying the limits of the water surface and dispensing with constraints of perspective, Monet produced highly sought-after works which are considered masterpieces of another style—impressionism. Post-impressionism developed as a reaction to the impressionists' concern with naturalistic depiction of color, leading to works like *Dune (Le Port Hue)* (1890) by Paul Signac

FIGURE 8.4 John Everett Millais, *Ophelia*, 1852. Tate Britain, London. Photo by DeAgostini/Getty Images.

(1863–1935) in the Pushkin Museum, Moscow. Painted in pointillist manner it shows a very accurately represented sea holly (*Eryngium maritimum*) in the foreground. In a quite different vein are the paintings by Henri Rousseau (1844–1910), e.g. *Le rêve* (1910) in the Museum of Modern Art, New York, with its bizarrely modified, "alien," sky-blue or grey representations of the flowers of sacred lotus (*Nelumbo nucifera*). Together with other less outlandish depictions of plants, the lotuses are set in a clearing dominated by a nude woman called Yadwigha. Rousseau's work is regarded by some as naïve art, by some as primitive art, while others regard him as standing at the beginning of surrealism.

The large canvases by Paul Gauguin (1848–1903) contain copious references to tropical plant life. His *Vaïraumati téi oa* (*Her Name was Vaïraumati*) (1892), for example, painted in Tahiti and now also in the Pushkin Museum, contains representations of mango fruits (*Mangifera indica*), leaves of the breadfruit tree (*Artocarpus altilis*), and flowers of the Tahitian gardenia (*Gardenia taitensis*). An icon of Art Nouveau or Jugendstil (see below) is Klimt's *Bauerngarten mit Sommerblumen* (1907) in the Belvedere; it shows a farmer's garden in full flower, with sunflower, phlox (*Phlox paniculata*), dahlias, and daisies clearly discernible. More difficult is the interpretation of a fruit shown in another key work—Schiele's *Selbstbildnis mit Lampionsfrüchten* (1912) in the Leopold Museum, Vienna. It contains elements characteristic of expressionism, a movement representing the world solely from a subjective perspective, often distorting it radically for emotional effect. In that year the artist, already known for his sexually very explicit portraits, had been taken into investigative custody for allegedly having kidnapped an underage girl. Although the charges were dropped, Schiele, deeply hurt, produced a self-portrait with one shoulder pulled up, the other lowered, and a twig of the Chinese lantern (*Alkekengi officinarum*) with its bright-red calices in the left part of the canvas. Why Schiele chose this plant is uncertain. In a sense the botanical watercolors and still lifes from Walberswick (1914–15) by Charles Rennie Mackintosh (1868–1928) in the Hunterian, Glasgow, also belong to the sphere of fine arts—many are identifiable to the specialist, but they show deliberately distorted, stylized, often incompletely colored "specimens."

In the nineteenth century the various styles mentioned above were truly global. An example exhibiting this phenomenon is a *c.* 9 × *c.* 14 cm (3.5 × 5.5 in) anonymous lithograph called "Blossoming plum tree by a river" published by Chūkyō zuan kai (Nagoya Design Association), now in the Museum of Fine Arts, Boston. It shows a landscape in early spring, is dated late Meiji era, i.e. *c.* 1900, and is comparable to similar, small format prints produced by Art Nouveau artists in Europe then.

The Representation of Plants in Sculpture

The intricate and complex structure of plants is often a challenge to the sculptor, irrespective of the type of material used, whether it be stone, metal, or wood, but world-famous buildings and statues erected during the nineteenth century include three-dimensional representations of plants. A striking example is found in the Arc de Triomphe du Carrousel in Paris (1806–8), built to commemorate Napoleon's diplomatic and military victories of 1805. Two winged female figures in Empire dress are shown as witnesses of the triumph—one carrying a palm frond and an olive branch (*Olea europea*) in her left hand, the other with a wreath of laurel (*Laurus nobilis*) leaves in her right, with all details carved as low reliefs in stone. Many similar winged female figures carrying similar symbols were to be erected in other countries to commemorate military or political success. The column in the center of Berlin's Mehringplatz is crowned with a Victoria

FIGURE 8.5 Paul Gauguin, *Vaïraumati téi oa*, 1892. Pushkin Museum, Moscow. Photo by DeAgostini/Getty Images.

(personification of victory, 1843) by Christian Daniel Rauch (1777–1857) of which two copies were given by Friedrich Wilhelm IV, King of Prussia, to his brother-in-law, Nicolas I, Tsar of Russia. The latter had them placed on columns in the Konnogvardeyskiy Bulvar in St. Petersburg. The Prussian victories over Denmark, Austria, and France in 1864–71

FIGURE 8.6 Egon Schiele, *Selbstbildnis mit Lampionsfrüchten*, 1912. Leopold Museum, Vienna. Photo by DEA/E. LESSING/De Agostini via Getty Images.

led to the building of Berlin's *Siegessäule*, which has a Victoria (1873) by Friedrich Drake (1805–82) at its top. By contrast no military triumph is commemorated by the Nike ("Victory" in Greek) or Victoria with laurels crowning the *Queen Victoria Monument* (1911), opposite Buckingham Palace in London, and the same applies to the four Victoria statues in cast bronze, two each flanking the main entrances of the Kunsthistorisches Museum and the Naturhistorisches Museum in Vienna. In the United States the Victory "fashion" continued in Manhattan's Grand Army Plaza where a monument (1902) by Augustus Saint-Gaudens (1828–1907) portrays the Civil War general William Tecumseh riding behind a gilt Nike carrying a palm frond.

An enormous hollow sphere, nicknamed Krauthappel (cabbage head), its surface made up of a network of representations of laurel twigs in gilt metal, crowns the Vienna Secession building (1897) by Joseph Maria Olbrich (1867–1908). It created a scandal as the architectural symbol or representation of the Secession Movement (see below) with its motto "Der Zeit ihre Kunst. Der Kunst ihre Freiheit" (To Every Age its Art. To Art its Freedom). Leaves of another tree, the ginkgo (*Ginkgo biloba*), were used as models for carvings in stone walls, as in the *Maison Ginkgo biloba* in Nancy, France. When the architect Otto Wagner (1868–1918) selected Koloman Moser (1868–1918) to decorate the house Linke Wienzeile 38, a block of flats in Vienna, the latter chose crossed palm fronds in stucco, functioning somewhere between sculpture and decoration, painted in gold for the walls between the windows on the top floor. The adjacent block of flats, Linke Wienzeile 40, the *Majolikahaus* (1899), also built by Wagner, has a polychrome façade consisting of glazed tiles forming rather abstract plant ornaments. The head of an Asian woman with ginkgo branches to the right and left of her helmet is rendered in low stucco relief in Prague's *Hlavní Nádraží* (Main Railway Station) (1901–9), which was built according to plans of the architect Josef Fanta (1856–1954), and documents the special attraction that this Chinese tree held in the Art Nouveau period. By contrast the caryatides in the domed Primátorský sál of Prague's *Obecní Dům* (1905–12) are flanked by laurel twigs worked in *bas-relief*. Other projects with botanical associations were not carried out, among them a colonnaded and domed tower for Hyde Park in Sydney projected by Henry Lucien (1850–96) with the apex imitating a gigantic flowerhead of the warratah (*Telopea speciosissima*), a common motif in the decorative arts of New South Wales, where it is the State Flower, once argued to be that of Australia as a whole.

FIGURE 8.7 Joseph Maria Olbrich, *Secession Building in Vienna*, 1898. Photo by Alan John Ainsworth/Heritage Images/Getty Images.

The Representation of Plants in Literature

Plants are omnipresent in nineteenth-century literature, in all its major forms, genres, and styles, so it would be otiose to present a list of quotations and authors referring to plants which can be identified botanically with certainty. It seems more appropriate to concentrate here on Symbolism, a nineteenth-century literary and artistic movement that rejected realism and was characterized by the usage of symbols, images, and tropes, as in the case of *Fleurs du mal* (1857) by Charles Baudelaire (1821–67). A good example is the poem *Un dahlia* (1866) by Paul Verlaine (1844–96), published in Paris in his *Poèmes saturniens*. Verlaine compares the dahlia's scentless "proud" flower-head to a young courtesan "au sein dur" (with a hard breast). More somber is the narrative of the poem *Die Magd* (1913) by Richard Dehmel (1863–1920). In spring a maidservant falls in love with a young man, becomes pregnant, and in winter is dismissed, when she gives birth to the child. From the very beginning there are forebodings expressed in the line "er sah mich trüb an und müd im Faulbaum rief die Nachtigall: die Blüte flieht!" (he looked at me, gloomy and tired, the nightingale called in the alder buckthorn: the blossom flees!). In German, Faulbaum (*Frangula alnus*) literally means "foul tree," a reference to the unpleasant smell of its bark and the suspicion of it being poisonous. The posthumously published poem *Träumerei am Abend* (1915) by Georg Trakl (1897–1914) has a similarly somber line "Dem einsam Sitzenden löst weißer Mohn die Glieder" (white poppy relaxes the limbs of the man sitting all alone). Here the symbolism is easy to understand—white poppy (*Papaver somniferum*) stands for opium, a scarcely surprising reference by an author well known for his addiction to psychotropic drugs.

In the nationalistic propaganda of the late nineteenth century several plants served as symbols, for example the oak (*Quercus* spp.) and the linden (*Tilia* spp.) representing the German-speaking people and the people speaking Slavonic languages respectively. The poem *Nachruf auf Preschérn* (i.e. France Prešeren, 1800–49) (1876) by Anastasius Grün, a nom de plume of Anton Alexander Graf Auersperg (1806–76), epitomizes such nationalistic sentiment. A rural feast in what is now Slovenia with people speaking either German or Slovenian symbolized by the oak and the linden is described in this text. Even today the linden is regarded as national symbol for no fewer than three modern European states: Croatia, where the lipa (linden) is today the smaller unit of the national currency kuna; Slovenia; and the Czech Republic. In *Gorski Vjenas* (Mountain Wreath, 1857) by Petar II Petrović Njeguš (1813–51), prince-bishop of Montenegro, the linden is, by contrast, associated with Christianity. This national epos has the need for Serbian Orthodox Christians to rise against their Ottoman oppressors as a broader message and is even today important for Serbian and Montenegrin nationalists (Roberts 2007). Here the "raskršće lipovo" (linden cross) is understood as counterpart to the "čelik" (steel, i.e. the sword of the Ottomans). Both the oak and the linden tree are able to create patriotic feelings also in other countries in Europe, although the two tree species are not restricted to any of them.

The Representation of Plants in Music

Plants rarely produce sound and, unsurprisingly, they are neither echoed nor imitated in instrumental music—even Richard Wagner (1813–83), the high priest of the *Leitmotiv* in music, did not produce a single strictly instrumental "phrase" that was obviously associated with a clearly identifiable plant species. However, when vocal and instrumental sounds are combined there is a wide range of musical phrases directly or indirectly associated with plants, sometimes even with given species. Perhaps most famous is the poem *Heiden Röslein* written by Goethe in 1770–1, subsequently set to music by several composers but notably Franz Schubert (1798–1829) as early as 1815. A young woman represented by a rose, its thorns causing a man to bleed, is in the focus of the verses, and the story is open to several interpretations, among them rape. The poem *Kennst du das Land wo die Citronen blühen*, written by Goethe in 1777–85, but published only in 1795, also known as *Lied der Mignon*, was also set to music—by Beethoven (1809), Schubert (1815), Franz Liszt (1848), Robert Schumann (1849), Hugo Wolf (1888), and Alban Berg (1907). Its first verse has references to three specific plants—lemon (*Citrus* × *limon*), laurel, and myrtle (*Myrtus communis*), whereas a fourth one, i.e. Goldorange, is less clear; Goethe was perhaps referring to the sour orange (*Citrus* × *aurantium*), which he may have seen in the orangery in the Belvedere, Weimar, or elsewhere. The poem *Сырен* (1878) by Ekaterina Andreyena Beketova (1855–92), set to music by Sergei Rachmanninoff (1873–1943) in 1902, refers to the lilac (*Syringa vulgaris*), while the Moravian folk song *Rozmarýn* formed the basis for Leoš Janaček's (1854–1928) homonymic song which is part of the cycle *Moravské lidová poezie v písních* published in 1922. It is centered on rosemary (*Rosmarinus officinalis* [i.e. *Salvia rosmarinus*]).

The text of *Der Lindenbaum* (1823) by Wilhelm Müller (1794–1827), soon made famous by Schubert as part of his cycle *Die Winterreise* (1828), is less botanically precise as it could refer to either the small-leaved linden (*Tilia cordata*), the large-leaved linden (*T. platyphyllos*), both wild in central Europe where Müller and Schubert lived, or the hybrid of the two. Much more complex is the interpretation of a plant name in *Der Abschied* forming the last song of the famous cycle *Das Lied von der Erde* by Gustav Mahler (1860–1911). The French sinologist Marie Jean Léon Le Coq, Baron d'Hervey had paraphrased three poems written respectively by Meng Hao-Ran and Wang Wei, both of the Tang period (618–917), in his poem *L'adieu* (1862). This was translated by Hans Bethge (1876–1946) into German with this version subsequently modified by Mahler (Lack 2010). In this sequence, the meaning of a plant name got lost: the Chinese word "song" meaning pine (*Pinus* sp.) became in Bethge and Mahler "Fichte," i.e. Norway spruce, a tree which is not native in China. As a consequence, the implied allusion in Chinese poetry of "song" being representative of life as a hermit was lost. Similarly, in the third *lied, Von der Schönheit*, the Chinese word "yang" used for poplar (*Populus* sp., Salicaceae) metamorphosed via "*saule pleureur*," i.e. weeping willow (*Salix babylonica*), to "Weide," i.e. any willow. In this case the connotation of "yáng" as spring and sexual desire disappeared.

No composer wrote more *lieder* with direct references to plants than did Richard Strauss (1864–1949), among them those based on poems by Felix Ludwig Dahn (1834–1912), namely the cycle (or group) *Mädchenblumen* (1891). The cycle comprises four songs, i.e. *Kornblumen* (cornflower, *Centaurea cyanus*), *Mohnblumen* (*Papaver rhoeas*), *Epheu* (*Hedera helix*), and *Wasserrose* (waterlily, either yellow water lily, *Nuphar lutea*, or white, *Nymphaea alba*). Richard Strauss' earlier cycle *Letzte Blätter*, based on poems by Herman von Gilm zu Rosenegg (1812–64), includes two more with botanical titles, namely *Georgine* (*Dahlia* sp.) and *Zeitlose* (*Colchicum autumnale*). At that time both dahlias from Mexico and florist's chrysanthemums (*Chrysanthemum* × *morifolium*) from China were widely cultivated in central Europe. Rainer Maria Rilke (1875–1926) referred to chrysanthemums in the poem *Das war der Tag der weißen Chrysanthemen*, which appeared in his collection *Traumgekrönt* (1897). In 1907 Alban Berg (1885–1935) based his *lied Traumgekrönt* on this somewhat erotic text forming part of his cycle *Sieben frühe Lieder*.

Sweet-smelling flowers are associated by some writers with longing, love, and desire, and composers have taken up these texts. The poem *Allerseelen* (1844) by Gilm zu Rosenegg, with the first line "Stell auf den Tisch die duftenden Reseden," was the basis of the song (1885) of the same name by Richard Strauss and combines the sentiment of All Souls Day and feelings of longing with the scent of mignonette (*Reseda odorata*). *Ich atmet einen linden Duft* (c. 1844) by Friedrich Rückert (1788–1866) is a quintessential poem of romanticism and attracted the attention of Mahler, who composed a song of the same title that premiered in Vienna in 1905. A very soft melody accompanies the lyrical text, which has a flowering twig of linden collected by a lover's hand as its thematic center and calls the sweet scent of the linden flower the fragrance of love.

What has been said about the texts of *lieder* can also be said about the librettos of operas. Wagner's text for *Die Meistersinger von Nürnberg* (1862) contains, for example, the line "Der Flieder war's—Johannisnacht," which is usually interpreted as elderflower (*Sambucus nigra*), a bush which indeed may still be in flower on June 24 (hence Johannisnacht) in Nürnberg. In Wagner's *Siegfried* (1876) the lead character is resting, as stated in the libretto, under a linden, while the identity of the "Welt-Esche" (world ash), the tree of the three norns in Wagner's *Die Götterdämmerung* (1876), is obscure. It may refer to Yggdrasil, the world tree of Nordic mythology (Heizmann 2002). However, in some operas there are botanical paradoxes: Francesco Maria Piave's libretto for *La Traviata* (1853) by Giuseppe Verdi (1813–1901) does not contain an explicit reference to a camellia (*Camellia japonica*) though the text is based on Alexander Dumas' novel *La dame aux camellias* (1848), and Hugo von Hofmannsthal's libretto for *Der Rosenkavalier* (1913) by Richard Strauss clearly states that Octavian Graf Rofrano, the title-giving figure, hands over an artifact to Sophie von Faninal—not a living rose, but one made of silver.

THE REPRESENTATION OF PLANTS IN THE APPLIED ARTS

The line between fine and applied arts is often difficult to draw; consequently, some of the above-cited examples, like the stone carvings in the *Maison Ginkgo biloba*, could easily be placed in this category, as could the exquisite Art Nouveau book-bindings, like those issued by Blackie's of Edinburgh. However, in the applied arts repetitive patterns are more widespread, as in wallpaper, textiles, and other large-scale artifacts, likewise the tendency to stylize representations of plants, as in ceramics and jewelry. More importantly, and increasingly so in the nineteenth century, items were mass-produced for a greatly widened market augmented by the aspirations of a newly growing middle class. No attempt is made here to systematize the various expressions of applied arts which extend to furniture, carpets, metalwork, and cutlery, making use of a wide variety of materials, but some of the more remarkable developments are highlighted below.

There seems to have been a general trend for the degree of stylization in the representation of plants to increase over time. This can be illustrated by considering two objects from Vienna: a drinking cup made of transparent glass by Gottlob Samuel Mohn (1789–1825), dated 1813 and now in the Wien Museum (Scholda 2002), and the prints based on expressionistic drawings by Oskar Kokoschka (1886–1980) intended to illustrate his *Die träumenden Knaben* (1908). Mohn decorated his cup with two types of flower garlands in transparent colors—one used as a border comprising a pink rose, wild daffodil (*Narcissus pseudonarcissus*), morning glory (*Ipomoea tricolor*), and other plants, while on the body of the glass appear garlands of flowers and leaves of forget-me-not (*Myosotis sylvatica*) forming the letters E and H, the initials of the dedicatee. Their botanical identity is unquestionable. By contrast, Kokoschka's color lithograph *Das Mädchen Li und ich*, taken from his book, has completely abstract plants that are therefore impossible to interpret (Wingler and Welz 1975).

The nineteenth century was dominated by several waves of revival—among them neo-classic, neo-renaissance, neo-baroque, neo-rococo, and neo-gothic styles, the latter also called Gothic revival, all five referring to styles of previous centuries. These neo-movements were mainstream and the current emphasis, if not craze, on the subsequent styles distinguished by their originality, such as Art Nouveau, distorts what applied arts were like for long periods of the nineteenth century. The Gothic revival culminated in highly stylized plant representations, often geometrically arranged in sophisticated repetitive patterns as in the book *Floriated ornament* (1849) by Augustus Pugin (1812–52), one of the architects of the Gothic revival Palace of Westminster in London (Atterbury 1995). More ornamental representations are found on the three *Screens with embroidered panels* (1885–1910) designed by John Henry Dearle (1859–1932) in the Victoria and Albert Museum (V&A), London, to be associated with the Arts and Crafts movement. In the same vein is *Artichoke wallpaper* (c. 1897) by John Henry Dearle and many similarly stylized plant representations like the tiles of William de Morgan (1839–1916) with its many blue flowers, with representative specimens of wallpaper and titles also in the V&A.

Art Nouveau or Jugendstil is best understood as an umbrella term for a broad spectrum of styles ranging from Wiener Secession (see above) to Stile Liberty in Italy, from Szecesszió in Hungary to Модерн in Russia, from Arte Joven in Spain to Jügendstila in Latvia. All these variants express themselves extensively in the applied arts, but it should be noted that they were basically elite trends, while historicism prevailed. Art Nouveau radiated into almost all parts of the world and to all types of artifact. Two pieces of furniture exemplify this—a tray decorated with representations of clematis flowers (1900) from the workshop of Emile Gallé (1846–1904) in Nancy and a pair of wooden tablets (1899) by one Carl Spindler in the Museum für angewandte Kunst (MAK) in Vienna decorated with representations of hops (*Humulus lupulus*), all three worked in intarsia. Famous objects of Art Nouveau include the *Lily lamp* (1900–10) by Louis Comfort Tiffany (1848–1933) and his glass window *Magnolias and irises* (1908–10), both in The Metropolitan Museum, New York. Other masterpieces in glass were created by the workshop of Gallé, which produced vases consisting of an internal layer of opal glass and outer layers of colored cameo glass with representations of various plants such as fuchsias, cut out and acid etched. From the same workshop also came tiny perfume bottles decorated with representations of flowers of clematis, of which a fine specimen is in the Pola Museum of Art in Hakore, Japan. Of a quite different size are the tall windows in High Victorian churches in England or the often broad, Art Nouveau stained glass windows in hotels, both including representations of plants. The latter embellish staircases built around 1900, one of the most striking being in the Grand Hotel in Lund, Sweden, another in the Danubius Gellért Hotel in Budapest, Hungary. The Hotel Villa Igiea in Palermo, Italy has a counterpart in fresco (1908)—one of the halls is decorated with a staggering profusion of irises, Madonna lilies, calla lilies, nymphaeas, and tree peonies (*Paeonia suffruticosa*) as a setting for a glorious garden party of young ladies painted by Ettore De Maria Bergler (1850–1938).

Otto Wagner (see above) was a fan of the sunflower—he decorated the portico of the *Stadtbahnstation Karlsplatz* (1899) in Vienna with a row of stylized representations of it. He also used this motif, albeit even more stylized, in his *Sonnenblumengitter* (1898–1901) originally painted beige, which motif survives as the dominant element of today's Stadtbahn network, an early example of corporate identity for a railway line. For the Paris Metro Hector Guimard (1867–1942) relied on more abstracted vegetal forms, among them lily-of-the-valley (*Convallaria majalis*) (Anon. 1978). His other buildings have contorted plants in iron, stone, wood, and glass as decorative elements, which are echoed in the works of Antoni Gaudi i Cornet (1852–1926), whose buildings in Barcelona, Spain, are topped with structures allegedly representing cypress cones.

At the other end of the spectrum is a wide range of small, everyday items, like letter openers or napkin rings decorated with representations of plants. Among these, ginkgo was particularly fashionable in the Art Nouveau period, with a gilt ceramic pitcher made in Paris in 1900 a good example (Schmoll, gen. Eisenwerth 1994). A specialized art form is what could be called botanical jewelry, the representation of plants as elements of brooches, necklaces, and other pieces, making use of gold, silver, bronze, precious and semiprecious stones, enamel, and glass. Flowers of the orchid genus *Cattleya*, twigs of mistletoe (*Viscum album*), and ginkgo leaves are among the subjects used in what today are heirlooms, collectors' items, and museum pieces. The representation of a dark blue violet dominates a gold and silver bracelet with diamonds manufactured in Munich in 1887 and now in the Musej Primenjene Umetnosti in Belgrade, Serbia. Another example of this kind of cabinet piece was produced in *c.* 1890 by Tiffany, New York: a 5-cm (about 2 in.) brooch, now in private hands, representing the orchid *Calanthe* × *veitchii* realistically modeled with an emerald stem, the lip worked as matt pink enamel with a cut diamond in the upper part.

In addition to luxury objects, the nineteenth century witnessed depiction of plants in architecture, including ironwork, and in textiles, crockery, and wallpaper. For example, the classical *anthemion* (Greek palmette) used in architectural features, was ubiquitous in Boston ironwork, notably railings, at the beginning of the nineteenth century, as it was in 1820s in Cheltenham and London ironwork in Britain, while New Orleans ironwork with its exquisite botanical patterns dates from the 1850s (Robertson and Robertson 1977). The capitals of architectural iron columns often exemplified great realism, as for example the chestnuts on the pillars at Great Malvern railway station (1860–1) in England (Robertson and Robertson 1977: fig. 82). The national emblems of the British Isles—Tudor rose, Scotch thistle, Irish shamrock, and Welsh leek—were used throughout the British Empire, and could be combined with local ones, as in the case of maple leaves in Canada. The Aesthetic Movement in London in the 1880s had ironwork with the ubiquitous sunflowers, which also figured in Australian work.

In wallpaper there was an array both of conventional and new designs, which was also true of other expressions of applied art of the period. Good examples of realism are some of the papers produced by Zuber & Cie. in Rixheim, France; such was the large Isola Bella panorama printed in 1843 from no fewer than 550 blocks in 85 colors, depicting an imaginary landscape including accurate representations of a palm, hummingbird fuchsia (*Fuchsia magellanica*), and lily magnolia (*Magnolia liliiflora*). A repetitive, conventional, and much more decorative approach was followed in a paper designed by Louis Marie Lemaire (1824–1910) and produced by the same company in 1856 with a pattern of

FIGURE 8.8 William Morris, Blackthorn wallpaper, *c.* 1890. Victoria and Albert Museum, London. Photo © Historical Picture Archive/CORBIS/Corbis via Getty Images.

roses, clematis, campanula, and acacia, an example now being in the Musée du papier peint in Rixheim, France. *Blackthorn* wallpaper designed by William Morris (1834–96) in the late nineteenth century is an example of botanically inspired wallpaper (Parry 2013) manufactured by Jeffrey & Co., London, it has representations of violets and fritillaries besides other unidentifiable plants. This type of wallpaper later became one of the specialities of the architect and textile designer Charley Voysey (1857–1941).

The Arts and Crafts movement radiated in many directions—one of them was Vienna, where the Wiener Werkstätte, a production community of visual artists established in 1903 by Koloman Moser and Josef Hoffmann (1870–1956), with Kokoschka among its temporary associates, produced extraordinary book-bindings with highly stylized flower composition in predominantly beige and greys. Graphic art in the broad sense, including sketches and preliminary drawings for paintings, is full of representations of plants. In the Israel Museum in Jerusalem a good example is a pen-and-ink drawing on paper by Ephraim Mose Lilien (1874–1925) intended to be the basis for a commemorative print for the Fifth Zionist Congress in Basle, Switzerland, in 1901. To the left is depicted an old Orthodox Jew seated behind a bush with thorny branches, while to the right is a field of mature wheat (Dimitrieva 2006). This symbolizes the Aliyah, the emigration to Israel or, more metaphorically, the route to *Altneuland*, which was the title of a utopian novel by Theodor Herzl (1860–1904), published in Leipzig, Germany, in 1902, which soon became emblematic of Zionism.

As regards textiles, there is both direct evidence in museums and indirect evidence in the form of paintings of them. For the neo-classical style, examples of the latter include a full-size portrait in oils of Empress Joséphine (1808) by Antoine Jean Gros (1771–1835) now in the Palais Masséna in Nice, France. She is shown in a long dress, the lower part of which is covered with representations of very numerous, stylized flowers in radiant colors embroidered (?) on white silk (Lack 2004). A watercolor (1826) by Johann Stephan Decker (1784–1844) in The Metropolitan Museum, New York, documents the wool carpet completely covering the floor of the sitting room in the apartment of Franz I, Emperor of Austria, in the Hofburg, Vienna. Long straight rows of groups of pink roses were selected as the pattern for this long-lost Biedermeier piece (Hanzl-Wachter 2002). By contrast Hermann Bollé (1845–1926) chose geometrically arranged and highly stylized representations of flowers in a gouache (1898), now in the MAK, to act as the pattern for a carpet intended for the cathedral of Zagreb, Croatia. For Jugendstil and its fine textiles the paintings of Klimt offer excellent source material, as in *Der Kuss* (Fig. 8.1) or *Bildnis Eugenia (Mäda) Primavesi* (1913–14) in the Municipal Museum of Art in Toyota, Japan. Here the sitter wears a colorful kimono, which includes highly stylized representations suggesting poppies and roses (Natter 2000).

THE REPRESENTATION OF PLANTS IN SCIENCE

Science is quite the opposite of the arts, since aesthetic considerations play no part, with the result that in science a representation of any plant is not valued for its beauty but for its accuracy, i.e. truth to nature. Indeed, most botanical illustrations are not produced in order to create or intensify a viewer's emotional response, but rather to document facts.

Botanical Illustration

A botanical illustration is best understood as an exact pictorial representation of a plant—precisely like a living plant before being pressed between sheets of paper to become an herbarium specimen. This kind of image, magnified or not, is normally decontextualized by being placed on a white background. In the nineteenth century such illustrations were usually not prepared by a botanist, though William Jackson Hooker (1785–1865) in

Britain was a striking exception, but by a specialist—the botanical illustrator or botanical artist. The latter generally worked under the instructions of the botanist who has always been responsible for the illustration's accuracy, the illustrator in a sense lending eyes and hand to the botanist. Illustrations provide an immediate impression of a given specimen, which was particularly important in the first half of the nineteenth century when botanical terminology was not yet standardized. In them, ephemeral features like flower colors, which are soon lost, or parts of plants, like juicy fruits, that are difficult to conserve, are readily recorded. Botanical illustrations are also less prone to insect attack than are herbarium specimens, at the same time being much more convenient to transport. Finally, the original illustration can easily be multiplied—by making copies by hand, printing or, via photography, by preparing double negatives (see below). However, there are also drawbacks: for the purpose of identification of a specimen, botanical illustrations can only provide the characteristics and parts of plants that have been drawn, the extraction of plant parts, like pollen or cells, not being possible, though this generally had little significance in this period.

Botanical Illustrations Made by Hand

Botanical illustrations in the conventional sense are prepared by hand in the form of a pencil or ink drawing, watercolor, or painting. Botanical illustrators generally approach their subjects by first preparing sketches, focusing on details, and only finally producing a finished image, often including the representation of magnified plant parts like stamens, or fruits transversally sectioned. Sometimes information noted on the sketches is later lost, as in the preliminary sketches documenting plants cultivated in c. 1810–24 in the imperial gardens in Vienna prepared by Mathias Schmutzler (1752–1824) and in parts now in the Biblioteka Jagiellońska in Cracow, Poland (Lack 2006). Delays in the production process were frequent, as with the color-coded pencil drawing of *Cyclamen persicum* prepared in Cyprus in 1787 by Ferdinand Bauer (1760–1826), the watercolor based on it being made by him in in Oxford c. 1792, yet the colored copper engraving by James Sowerby (1757–1822), having the watercolor as its basis, was published in London only in 1825 (Lack with Mabberley 1998).

Most botanical illustrations prepared in the nineteenth century—and, considering print run, millions went into circulation—are on paper, a relatively cheap material. By contrast, vellum was an extremely expensive, durable material used only occasionally for botanical illustrations, and generally only for imperial and royal patrons; in the second half of the nineteenth century it fell out of use. The "Collection des vélins" in the Bibliothèque Centrale of the Muséum National d'Histoire Naturelle in Paris has many such illustrations of superb quality (Heurtel and Lenoir 2016). The period from 1789 to 1832 fittingly has been called the Golden Age of this collection (Heurtel and Lenoir 2016), when key figures like Pierre-Joseph Redouté (1759–1840) made fine botanical illustrations on vellum. Several of those prepared by him for Empress Joséphine (see above) ended up in the Fitzwilliam Museum in Cambridge, England, after having been sold by her heirs, though others are still on the market.

Some large-sized botanical illustrations were prepared as "swagger prints" in the nineteenth century, but more familiar are those on card, or backed with linen so that they could be rolled up like a scroll, playing an important role in instruction, both in schools and at universities (Noltie 2014). These botanical wall charts were mainly produced in Europe and Japan, an early example being *Prof. Henslow's Botanical Drawings* by Walter

FIGURE 8.9 Ferdinand Bauer, *Cyclamen persicum*, c. 1792. Courtesy of Sherardian Library, University of Oxford.

Hood Fitch (1817–92), which were printed in London in 1857 (Laurent 2016). Such charts were often issued as series of lithographs, the *Botanische Wandtafeln* by Leopold Kny (1841–1916) published in Berlin between 1874 and 1911 forming the high point of this development. This series consisted of 120 charts that were accompanied by a detailed explanatory textbook covering plant systematics, morphology, and anatomy. Arnold and Carolina Dodel aptly wrote, "Natural, scientifically reliable wall charts can replace a natural object in classroom teaching and in lectures; they are more enlightening than the spoken word" (Dodel and Dodel-Port 1883). Although spurned and often destroyed in the twentieth century, collections of wall charts still survive in several botanical institutions, a particularly extensive one being in the Botanisches Museum Berlin; they are now valuable collectors' items. It should be noted in this context that wall charts were also specifically prepared for university teaching as single items that often remained unpublished.

While lithographic stones were re-polished and reused, engraved copper plates with botanical illustrations were usually melted down after printing so that the metal could be reused. Only rarely were engraved plates preserved, as in the case of those forming the series now in the Statens Naturhistorske Museum in Copenhagen. These were used to print the 3,060 copper engravings of *Flora Danica*, which started to appear under the orders of Christian VII, King of Denmark in 1766 but was completed only under Christian IX in 1883. Similarly, the copper plates engraved by Ferdinand Bauer for his *Illustrationes florae Novae Hollandiae* (1813–16) are now in the Natural History Museum, London. Both sets have recently been used to produce genuine reprints. With regard to accuracy the engravings of *Illustrationes* are of the highest quality and were produced for wealthy patrons, while the very numerous, useful plant illustrations published in gardeners' magazines such as those in the *Gardeners' Chronicle* (1841–), nursery catalogs, seed lists, and popular encyclopedias, often published on cheap paper and with large print runs for the general public, are at the lower end.

Die Kunstformen der Natur by Ernst Haeckel (1834–1919) published in Leipzig between 1899 and 1903 contain mainly zoological illustrations which are of no interest here; however, he also included some botanical plates which, though not inaccurate, are on the fringes of the applied arts—plants are identifiable, but the illustrator has used a hardly acceptable amount of freedom to add ornamental elegance to his representations of cycads, orchids, and mosses while omitting the conventional detail.

Though largely a Western phenomenon, botanical illustration in the nineteenth century had some remarkable practitioners in Japan, India, and China, the latter largely anonymous. Often the plant illustrators worked in more or less close association with Western botanists with, for example, Kawahara Keiga (1786–after 1860) in Nagasaki strongly influenced by Philipp Franz von Siebold (1796–1866) and following "yōfūga," that is, the realistic Western style. His undated watercolors are mostly in the Botanical Institute of the Russian Academy of Sciences in St. Petersburg, Russia. Another early example is Kane'en Iwaski's *Honzō Zufu,* published in Tokyo in 1828, with hand-colored wood-block prints depicting some two thousand Japanese medicinal plants in some ninety-six volumes (Loudon 2015: 96). In Calcutta, Rungiah and Govindoo painted with Indian pigments for the prolific Robert Wight (1796–1872) and had their illustrations published as colored lithographs in his *Icones plantarum Indiae orientalis*, which appeared in Madras in 1838–53 (Noltie 2007); the undated originals are now in four British institutions including the Royal Botanic Gardens, Kew. However, very many other such Indian illustrations remain unpublished, though the anonymous, undated Dapuri drawings, now in the Royal Botanic Garden Edinburgh, have recently been reproduced, as have some of those in the Cleghorn Collection, 1845–60 (Noltie 2002, 2016).

Only rarely do botanical illustrations form elements of oil paintings (but see the portrait of Empress Joséphine above), since oils are generally unsuited for the required detail in scientific documentation. Portraits, in particular of botanists, sometimes comprise botanical illustrations representing associations with the person depicted. A good example is that of José Celestino Mutis (1732–1802) of *c*. 1800, attributed to Salvador Rízo (1760–1816) and now in the Real Academia Nacional de Medicina in Madrid. It shows the botanist seated with a lens in his right hand and a head of the scarlet-flowering *Mutisia clematis* in his left; this specimen is so accurately documented that it can be identified with certainty. In Bogotá Mutis had entertained Alexander von Humboldt (1769–1859), who was to be portrayed in Berlin in 1806 by Friedrich Georg Weitsch (1758–1828) in the oil now in the Nationalgalerie in Berlin. Humboldt is shown holding in his right hand a flowering specimen of *Merania speciosa*, while a twig of *Passiflora emarginata* is near his left. These two illustrations were accurately copied from copper engravings hot off the press—they had been published in Paris in the first instalments of Humboldt's travel account *Voyage aux regions équinoxiales du nouveau continent* appearing in that very year.

Porcelain, like vellum, is a luxury material for botanical illustration and, likewise, largely royalty and aristocracy were the commissioners for very correct representations of plants in this medium, annotated with the scientific names on the reverse (Baer 1999). The pinnacle of botanical illustration on porcelain is the service of *c*. 1500 pieces decorated in 1790–1802 with images copied from *Flora Danica*, probably intended by Christian VII as a gift to Catherine II, Empress of Russia. However, due to her early death it was never sent to St. Petersburg, but remained in Copenhagen. Today the Flora Danica service, still in use on state occasions, is in several palaces of the royal family and in museums there. Pieces from other porcelain services with botanical illustrations are precious museum items, e.g. those made for the imperial courts in St. Petersburg and Vienna, now in the Hermitage, St. Petersburg, and the Hoftafel- und Silberkammer, Vienna respectively. With services from the Staffordshire Porcelain Factory produced in England in *c*. 1835, this fashion came to an end.

Japan is the only country where polished wood blocks—not to be confused with wooden printing blocks—were decorated with botanical illustrations, in a sense offering a pictorial explanation of the tree species they originally came from. During his famous "hofreis" (court journey) to Edo (Tokyo) undertaken in 1828, von Siebold acquired such blocks, now in the Naturalis Biodiversity Center in Leiden, Netherlands (Baas et al. 2021). A much larger collection of 156 illustrated woodblocks, each with four cylindrical slices from branches and four rectangular pieces of bark affixed to them, is conserved in the Botanisches Museum Berlin. It has convincingly been shown that the blocks, bearing the red stamp of Chikusai Kato originated in Meiji 11, i.e. 1878 (Lack 1999), and served educational purposes in the newly founded Tokyo University.

Nature Printing

Using inked herbarium specimens for printing has a long tradition with the oldest example being found in a thirteenth-century Syrian Dioscorides manuscript now in the Topkapi Saray Museum, Istanbul (Geus 1995). However, this method has a serious disadvantage: the printing process quickly destroys the fragile specimens with the result that only small print runs are possible. Nevertheless, the method had a few followers in the nineteenth century, among them Humboldt and Aimé Bonpland (1773–1868), who in 1801–2 recorded some of their most precious herbarium specimens collected in tropical America as nature prints, now in the Institut de France in Paris (Lack 2001). These crude images had a better chance of survival under tropical conditions than did the specimens prone to destruction by insects and mold. In the middle of the nineteenth century a major breakthrough was made in the K.K. Hof- und Staatsdruckerei in Vienna by an anonymous technician who had the brilliant idea of laying "a [pressed and dried] plant ... between a copper and a lead plate" and "letting both run between two rollers firmly screwed together" with the impressed lead plate after hardening used for printing. The result was the *Physiotypia plantarum austriacarum* published by Constantin von Ettingshausen (1828–97) and Alois Pokorny (1826–86) in ten volumes in 1856–73 (Lack 2016). Less well understood are the techniques used for the production of the sumptuous *The ferns of Great Britain* (1855) by Thomas Moore (1821–77) and *The nature-printed British seaweeds* (1859–60) by William Grosart Johnstone and collaborators. Although large print runs were now possible, this technique did not become widespread due to cost and could not compete with the then burgeoning use of photography.

Botanical Illustrations Made Using Light

Using light to produce an image was a major breakthrough in cultural history, and in the very early phases of photography, plants were important subjects. In November, 1838, Henry Fox Talbot (1800–77) managed to make what was then called a "photogenetic drawing" of a flowering lesser masterwort (*Astrantia minor*). He simply placed the plant specimen on what was then called salted paper in the sun and subsequently made the image permanent (Schaaf 2000). The result is one of the most famous of botanical illustrations. An industrious follower was Anna Atkins (1799–1871) who used Herschel's cyanotype process, whereby the images end up on a blue background (hence the term "blueprint") in her monumental *Photographs of British algae: Cyanotype impressions* (three volumes 1843–53) and later works on vascular plants (Schaaf 1985); today her images are more highly valued as fine art rather than as the scientifically oriented images that they were originally intended (Loudon 2015).

A number of technical developments, among them new light-sensitive substances, new carriers for the latent and subsequently fixed image, and new apparatuses called "cameras" with new lenses to direct the light on to the carrier of the image, made photography a method for recording plants in a way that is very true to nature and without the need for an illustrator. In addition, copies of images could be produced easily, cheaply, and quickly, opening the gates for a speedy distribution of images which could be reproduced by conventional printing methods as well. Two works on conifers by Franz Antoine, Jr. (1815–86) illustrate this technological shift: his *Die Coniferen des Cilicischen Taurus* (1855) was illustrated with conventional lithographs, his *Die Cupressinen-Gattungen* (1857) with black-and-white photographs on paper pasted into the book. Later improvements of lenses led to the opportunity to record vegetation in situ in a way undreamt of before, the results being printed as illustrations in books and journals.

Botanical Diagrams

According to the *Oxford English Dictionary* a diagram is "an illustrative figure which, without representing the exact appearance of an object, gives an outline of the general scheme of it, so as to exhibit the shape and the relations to its various parts." This is what became the standard approach in botany, in particular plant morphology, namely drawing only the outlines of complex structures. Transversal sections of flowers and fruits, the development of pollen and egg cells, and histological preparations are often depicted in this simplified form. Perhaps the most spectacular botanical diagrams ever were those

FIGURE 8.10 Henry Fox Talbot, *Astrantia minor*, 1838. Photo by The Royal Photographic Society Collection/Victoria and Albert Museum, London/Getty Images.

published in Leipzig in 1849 in the first volume of *Historia Naturalis Palmarum* edited by Carl Ludwig Philipp von Martius (1794–1868). They are based on the drawings by an anonymous illustrator and show the complex structure of palm inflorescences as diagrams printed as colored lithographs of exceptionally large format (60 × 44 cm, about 24 × 17 in.). No doubt, the sober *Die Blüthendiagramme* by August Wilhelm Eichler (1839–87) in conventional format published in 1875–8 marked the high point of this approach, since an extremely broad spectrum of families of flowering plants was covered with the results discussed in a comparative way.

Botanical Models

In essence, botanical models are three-dimensional botanical illustrations made, in the nineteenth century, from a range of materials, wax, wood, bronze, glass, plaster, and papier-mâché or even ivory, though these last were largely decorative rather than didactic, like a Meiji Era banana from Japan (Grotz 2015; Loudon 2015). Usually the models were produced for educational purposes and formed the backbone of botanical exhibitions in many natural history museums. In comparison with the plants or plant parts that they represent, botanical models are usually greatly enlarged and may be either very true to nature or more or less stylized and then become almost three-dimensional botanical diagrams. Like botanical wall charts, such models were commercially produced by specialist companies, with Brendel Verlagsanstalt für Lehrmittel in Berlin being a particularly successful manufacturer, active *c.* 1880–1920. Made largely of papier-mâché fixed on a wooden base, Brendel models are widespread, with a particularly complete series in the Museu de História Natural of Coimbra University, Portugal. Similar are the *c.* 4,300 botanical glass models made in Dresden from 1887 until 1936 by Rudolf Blaschka (1857–1939) for the Harvard Museum of Natural History in Cambridge, Massachusetts. Models of apple and other fruits made of papier-mâché covered with a layer of stucco, painted and covered with a thin layer of wax are similar artifacts, of which a particularly good collection made in Gotha, Germany, is in the Santos Museum of Economic Botany in the Adelaide Botanic Garden, South Australia. In Italy, those made of wax even included diseased fruits like the lemon and peach models, among them those made by Francesco Garnier-Valletti (1808–89) and now in the Museo della Frutta, Turin (Loudon 2015).

EPILOGUE

It is the parallel progression of the arts and sciences in the "long" nineteenth century that is so very striking. All this could have meant harmony or antagonism—symbiosis or competition and conflict. Ultimately, this parallelism helped to enrich both the arts and sciences. Another factor in the nineteenth century was the rise of global trends with increasing interaction—as between the West and India, China, and Japan, still only partly understood and therefore a desideratum for future research.

ACKNOWLEDGEMENTS

Thanks are due to Andrew Drummond (Sydney) who made several valuable suggestions for topics to be included in the section on applied arts. Preliminary versions of this text were kindly read and commented on by James Compton (Tilbury, United Kingdom) and Eva Lack (Berlin).

NOTES

Introduction

1. Erected in Hyde Park, this prefabricated structure, three times the size of St Paul's Cathedral in London, had 293,000 panes of glass and was moved to Sydenham, south London, after the Exhibition, but was burnt down in 1936; its name was given to an extant soccer team.
2. As translated by J. F. W. Johnston in *Edinburgh Journal of Science* n.s. 4 (1831), 189.
3. The Belgian botanist, Charles François Antoine Morren (1807–58) coined the term "phenology" in 1849.
4. Darwin to Hooker, February 10, 1845 (Darwin Correspondence 3: 140).
5. Bowler (1988). Bowler plays down the role of the zoologist Huxley, a "pseudo-Darwinian," with the botanist Hooker emerging as the true Darwinist.

Chapter 1

1. The paper "Biological Globalization: The Other Grain Invasion" (2006) was cited with permission from the authors, A. L. Olmstead and P. W. Rhode.

Chapter 2

1. As quoted in Bickham (2008: 71).
2. The many sources on topic include Burnett (1999); Clarke (2015); Lukacs (2005, 2013); Robinson ([1994] 2015); Stewart (2013); and Taber (2006).
3. See, for example, Bender and Bender (1995: 20); Edgar (2010); Hunt (2013b); Kelly (2005: 486–8); and Offernans and Rosenthal (2008: 1219–22).
4. Note that West Africa has replaced South America as the dominant producer of chocolate in the twenty-first century but it is also *cacao forestera* that is mostly produced.
5. "… Vos confiseurs de Paris s'opposent à ce commerce … un pot de marmalade." Lettre 395, M. de Voltaire à Madame La Marquise du Deffand. *Correspondance complète de la marquise Du Deffand avec ses amis le Président Hénault, Montesquieu, D'Alembert, Voltaire, Horace Walpole, Classée dans l'ordre chronologique et sans suppression*, tome second. M. de Lescure, ed. (Paris: Henri Plon, 1865), 181.
6. See: "History of Jam," Museu de la Confitura, Torrent, Girona, Catalonia, Spain (https://www.museuconfitura.com/en [accessed May 14, 2021]).
7. See National Library of Scotland, Library Reference Acc. 10708/1.
8. British Museum Research Collection Database. Sheila O'Connell, "London 1753", BM 2003, cat. nos. 5.86, 5.87 (http://www.britishmuseum.org/research/search_the_collection_database/term_details.aspx?bioId=15164).
9. See "The Valley of Apricots." Swiss Food and Facts, Confederation Suisse, Expo Milano 2015, Padiglione Svizzero CH (https://www.eggandplant.ch/apricots-from-the-valais-region [accessed May 20, 2021]).

10 Library and Archives, Canada, record 1883-05-12, 458, *Indians of North America*, vol. XXVII, no. 19, 296.
11 In fact, the price of saffron may have increased since the demise of France's saffron production in the late nineteenth century, now having recovered in places like Quercy, but still small relative to Iran's dominant supply that has remained under some market sanctions in the last few decades despite periodic diplomatic thaws.
12 This is according to the *Republican Campaign Textbook*, 1880, Statistical Tables, 207.

Chapter 6

1 Plant-lore Archive—a collection of material relating to the folklore of plants, in the author's possession; see www.plant-lore.com (accessed May 20, 2021).

BIBLIOGRAPHY

Abel, L. 1876. *Garten-Architektur*. Wien: Lehmann and Wentzel.
Achan, J., A. O. Talisuna, A. Erhart, A. Yeka, J. K. Tibenderana, F. N. Baliraine, P. J. Rosenthal, and U. D'Alessandro. 2011. "Quinine, an old anti-malarial drug in a modern world: Role in the treatment of malaria." *Malaria Journal*, 10: 144.
Agnoun, Y., S. S. H. Biaou, M. Sié, R. S. Vodouhè, and A. Ahanchédé. 2012. "The African Rice Oryza glaberrima Steud.: Knowledge Distribution and Prospects." *International Journal of Biology*, 4: 158–80.
Aidoo, K. 1992. "Lesser-known fermented plant foods." *National Center for Biotechnology Information Press, U. S. National Library of Medicine*. National Academy of Science.
Alcorn, K. 2020. "From specimens to commodities: The London nursery trade and the introduction of exotic plants in the early nineteenth century." *Historical Research*, 93: 715–33.
Alexander, M. 2006. *A Surrey garland*. Newbury: Countryside Books.
Allan, M. 1977. *Darwin and his Flowers: The Key to Natural Selection*. London: Faber and Faber.
Allingham, E. G. 1924. *A romance of the rostrum being the business life of Henry Stevens and the history of thirty-eight King Street, together with some account of famous sales held there during the last hundred years*. London: Witherby.
Allston, R. F. W. 1846. "Memoir of the introduction and planting of rice in South Carolina." *DeBow's Review*, 39: 521–37.
Amery, P. F. S. 1905. "Twenty-second report of the Committee on Devonshire folk-lore." *Report and Transactions of the Devonshire Association for the Advancement of Science*, 37: 111–21.
Amphlett, J., and C. Rea. 1909. *The botany of Worcestershire*. Birmingham: Cornish.
Anderson, E. 1903. *The book of corn, a complete treatise upon the culture, marketing and use of maize in America and elsewhere*. New York: Orange Judd.
Anderson, S. C. 2010. "Pharmacy and empire: The British pharmacopoeia as an instrument of Imperialism 1864 to 1932." *Pharmacy in History*, 52: 112–21.
Anderson, W. 1992. "Climates of opinion: Acclimatization in nineteenth-century France and England." *Victorian Studies*, 35: 135–57.
André, É. 1879. *L'art des jardins: Traité général de la composition des parcs et jardins*. Paris: Masson.
Angus, W. J. 2001. "United Kingdom wheat pool." In A. P. Bonjean and W. J. Angus (eds.), *The World Wheat Book: A History of Wheat Breeding*, 103–26. Paris: Lavoisier Publishing.
Anon. 1978. *Hector Guimard*. London: Academy Editions.
Anon. 1992. "Alexandra Rose Day." *Reader's Digest*, June: 114.
Anon. n.d. *The language of flowers*. London: Warne's Bijou Books.
Aragão, F. J. L., R. P. V. Brondani, and M. L. Burle. 2011. "*Phaseolus*." In C. Kole (ed.), *Wild Crop Relatives: Genomic and Breeding Resources*, 223–36. Berlin and Heidelberg: Springer.
Arber, A. 1920. *Water plants: A study of aquatic angiosperms*. Cambridge: Cambridge University Press.
Arnold, D. 2006. *The Tropics and the Traveling Gaze: India, Landscape, and Science, 1800–1856*. Seattle, WA: University of Washington Press.

Arslan, M., S. Şar, and B. Sözen-Şahne. 2017. "The first military and non-official pharmacopoeias of the Ottoman Empire." *Journal of History Culture and Art Research*, 6: 82–91.

Atanasov, A. G., B. Waltenberger, E. M. Pferschy-Wenzig, T. Linder, C. C. Wawrosch, P. Uhrin, V. Temml, L. Wang, S. Schwaiger, E. H. Heiss, J. M. Rollinger, D. D. Schuster, J. M. Breuss, V. Bochkov, M. D. Mihovilovic, B. Kopp, R. Bauer, V. M. Dirsch, and H. Stuppner. 2015. "Discovery and resupply of pharmacologically active plant-derived natural products: A review." *Biotechnology Advances*, 33: 1582–1614.

Atterbury, P. (ed.). *A. W. N. Pugin: Master of Gothic Revival*. New Haven: Yale University Press.

Austin, H. A. 1928. *History and development of the sugar beet industry*. Washington, DC: National Press Building.

AWB Global Technical Services Group Australia Wheat Industry. 2007. Available online: http://muehlenchemie.de/downloads-future-of-flour/FoF_Kap_07.pdf (accessed February 11, 2018).

Baas, P., T. Fujii, N. Kato, M. Mertz, S. Noshiro, and G. Thijsse. 2021. "Mogami Tokunai's wood collection from Hokkaido, Japan: An early record of Ainu wood culture." *IAWA Journal*, 42: 1–16.

Babayan, É. A., M. D. Mashkovskii, and A. N. Oboimakova. 1978. "From the first Russian State Phamacopoeia to the eleventh edition of the State Pharmacopoeia of the USSR (the 200th anniversary of the first Russian State Pharmacopoeia)." *Pharmaceutical Chemistry Journal*, 12: 1543–7 (Translated from *Khimiko-Farmatsevticheskii Zhurnal*, 12 (12) (1978): 3–8.).

Badu-Apraku, B., and M. A. B. Fakorede. 2006. "*Zea mays* L." In M. Brink and G. E. Belay (eds.), *Cereals and pulses*, 229–37. Plant Resources of Tropical Africa (PROTA). Wageningen: Backhuys Publishers/CTA.

Baer, W. 1999. "Zur Entwicklung der botanischen Malerei beim Porzellan." In Anon., *Das Flora Danica—Service 1790–1802. Höhepunkt der Botanischen Porzellanmalerei*, 187–247. Copenhagen: Kongelige Udstillingsfond.

Baker, A. E. 1854. *Glossary of Northamptonshire words and phrases*, vol. 1. London: Russell Smith.

Baker, H. G. 1970. *Plants and civilization*, 2nd edn. Belmont: Wadsworth.

Baker, M. 1974. *Discovering the folklore and customs of love and marriage*. Princes Risborough: Shire.

Balole, T. V., and G. M. Legwaila. 2006. "*Sorghum bicolor* (L.) Moench." In M. Brink and G. E. Belay (eds.), *Cereals and pulses*, 169–75. Plant Resources of Tropical Africa (PROTA) Foundation. Wageningen: Backhuys Publishers/CTA.

Barbour, O. L. 1872. *Reports of cases in law and equity in the Supreme Court of the State of New York*, vol. 61. Albany: Little.

Barnes, C. R. 1893. "On the food of green plants." *Botanical Gazette*, 18: 403–11.

Barr and Sugden. 1862. *Barr and Sugden's Spring seed catalogue and guide to the flower and kitchen garden*. London: Barr and Sugden.

Bateson, W. 1905. Letter to A. Sedgwick, April 18, 1905. Facsimile. Norwich: John Innes Historical Collections, John Innes Centre.

Batsaki, Y., et al. 2016. *The botany of empire in the long eighteenth century*. Washington: Dumbarton Oaks.

Battiscombe, G. 1969. *Queen Alexandra*. London: Constable.

Beale, R. A. 1927. "A scheme of classification of the varieties of rice found in Burma." Pune: *Agricultural Research Institute Bulletin*, 167.

Bean, W. J., and G. Taylor. 1970. *Trees and shrubs hardy in the British Isles*, 8th edn., vol. I. London: Murray.

Beckert, S. 2014. *Empire of cotton: A new history of global capitalism*. London: Allen Lane.

Behçet, A. 2014. "The source-synthesis-history and use of atropine." *Journal of Academic Emergency Medicine*, 13: 2–3.

Belay, G. 2006a. "*Triticum turgidum* L." In M. Brink and G. E. Belay (eds.), *Cereals and pulses*, 183–7. Plant Resources of Tropical Africa (PROTA) Foundation. Wageningen: Backhuys Publishers/CTA.

Belay, G. 2006b. "*Triticum aestivum* L." In M. Brink and G. E. Belay (eds.), *Cereals and pulses*, 176–82. Plant Resources of Tropical Africa (PROTA). Wageningen: Backhuys/CTA.

Bender, D. A., and A. E. Bender. 1995. *A dictionary of food and nutrition*. Oxford: Oxford University Press.

Bender, D. 2009. *Oxford dictionary of food and nutrition*. Oxford: Oxford University Press.

Berger, R. 2008. "Ayurveda and the Making of the Urban Middle Class." In D. Wujastyk and F. M. Smith (eds.), *Modern and global ayurveda: Pluralism and paradigms*, 101–15. Albany: State University of New York Press.

Berkeley, M. J. 1846. "Observations, botanical and physiological on the potato murain." *Journal of the Horticultural Society of London*, 1: 9–34.

Berkeley, M. J. 1869. *Gardeners' Chronicle* [News item], 45: 1157.

Berra, T. M. 2015. "Darwin's harbingers." *Linnean*, 31: 11–19.

Bertomeu-Sánchez, J. R., and A. Nieto-Galan. 2006. *Chemistry, medicine and crime: Mateu J B Orfila (1787–1853) and his times*. Sagamore Beach, MA: Science History Publications.

Bezançon, G. 1993. "Le riz cultivé d'origine Africaine *Oryza glaberrima* Steud. et les formes sauvages et adventices apparentées: diversité, relations génétiques et domestications." Thèse de Doctorat, Université de Paris Sud, Centre D'Orsay, France.

Bezançon, G., and S. Diallo 2006. "*Oryza glaberrima* Steud." In M. Brink and G. E. Belay (eds.), *Cereals and pulses*, 106–10. Plant Resources of Tropical Africa (PROTA). Wageningen: Backhuys/CTA.

Bhattacharya, N. 2016. "From materia medica to the pharmacopoeia: Challenges of writing the history of drugs in India." *History Compass*, 14: 131–9.

Bickham, T. 2008. "Eating the Empire: Intersections of food, cookery and imperialism in eighteenth-century Britain." *Past and Present*, 198: 71–109.

Bilston, S. 2008. "Queens of the garden: Victorian women gardeners and the rise of the gardening advice text." *Victorian Literature and Culture*, 36: 1–19.

Bisgrove, R. 2008. *William Robinson: The wild gardener*. London: Frances Lincoln.

Blackman, F. F. 1905. "Optima and limiting factors." *Annals of Botany [London]*, 19: 281–95.

Blackman, F. F., and G. L. C. Matthaei. 1905. "Experimental researches on vegetable assimilation and respiration. IV. A quantitative study of carbon dioxide assimilation and leaf temperature in natural illumination." *Proceedings of the Royal Society of London B*, 76: 402–60.

Blanning, T. 2007. *The pursuit of glory: Europe 1648–1815*. London: Viking.

Bleichmar, D. 2008. "Atlantic competitions: Botany in the eighteenth-century Spanish empire." In James Delbourgo and Nicholas Dew (eds.), *Science and empire in the Atlantic world*. New York: Routledge.

Blevins, S. M., and M. S. Bronze 2010. "Robert Koch and the 'golden age' of bacteriology." *International Journal of Infectious Diseases*, 14: e744–e751.

Blomfield, R. T. 1892. *The formal garden in England*. London: MacMillan.

Boitard, Pierre. 1833. *Les instrumens aratoires*. Paris: A. Ledoux.

Le Bon Jardinier. 1844. *Almanach pour l'an*. Paris: Audot.

Bonjean, A. P., G. Doussinault, and J. Stragliati. 2001. "French wheat pool." In A. P. Bonjean and W. J. Angus (eds.), *The world wheat book: A history of wheat breeding*, 128–65. Paris: Lavoisier.

Bonnemain, B. n.d. "Pharmacopoeias in France after the French Revolution." Available online: www.histpharm.org/ISHPWG%20France2.pdf (accessed August 12, 2017).

Bosse, J. F. W. 1829. *Vollständiges Handbuch der Blumen-Gärtnerei*. Hannover: Hahn.

Bosse, J. F. W. 1861. *Vollständiges Handbuch der Blumen-Gärtnerei*, 3rd rev. edn. Hannover: Hahn.

Boulger, G. S. 1878. "The merit of carpet bedding." *Garden*, 14: 4.

Bourke, A. 1991. "Potato blight in Europe in 1845: The scientific controversy." In J. A. Lucas, R. C. Shattock, D. S. Shaw, and L. R. Cooke (eds.), *Phytophthora*, 12–24. Cambridge: Cambridge University Press.

Boussingault, J. B. 1864. "De la végétation dans l'obscurité." *Annales des Scences Naturelles (Paris)*, 1: 314–24.

Bowler, P. J. 1988. *The non-Darwinian revolution: Reinterpreting a historical myth*. Baltimore: John Hopkins University Press.

Boysen-Jensen, P. 1910. "Über die Leitung des phototropischen Reizes in Avena–keimpflanzen." *Berichte der Deutschen Botanischen Gesellschaft*, 28: 118–20.

Bradford, C. A. 1939. *Hugh Morgan, Apothecary-in-Ordinary to Queen Elizabeth*. London: Heron.

Bradshaw, J. E. 2016. *Plant breeding: Past, present and future*. Dordrecht: Springer.

Brady, D. S. 1964. "Relative prices in the nineteenth century." *Journal of Economic History* 24: 145–203.

Braithwaite, M. E. 2012. *The wildflowers of a Berwickshire bard – George Henderson of Chirnside (1800–64)*. Kelso: Berwickshire Naturalists Club.

Bratránek, F. T. 1853. *Beiträge zu einer Aesthetik der Pflanzenwelt*. Leipzig: Brockhaus.

Braun, A. C. 1951. "Recovery of tumor cells from effects of the tumor-inducing principle in crown gall." *Science*, 113: 651–3.

Braun, A. C., and R. J. Mandle. 1948. "Studies on the inactivation of the tumor-inducing principle in crown gall." *Growth*, 12: 255–69.

Brennan, E., L. -A. Harris, and M. Nesbitt. 2013. "Jamaican lace-bark: Its history and uncertain future." *Textile History*, 44: 235–53.

Brightman, R. 1956. "Perkin and the dyestuffs industry in Britain." *Nature*, 177: 815–21.

Britten, J. 1878. "Plant-lore notes to Mrs Latham's West Sussex superstitions." *Folk-lore Record*, 1: 155–9.

Britten, J., and R. Holland. 1886. *A dictionary of English plant-names*. London: Trübner, for the English Dialect Society.

Broadbent, M. 1991. *The new great vintage wine book*. New York: Knopf.

Brock, E. K. 2014. "New patterns in old places: Forest history for the global present." In A. C. Isenberg (ed.), *The Oxford handbook of environmental history*, 154–77. Oxford: Oxford University Press.

Brock, T. D. 1999. *Robert Koch: A life in medicine and bacteriology*. Washington, DC: American Society of Medicine.

Brockway, L. 1979. *Science and colonial expansion. The role of Kew Royal Botanic Gardens*. New York: Academic Press.

Brown, M. 2007. "Medicine, quackery and the free market: The 'war' against Morison's pills and the construction of the medical profession, *c*.1830–*c*.1850." In M. S. R. Jenner and P. Wallis (eds.), *Medicine and the market in England and its colonies, c.1450–1850*, 238–61. Basingstoke: Palgrave Macmillan.

Brown, R. 1831. *Observations on the organs and mode of fecundation in Asclepiadeae and Orchideae*. London: Taylor.
Browne, J. 1995. *Charles Darwin: Voyaging*. London: Cape.
Browne, J. 1997. "Biogeography and empire." In N. Jardine, J. A. Secord, and E. C. Spary (eds.), *Cultures of natural history*, 305–20. Cambridge: Cambridge University Press.
Browne, J. 2002. *Charles Darwin: The power of place*. London: Cape.
Buchner, A. 1828. "Uber das Rigatellishe Fiebermittel and über eine in der Weidenrinde entdeckte alcaloidische Substanz" (On Rigatelli's antipyretic [i.e. anti-fever drug] and on an alkaloid substance discovered in willow bark). *Repertorium für die Pharmacie*, 29: 405–20.
Buller, A. H. R. 1919. *Essays on wheat*. New York: MacMillan.
Burbidge, F. W. 1874. *Domestic floriculture*. Edinburgh: Blackwood.
Burbidge, F. W. 1885. *The chrysanthemum: Its culture, classification, and nomenclature*, 2nd edn. London: The Garden.
Burne, C. S. 1883. *Shropshire folk-lore*. London: Trübner.
Burnett, John. 1999. *Liquid pleasures: A social history of drinks in modern Britain*. London: Routledge.
Burns, E. 2007. *The smoke of the gods: A social history of tobacco*. Philadelphia: Temple University Press.
Cain, P. J. 1999. "Economics and empire: The metropolitan context." In A. Porter (ed.), *The Oxford history of the British Empire: The nineteenth century*, 31–52. Oxford: Oxford University Press.
Cain, P. J., and A. G. Hopkins. 2016. *British imperialism 1688–2015*. Abingdon: Routledge.
Campbell, C. G. 1995. "Buckwheat. *Fagopyrum esculentum* (Polygonaceae)." In J. Smartt and N. W. Simmonds (eds.), *Evolution of crop plants*, 2nd edn., 409–12. Singapore: Longman.
Campbell, C. G. 1997. *Buckwheat. Fagopyrum esculentum Moench. Promoting the conservation and use of underutilized and neglected crops*. Gatersleben: International Plant Genetic Resources Institute.
de Candolle, A. P., and A. de Candolle. 1824–73. *Prodromus systematis naturalis regni vegetabilis*. 17 vols. Paris: Treuttel et Würtz.
Carleton, M. A. 1900. *The basis for the improvement of American wheats*. United States Department of Agriculture Bulletin No. 24. Washington DC: Government Printing Office.
Carney, Judith A. 2001. *Black rice: The African origins of rice cultivation in the Americas*. Cambridge, MA: Harvard University Press.
Carter, N. C. 2007. "San Diego olives: Origins of a California industry." *Journal of San Diego History*, 54: 137–61.
Cavara, F. 1897a. "Eziologia di alcune malattie di piante coltivate." *Le Stazioni Sperimentale Agrarie Italiane*, 30: 482–509.
Cavara, F. 1897b. "Tuberculosi della vite. Intorno alla eziologia di alcune malattie di piante coltivate, nota." *Le Stazioni Sperimentale Agrarie Italiane*, 30: 483–7.
Chambers, R. 1850. *Vestiges of the natural history of creation*, 8th edn. London: Churchill.
Chaplin, J. E. 1993. *An anxious pursuit: Agricultural innovations and modernity in the lower south 1730–1815*. Chapel Hill: University of North Carolina Press.
Chodubski, A. 1982. "Ludwik Młokosiewicz (1831–1909)—pioneer badań flory i fauny Kaukazu." *Kwartalnik Historii Nauki i Techniki*, 27: 421–8.
Christison, R. 1829. *A treatise on poisons: In relation to medical jurisprudence, physiology, and the practise of physic*. Edinburgh: Black.
Christnacher, F., A. Hartmann, and A. Baumann. 2015. *L'épopée des pépinières Baumann de Bollwiller (1730–1990) et leur influence sur les parcs et jardins à travers le monde*. Available

online: http://arbo-haut-sundgau.com/articlesFA/Article_complet-Pepiniere_Baumann_Bollwiller_janv2015.pdf (accessed May 25, 2017).

Clarence-Smith, W. G. 2000. *Cocoa and chocolate 1765–1914*. London: Routledge.

Clarke, O. 2015. *The history of wine in 100 bottles: From Bacchus to Bordeaux and beyond.* London: Pavilion.

Cloudsley, P. 1999. "The art of the shaman: Healing in the Peruvian Andes." *Journal of Museum Ethnography*, 11: 63–78.

Cleene, M. de, and M. C. Lejeune. 2003. *Compendium of symbolic and ritual plants in Europe*. 2 vols. Ghent: Man and Culture.

Clowse, C. D. 1971. *Economic beginnings in colonial South Carolina, 1670–1730*. Columbia: University of South Carolina Press.

Coates, C. 2004. *The wines of Bordeaux: Vintages and tasting notes, 1952–2003*. Berkeley: University of California Press.

Coats, A. M. 1969. *The quest for plants: A history of the horticultural explorers*. London: Studio Vista.

Coats, A. M. 1992. *Garden shrubs and their histories*. New York: Simon and Schuster.

Coclanis, P. A. 2010. "The rice industry of the United States." In S. D. Sharma (ed.), *Rice: Origin, antiquity and history*, 411–31. Boca Raton: CRC Press.

Coffin, A. I. 1845. *A botanic guide to health, and the natural pathology of disease*. Leeds: Moxon.

Coleman, W. 1977. *Biology in the nineteenth century: Problems of form, function, transformation*. Cambridge: Cambridge University Press.

Conn, H. J. 1942. "Validity of the genus Alcaligenes." *Journal of Bacteriology*, 44: 342–60.

Conrad, Barnaby. 1997. *Absinthe: History in a bottle*. San Francisco: Chronicle Books.

Coote, A., A. Haynes, J. Philp, and S. Ville. 2017. "When commerce, science, and leisure collaborated: The nineteenth-century global trade boom in natural history collections." *Journal of Global History*, 12: 319–39.

Cornish, C. 2013. "Curating science in an age of empire: Kew's Museum of Economic Botany." PhD thesis, Royal Holloway and Bedford New College, University of London.

Cornish, C. 2015. "Nineteenth-century museums and the shaping of disciplines: Potentialities and limitations at Kew's Museum of Economic Botany." *Museum History Journal*, 8: 8–27.

Cornish, C. 2017. "Botany behind glass: The vegetable kingdom on display at Kew's Museum of Economic Botany." In C. Berkowitz and B. Lightman (eds.), *Science museums in transition: Cultures of display in nineteenth-century Britain and America*, 188–214. Pittsburgh: University of Pittsburgh Press.

Cornish, C., and Driver, F. 2019. "'Specimens distributed': The circulation of objects from Kew's Museum of Economic Botany, 1847–1914." *Journal of the History of Collections,* 32: 327–40. DOI: 10.1093/jhc/fhz008.

Cornish, C., and M. Nesbitt. 2014. "Chapter 20. Historical perspectives on western ethnobotanical collections." In J. Salick, K. Konchar, and M. Nesbitt (eds.), *Curating biocultural collections: A handbook*, 271–93. Kew: Royal Botanic Gardens, Kew.

Cowen, D. L. 2001. *Pharmacopoeias and related literature in Britain and America, 1618 to 1847*. Aldershot: Ashgate Variorum.

Critz, J. M., A. Olmstead, and P. Rhode. 1999. "'Horn of Plenty': The globalization of Mediterranean horticulture and the economic development of southern Europe, 1880–1930." *Journal of Economic History*, 59: 316–52.

Crockett, R. (ed.). 1947. *Leaves from the life of a country doctor*. Edinburgh: Ettrick.

Cronk, Q. C. B. 1989. "The past and present vegetation of St Helena." *Journal of Biogeography*, 16: 47–64.
Crosby, A. W. 1986. *Ecological imperialism: The biological expansion of Europe, 900–1900*. Cambridge: Cambridge University Press.
Crow, J. F. 1998. "Perspectives anecdotal, historical and critical commentaries on genetics, 90 years ago: The beginning of hybrid maize." *Genetics*, 148: 923–8.
Cruthers, N., D. Carr, and R. Laing. 2009. "The New Zealand flax fibre industry." *Textile History*, 40: 103–11.
Curry, H. A. 2016. *Evolution made to order. Plant breeding and technological innovation in twentieth-century America*. Chicago and London: University of Chicago Press.
Curtis, H. A. 1924. "Fertilizers: The world supply." *Foreign Affairs*, 2: 436–45.
Cuveland, H. de. 2006. *Natur im Aquarell. Meisterwerke der Wiener Hofmaler Johann und Joseph Knapp*. Munich: Prestel.
Dacombe, M. (ed.) 1951. *Dorset up along and down along*, 3rd edn. Dorchester: Dorset Federation of Women's Institutes.
Darlington, C. D. 1961. *Darwin's place in history*. New York: Macmillan.
Dartnell, G. E., and E. H. Goddard. 1894. *A glossary of words used in the county of Wiltshire*. London: English Dialect Society.
Darwin, C. 1859. *On the origin of species or the preservation of favoured races in the struggle for life*. London: Murray.
Darwin, C. 1868. *The variation of animals and plants under domestication*, 2 vols. London: Murray.
Darwin, C. 1876. *The effects of cross and self-fertilisation in the vegetable kingdom*. London: Murray.
Darwin, C. 1880. *The power of movement in plants*. London: Murray.
Darwin, E. 1800. *Phytologia, or the philosophy of agriculture and gardening*. London: Johnson.
Davidson, A. 1999. *The Oxford companion to food*. Oxford: Oxford University Press.
Davidson, A. 2006. *The Oxford companion to food*. 2nd edn. Oxford: Oxford University Press.
Davies, J. 2000. *Saying it with flowers: The story of the flower shop*. London: Headline.
Daws, M. I., J. Davies, E. Vaes, R. van Gelder, and H. W. Pritchard. 2007. "Two-hundred-year seed survival of *Leucospermum* and two other woody species from the Cape Floristic region, South Africa." *Seed Science Research*, 17: 73–9.
Dean, R. 1903. *The dahlia: Its history and cultivation*. London: MacMillan.
Dean, W. 1987. *Brazil and the struggle for rubber: A study in environmental history*. Cambridge: Cambridge University Press.
Deane, T., and T. Shaw 1975. *The folklore of Cornwall*. London: Batsford.
DeBary, A. 1876. "Researches into the nature of the potato-fungus—*Phytophthora infestans*." *Journal of the Royal Agricultural Society*, 12: 239–68.
Debouck, D. G., and J. Smartt. 1995. "*Phaseolus* spp. (Leguminosae-Papilionoideae)." In J. Smartt and N. W. Simmonds (eds.), *Evolution of crop plants*, 2nd edn., 287–93. Singapore: Longman.
Dehnen-Schmutz, K., J. Touza, C. Perrings, and M. Williamson. 2007a. "The horticultural trade and ornamental plant invasions in Britain." *Conservation Biology*, 21: 224–31.
Dehnen-Schmutz, K., J. Touza, C. Perrings, and M. Williamson. 2007b. "A century of the ornamental plant trade and its impact on invasion success." *Diversity and Distributions*, 13: 527–34.
Dehnen-Schmutz, K., and M. Williamson. 2006. "*Rhododendron ponticum* in Britain and Ireland: Social, economic and ecological factors in its successful invasion." *Environment and History*, 12: 325–50.

Denham, A. 2013. "Herbal medicine in nineteenth century England: The career of John Skelton." MA thesis, University of York, UK.
Desmond, R. 1979. "The problems of transporting plants." In J. Harris (ed.), *The garden: A celebration of one thousand years of British gardening*, 99–104. London: Mitchell Beazley.
Desmond, R. 1992. *The European discovery of the Indian flora*. Oxford: Oxford University Press/Royal Botanic Gardens.
Desmond, R. 2007. *History of the Royal Botanic Gardens Kew*, 2nd edn. Kew: Royal Botanic Gardens, Kew.
Destler, C. M. 1968. "'Forward Wheat' for New England: The correspondence of John Taylor of Caroline with Jeremiah Wadsworth, in 1795." *Agricultural History*, 42: 201–10.
Diagre, D. 2002. "La naissance du jardin botanique de la Societe Royale d'Horticulture des Pays-Bas: Attendue et placée sous les meilleurs auspices." *Scientiarum Historia*, 28: 63–94.
Dickerson, B. C. 1992. *The old rose advisor*. Portland: Timber Press.
Dierig, D., H. Blackburn, D. Ellis, and M. Nesbitt. 2014. "Chapter 8. Curating seeds and other genetic resources for ethnobiology." In J. Salick, K. Konchar, and M. Nesbitt (eds.), *Curating biocultural collections: A handbook*. Kew: Royal Botanic Gardens.
Dimitrieva, M. 2006. 138. Ephraim Mose Lilien. In B. Lange (ed.), *Geschichte der Bildenden Kunst in Deutschland* 8, 138. Darmstadt: Wissenschaftliche Buchgesellschaft.
Dodel, A., and Dodel-Port, C. 1883. *Erläuternder Text zum anatomisch-physiologischen Atlas der Botanik für Hoch- und Mittelschulen*. Esslingen am Neckar: Schreiber.
Doggett, H., and K. E. P. Rao. 1995. "Sorghum (*Sorghum bicolor*) (Gramineae-Andropogoneae)." In J. Smartt and N. W. Simmonds (eds.), *Evolution of crop plants*, 2nd edn. Singapore: Longman.
Dondlinger, P. T. 1908. *The book of wheat*. London: Orange Judd.
Doughty, R. W. 2000. *The eucalyptus: A natural and commercial history of the gum tree*. Baltimore: Johns Hopkins University Press.
Dowling, J. E. 2001. *Neurons and networks: An introduction to behavioral neuroscience*. Cambridge, MA: Belknap.
Downing, A. J. 1841. *A treatise on the theory and practice of landscape gardening, adapted to North America*. New York: Wiley and Putnam.
Drayton, R. 2000. *Nature's government: Science, imperial Britain, and the "improvement" of the world*. New Haven, CT: Yale University Press.
Dritsas, L. 2006. "Civilising missions, natural history and British industry: Livingstone in the Zambezi." *Endeavour*, 30: 50–4.
Dritsas, L. 2010. *Zambesi: David Livingstone and expeditionary science in Africa*. London: Tauris.
Driver, F., and L. Martins (eds.). 2005. *Tropical visions in an age of empire*. Chicago: Chicago University Press.
Duncan, A. 1818. "Discourse, 1816." *Memoirs of the Caledonian Horticultural Society*, 2: 344–54.
Duncan, L. L. 1896. "Fairy beliefs and other folklore notes from County Leitrim." *Folk-lore*, 7, 161–83.
Dunlop, D. M., and T. C. Denston. 1958. "The history and development of the British pharmacopoeia." *British Medical Journal*, 2 (5107): 1250–2.
Durand, A. A. B. 1851. "On Hydrastis Canadensis." *American Journal of Pharmacy*, 18: 112–18.
Duthie, R. 1988. *Florists' flowers and societies*. Princes Risborough: Shire.

Dutrochet, [R. J. H.] 1832. "'Mémoire sur les organes aérifères des végétaux, et sur l'usage de l'air que contiennent ces organes." *Annales des sciences naturelles*, 25: 242–59.

Dwarswaard, A., and M. Timmer. 2010. *Van windhandel tot wereldhandel: Canon van de bloembollen*. Houten: Hes and De Graaf.

Dyke, G. V. 1993. *John Lawes of Rothamsted. Pioneer of science, farming and industry*. Harpenden: Hoos.

Edgar, B. 2010. "The power of chocolate." *Archaeology*, 63 (6): 20–5.

El Azeem, A., T. Badawi, M. A. Maximos, I. R. Aidy, R. A. Olaoye, and S. D. Sharma. 2010. "History of rice in Africa." In S. D. Sharma (ed.), *Rice: Origin, antiquity and history*, 373–410. Boca Raton: CRC Press.

Elgood, G. S., and G. Jekyll. 1904. *Some English gardens*. London: Longmans, Green.

Elliott, B. 1981. "Mosaiculture: Its origins and significance." *Garden History*, 9: 76–98.

Elliott, B. 1986. *Victorian gardens*. London: Batsford.

Elliott, B. 2004. *The Royal Horticultural Society: A history 1804–2004*. Chichester: Phillimore.

Elliott, P., C. Watkins, and S. Daniels. 2007. "Combining science with recreation and pleasure: Cultural geographies of nineteenth-century arboretums." *Garden History*, 35, Suppl. 2: 6–27.

Elliott, P., C. Watkins, and S. Daniels. 2011. *The British arboretum: Trees, science and culture in the nineteenth century*. London: Pickering and Chatto.

Emdad-ul Haq, M. 2000. *Drugs in south Asia: From the opium trade to the present day*. London: Palgrave.

Endersby, J. 2008. *Imperial nature: Joseph Hooker and the practices of Victorian science*. Chicago: Chicago University Press.

Endersby, J. 2016. *Orchid: A cultural history*. Kew: Kew Publishing.

Endersby, J. 2017. "Review of John F. Mcdiarmid Clark, *Bugs and the Victorians*. Yale University Press, 2009." *Reviews in History*, Institute of Historical Research, University of London, 2017. DOI: 10.14296/RiH/issn.1749.8155. Available online: http://www.history.ac.uk/reviews/review/924 (accessed May 20, 2021).

Englemann, T. W. 1882. "Uber Sauerstoffausscheidung von Pflanzenzellen im Mikrospectrum." *Botanische Zeitung*, 40: 419–26.

Englemann, T. W. 1884. "Untersuchungen über die quantitativen Beziehungen zwischen Absorption des Lichtes und Assimilation in Pflanzenzellen." *Botanische Zeitung*, 44: 43–52, 64–9.

Fabre, E., and Dunal, F. 1853. "Observations sur les maladies régnantes de la vigne." *Bulletin de la Société Centrale d'Agriculture du Département de l'Hérault*, 40: 11–75.

Falkus, M. E. 1966. "Russia and the international wheat trade, 1861–1914." *Economica*, 33 (132): 416–29.

Fan, F. -T. 2004. *British naturalists in Qing China: Science, empire and cultural encounter*. Cambridge, MA: Harvard University Press.

Fara, P. 2003. *Sex, botany and empire*. Cambridge: Icon Books.

Farooq, S., R. U. Rehman, T. B. Pirzadah, B. Malik, F. A. Dar, and I. Tahir. 2016. "Cultivation, agronomic practices, and growth performance of buckwheat." In M. Zhou, I. Kreft, S. -H. Woo, N. Chrungoo, and G. Wieslander (eds.), *Molecular breeding and nutritional aspects of buckwheat*, 299–319. London: Academic Press.

Federico, G., and A. T. Junguito. 2016. *World trade, 1800–1938: A new data-set*. European Historical Economics Society Working Paper 93. Available online: www.ehes.org/EHES_93.pdf (accessed May 20, 2021).

Feldman, M. 2001. "Origin of cultivated wheat." In A. P. Bonjean and W. J. Angus (eds.), *The world wheat book: A history of wheat breeding*, 3–56. Paris: Lavoisier.

Feldman, M., F. G. H. Lupton, and T. E. Miller. 1995. "Wheats *Triticum* spp. (Gramineae-Triticineae)." In J. Smartt and N. W. Simmonds (eds.), *Evolution of crop plants*, 2nd edn., 184–92. Singapore: Longman.

Fesenko, A. N., N. N. Fesenko, O. I. Romanova, and I. N. Fesenko. 2016. "Crop evolution of buckwheat in Eastern Europe: microevolutionary trends in the secondary center of buckwheat genetic diversity." In M. Zhou, I. Kreft, S. -H. Woo, N. Chrungoo, and G. Wieslander (eds.), *Molecular breeding and nutritional aspects of buckwheat*, 99–107. Oxford: Academic Press.

Ffennell, M. C. 1898. "The shrew ash in Richmond Park." *Folk-lore*, 9: 330–6.

Fiala, J. L. 1988. *Lilacs: The genus Syringa*. Portland: Timber Press.

Fintelmann, G. A. 1834. "Ueber Anwendung und Behandlung von Blattzierpflanzen und deren Verbindung mit Rankgewächsen für Schmuckgruppen." *Verhandlungen des Vereins zur Beförderung des Gartenbaues in den Königlich Preußischen Staaten*, 10: 359–71.

Fleming, J. 1865. "The royal gardens of Sans Souci, at Potsdam." *Gardeners' Chronicle*, 25: 1159.

Flemming, W. 1882. *Zellsubstanz, Kem und Zelltheilung*. Leipzig: Vogel.

Foerster, K. 1911. *Winterharte Blütenstauden und Sträucher der Neuzeit*. Leipzig: Weber.

Foerster, K. 1924. *Winterharte Blütenstauden und Sträucher der Neuzeit*, 3rd edn. Leipzig: Weber.

Foerster, K., and C. Schneider. 1927. *Das Dahlienbuch*. Berlin: Verl. der Gartenschönheit.

Fraser, A. S. 1973. *The hills of home*. London: Routledge and Kegan Paul.

Frazer, J. G. 1922. *The Golden Bough* (abridged edn.). London: Macmillan.

Freedman, P. 2012. "The medieval spice trade." In J. Pilcher (ed.). *The Oxford handbook of food history*, 324–40. Oxford: Oxford University Press.

Fried, A., and R. Elman. 1969. *Charles Booth's London*. London: Hutchinson.

Friedrich, C. n.d. "German pharmacopoeias." Available online: www.histpharm.org/ISHPWG%20Germany.pdf (accessed August 12, 2017).

Friedrich-Sander, S. 2006. "Erasmus Ritter von Engert." In S. Schulze (ed.), *Gärten. Ordnung Inspiration Glück*, 158–9. Ostfildern: Hatje Cantz.

Friend, H. 1884. *Flowers and flower lore*. London: Sonnenschein.

Fussell, B. 1992. *The story of corn*. New York: Knopf.

Gager, C. S. 1940. "Elizabeth G. Britton and the movement for the preservation of native American wildflowers." *Journal of the New York Botanical Garden*, 41: 137–42.

Galloway, J. N., A. M. Leach, A. Bleeker, and J. W. Erisman. 2017. "A chronology of human understanding of the nitrogen cycle." *Philosophical Transactions of the Royal Society B*, 368: 20130120.

Gamble, R. 1979. *Chelsea child*. London: British Broadcasting Corporation.

Ganeshram, Ramin. "Cracking coconut's history." *AramcoWorld*, January/February: 27–9.

Ganzinger, K. 1961. "Die Entwicklung der Arzneimittelprüfung im Spiegel der österreichischen Pharmakopöen zwischen 1855 und 1906." *Osterreichische Apotheker-Zeitung*, 15: 472–7.

Gardener, W. 1971. "Robert Fortune and the cultivation of tea in the United States." *Arnoldia*, 31: 1–18.

Ganzel, B. 2007. "Harvest technology corn combines, the Wessels Living History Farm." Available online: https://livinghistoryfarm.org/farminginthe50s/machines_13.html (accessed February 18, 2018).

Garris, A. J., T. H. Tai, J. Coburn, S. Kresovich, and S. McCouch. 2005. "Genetic structure and diversity in *Oryza sativa* L." *Genetics*, 169: 1631–8.

Gayon, J. A., and D. T. A. Zallen. 1998. "The role of the Vilmorin Company in the promotion and diffusion of the experimental science of heredity in France, 1840–1920." *Journal of the History of Biology*, 31: 241–62.

Gest, H. 1991. "The legacy of Hans Molisch (1856–1937), photosynthesis savant." *Photosynthesis Research*, 30: 49–59.

Gest, H. 2002. "History of the word photosynthesis and evolution of its definition." *Photosynthesis Research*, 73: 7–10.

Geus, A. 1995. "Natur im Druck—Geschichte und Technik des Naturselbstdrucks." In A. Geus (ed.), *Eine Ausstellung zur Geschichte und Technik des Naturselbstdrucks*, 9–28. Marburg an der Lahn: Basilisken-Presse.

Gilbert, R. M. 1984. "Caffeine consumption." In G. Spiller (ed.), *The methylxanthine beverage and foods: Chemistry, consumption and health effects*, 185–213. New York: Liss.

Gillbank, L. 1986. "The origins of the Acclimatisation Society of Victoria: Practical science in the wake of the Gold Rush." *Historical Records of Australian Science*, 6: 359–74.

Gilpin, W. S. 1832. *Practical hints upon landscape gardening: With some remarks on domestic architecture as connected with scenery*. London: Cadell.

Gisel, B. J. 2008. *Nature's beloved son: Rediscovering John Muir's botanical legacy*. Berkeley, CA: Heyday.

Glendinning, D. R. 1983. "Potato introductions and breeding up to the early 20th century." *New Phytologist*, 94: 479–505.

Glenny, G. 1847. *The standard of perfection for the properties of flowers and plants*. London: Houlston and Stoneman.

Goldstein, D. (ed.). 2015. *Oxford companion to sugar and sweets*. Oxford: Oxford University Press.

Gondola, I., and P. P. Papp. 2010. "Origin, geographical distribution and phylogenetic relationships of common buckwheat (*Fagopyrum esculentum* Moench.)." In J. Dobranszki (ed.), Buckwheat 2. *European Journal of Plant Sciences and Biotechnology*, 4, Special Issue 2: 17–32.

Goeschke, F. 1899. *Die Staudengewächse*. Berlin: Siegismund.

Goldthorp, W. O. 2009. "Medical classics: An account of the foxglove and some of its medicinal uses by William Withering, published 1785." *British Medical Journal*, 338: b2189.

González, A. n.d. "Spain: An account on the history of the Spanish pharmacopoeias." Available online: www.histpharm.org/ISHPWG%20Spain.pdf (accessed August 12, 2017).

Goodman, M. M. 1995. "*Zea mays* (Gramineae-Maydeae)." In J. Smartt and N. W. Simmonds (eds.), *Evolution of Crop Plants*, 2nd edn., 192–202. Singapore: Longman.

Goody, J. 1993. *The culture of flowers*. Cambridge: Cambridge University Press.

Gorer, R. 1976. "Nurseries, hybrids and plant collectors." *Archives (London)*, 12: 126–30.

Govindjee, and D. Krogmann. 2004. "Discoveries in oxygenic photosynthesis (1727–2003)." *Photosynthesis Research*, 80: 15–57.

Grabbe, H. 1897. *Unsere Staudengewächse: Kultur, Verwendung und Beschreibung derselben*. Stuttgart: Ulmer.

Granett, J., M. A. Walker, L. Kocsis, and A. D. Omer. 2001. "Biology and management of grape phylloxera." *Annual Review of Entomology*, 46: 387–412.

Grantham, G. 1984. "The shifting locus of agricultural innovation in nineteenth century Europe: The case of the agricultural experiment stations." *Research in Economic History*, suppl. 3: 191–214.

Grasse, S. 2016. *Colonial spirits: A toast to our drunken history*. New York: Abrams.

Greenstone, G. 2010. "The history of bloodletting." *British Columbia Medical Journal*, 52: 12–14.

Greenwood, D. 2008. *Antimicrobial drugs: Chronicle of a twentieth century triumph*. Oxford: Oxford University Press.

Grimspoon, L. 2005. "History of cannabis as a medicine." DEA Statement for Administrative Law Judge.

Grollman, A. P., and Z. Jarkovsky. 1975. "Emetine related alkaloids." In J. W. Corcoran and F. E. Hahn (eds.), *Mechanism of action of antimicrobial and antitumor agents. Antibiotics*, vol. 3, 420–35. Berlin: Springer.

Gross, B. L., and Z. Zhao. 2014. "Archaeological and genetic insights into the origins of domesticated rice." *Proceedings of the National Academy of Sciences*, 111: 6190–7.

Grotz, K. 2015. *modellSCHAU / modelSHOW*. Berlin: Botanischer Garten und Botanisches Museum Berlin.

Grove, R. H. 1995. *Green imperialism: Colonial expansion, tropical island Edens and the origins of environmentalism, 1600–1860*. Cambridge: Cambridge University Press.

Guardian. 2007. "Britain is built on sugar: Our sweet tooth defines us." October 12.

Haberlandt, G. 1884. *Physiologische Pflanzenanatomie im Grundriss dargestellt*. Leipzig: Engelmann.

Haberlandt, G. 1902. "Kulturversuche mit isolierten Pflanzenzellen." *Sitzungsberichte der Kaiserlichen Akademie der Wissenschaften. Mathematisch-Naturwissenschaftliche Classe*, 111: 69–92.

Habrich, C. (2013). "Über Goethes *Bryophyllum calycinum* in der Leopoldina-Ausgabe und seine Bedeutung als Arzneimittel in der anthroposophischen Medizin." *Acta Historica Leopoldina*, 62: 151–76.

Hacker, J., and F. Blum-Oehler. 2007. "In appreciation of Theodor Escherich." *Nature Reviews Microbiology*, 5: 902.

Hales, S. 1738. *Statical essays: Containing vegetable staticks; or an account of some statical experiments on the sap in vegetables*, vol. 1. London: Innys and Manby.

Halford, D. G. 1993. *Old lawnmowers*. Princes Risborough: Shire.

Hall, M. B. 1984. *All scientists now: The Royal Society in the nineteenth century*. Cambridge: Cambridge University Press.

Hallier, E. 1891. *Grundzüge der landschaftlichen Gartenkunst*. Leipzig: Haessel.

Hamm, W. 1845. *Die landwirthschaftlichen Geräthe und Maschinen Englands*. Braunschweig: Vieweg.

Hampel, C. 1893. *Gartenbeete und Gruppen*. Berlin: Parey.

Hancock, J. F. 2012. *Plant evolution and the origin of crop species*, 3rd edn. Wallingford: CABI.

Hanzl-Wachter, L. 2002. "254. Sitzzimmer im Appartement von Kaiser Franz II. (I.) in der Wiener Hofburg." In G. Frodl (ed.), *Das 19. Jahrhundert*, 554. Munich: Prestel.

Harris, D. R. 2005. "Origins and spread of Agriculture." In G. Prance and M. Nesbitt (eds.), *The Cultural History of Plants*, 13–26. Oxon: Routledge.

Harte, J. 2004. *Explore fairy traditions*. Loughborough: Explore Books.

Hartwig, J. 1861. *Die Anlage von Lustgebieten und Blumengärten*. Weimar: Voigt.

Haydak, M. H. 1957. "Petro Prokopovich—Ukrainian beekeeper, teacher and scientist." *American Bee Journal*, 97: 474–5.

Hayter, A. 1968. *Opium and the romantic imagination*. Berkeley CA: University of California Press.

Hazareesingh, S. 2012. "Cotton, climate and colonialism in Dharwar, western India, 1840–1880." *Journal of Historical Geography*, 38: 1–17.

Headrick, D. R. 1981. *The tools of empire: Technology and European imperialism in the nineteenth century*. Oxford: Oxford University Press.

Hedgcock, G. G. 1905. "Some of the results of three years' experiments with crown gall." *Science*, 22: 120–2.

Heine, H. H. 1967. "Ave Caesar, botanici te salutant". L'épopée napoléonienne dans la botanique. *Adansonia*, sér. 2, 7: 115–40.

Heiser, C. B. 1981. *Seed to civilization, the story of food*, 2nd edn. San Francisco: Freeman.

Heizmann, W. 2002. "Die Esche als Weltenbaum in der mythischen Uberlieferung der Nordgermanen." *Berichte aus der Bayerischen Landesanstalt für Wald- und Forstwirtschaft*, 34: 62–70.

Hemming, J. 2015. *Naturalists in paradise: Wallace, Bates and Spruce in the Amazon*. London: Thames and Hudson.

Hennig, M. 2006. "Johan Fredrick Lewis, Childe Hassam." In S. Schulze (ed.), *Gärten. Ordnung Inspiration Glück*, 172–3, 182–3. Ostfildern: Hatje Cantz.

Hesseltine, C. W. 1965. "A millennium of fungi, food and fermentation." *Mycologia*, 57: 149–97.

Heurtel, P., and M. Lenoir (eds.). 2016. *Les vélins du Muséum national d'histoire naturelle*. Paris: Citadelles and Mazenod.

Heyden, T. 2014. "Was there a time British people couldn't buy olive oil?" *Magazine Monitor of BBC News*, September 16.

Heynhold, G. 1840. *Nomenclator botanicus hortensis oder alphabetische und synonymische Aufzählung der in den Gärten Europa's cultivierten Gewächse*. Dresden: Arnold.

Heynhold, G. 1846. *Nomenclator botanicus hortensis oder alphabetische und synonymische Aufzählung der in den Gärten Europa's cultivierten Gewächse*, vol. 2. Dresden: Arnold.

Hibberd, S. 1857. *Rustic adornments for homes of taste*, 2nd edn. London: Groombridge.

Hibberd, S. 1858. *Garden favourites: Their history, properties, cultivation, propagation, and general management in all seasons*. London: Groombridge.

Hill, C. S. 1863. *Wild-flowers and their uses: A book for children*. London and Edinburgh: Chambers.

Hobhouse, H. 1985. *Seeds of change: Five plants that transformed mankind*. London: Sidgwick and Jackson.

Hodgson, B. 2001. *In the arms of Morpheus: The tragic history of morphine, laudanum and opium and patent medicines*. Vancouver, Canada: Greystone.

Hoffenberg, P. H. 2001. *An empire on display: English, Indian, and Australian exhibitions from the Crystal Palace to the Great War*. Berkeley, CA: University of California Press.

Hogg, J. 1911. *The microscope, its history, construction and application*. London and New York: Routledge.

Hooke, R. 1665. *Micrographia: Or some physiological descriptions of minute bodies made by magnifying glasses, with observations and inquiries thereupon*. London: Martyn and Allestry.

Hooker, W. 1855. *Museum of Economic Botany: Or, a popular guide to the useful and remarkable vegetable products of the Museum of the Royal Gardens of Kew*. London: Longman.

Hooper, R. 1820. *Medical dictionary*, 4th edn. London: Longman, Hurst, Rees, Orme.

Horn, J. 2011. "Transformation of Virginia: Tobacco, slavery and empire: introduction." *The William and Mary Quarterly*, 68: 327–31.

Horn, T. 2005. *Bees in America: How the honeybee shaped America*. Lexington: University Press of Kentucky.

Howarth, O. J. R. 1931. *The British Association for the advancement of Science: A retrospect 1831–1931*. Centenary edn. London: BAAS.

Humboldt, A. von. 1847. *Kosmos*, vol. 2. Stuttgart: Kosmos.
Hunt, P. 2013a. "Maya and Aztec chocolate history and antecedents." *Electrum Magazine*. Available online: http://www.electrummagazine.com/2013/04/maya-and-aztec-chocolate-history-and-antecedents/ (accessed May 20, 2021).
Hunt, P. 2013b. *Wine journeys: Myth and history*. San Diego: Cognella Academic Press.
Hunt, T. F. 1915. *The cereals in America*. New York: Orange Judd.
Hutton, R. 1996. *The stations of the sun: A history of the ritual year in Britain*. Oxford: Oxford University Press.
Illingworth, J. 1994. "Ruskin and gardening." *Garden History*, 22: 218–33.
Ingen-Housz, J. 1779. *Experiments upon vegetables, discovering their great power of purifying the common air in the sunshine and of inuring it in the shade and at night: To which is joined A new method of examining the accurate degree of salubrity of the atmosphere*. London: Elmsley and Payne.
Ingen-Housz, J. 1780. *Expériences sur les végétaux*. Paris: Didot.
Ingen-Housz, J. 1796. "Food of plants and the renovation of the soil." Appendix to the *Outlines of the Fifteenth Chapter of the Proposed General Report from the Board of Agriculture*. London: Elmsley and Payne.
Inglis, L. 2013. *Georgian London: Into the streets*. London: Penguin.
Irvine, F. R. 1974. *West African crops*. London: Oxford University Press.
Isenberg, A. C. 2014. "Seas of grass: Grasslands in world environmental history." In A. C. Isenberg (ed.), *The Oxford handbook of environmental history*, 133–54. Oxford: Oxford University Press.
Jackson, M. E. 1816. *The florist's manual: Hints for the construction of a gay flower garden*. London: Colburn.
Jäger, H. 1845. *Ideenmagazin zur zweckmässigsten Anlegung und Ausstattung geschmackvoller Hausgärten und anderer kleiner Gartenanlagen sowohl für den Luxus als zur Nutzung*. Weimar: Voigt.
Jäger, H. 1858. *Die Verwendung der Pflanzen in der Gartenkunst, oder Gehölz, Blumen und Rasen*. Gotha: Scheube.
Jäger, H. 1864. *Illustrirtes allgemeines Gartenbuch*. Leipzig: Spamer.
Janot, M. M. 1893. "The ipecac alkaloids." Cited in R. H. F. Manske and H. L. Holmes (eds., 1953), *The alkaloids: Chemistry and physiology*, vol. III, 363–94. New York: Academic Press.
Jansen, P. C. M. 2006. "*Fagopyrum esculentum* Moench." In M. Brink and G. E. Belay (eds.), *Cereals and pulses*, 72–6. Plant Resources of Tropical Africa (PROTA). Wageningen: Backhuys/CTA.
Jarvis, P. J. 1979. "Plant introductions to England and their role in horticultural and silvicultural innovation, 1500–1900." In H. S. A. Fox and R. A. Butlin (eds.), *Change in the countryside: Essays on rural England, 1500–1900*, 145–64. London: Institute of British Geographers (Special Publication 10).
Jeffrey, E. C. 1917. *The anatomy of woody plants*. Chicago: University of Chicago Press.
Jeffreys, D. 2005. *Aspirin: The remarkable story of a wonder drug*. New York: Bloomsbury Press.
Jekyll, G. 1882. "Colour in the flower garden." *Garden*, 22 (562): 177.
Jekyll, G. 1908. *Colour in the flower garden*. London: Country Life.
Jekyll, G. 1914. *Gardens for small country houses*, 3rd edn. London. Country Life.
Jensen, V. G., and A. Schæffer. 1960. *The history of pharmacy in Denmark. A survey* (Theriaca vol. 6). Copenhagen: Dansk Farmacihistorisk Selskab.

Ji, H. -F., X. -J. Li, and H. -Y. Zhang. 2009. "Natural products and drug discovery." *EMBO Reports*, 10: 194–200.

Jiggens, J. 2012. *Sir Joseph Banks and the question of hemp*. [Indooroopilly, Queensland]: Jayjay.

Johannsen, W. L. 1903. "Über Erblichkeit in Populationen und reinen Linien. Eine Beitrag zur Beleuchtung schwebender Selektionsfragen" [On heredity in pure lines and populations. A contribution to pending questions of selection]. Jena: Fischer.

Johnson, H. 1985. *The world atlas of wine*. New York: Simon and Schuster.

Johnson, R. 2003. *British Imperialism*. Basingstoke: Palgrave Macmillan.

Johnston, G. 1853. *The botany of the eastern borders*. London: Van Voorst.

Jones, B. 1967. *Design for death*. London: Deutsch.

Jones, D. F. 1918. "The effect of inbreeding and crossbreeding upon development." *Proceedings of the National Academy of Sciences of the United States of America*, 4: 246–50.

Joshi, B. D., and R. S. Rana. 1995. "Buckwheat (*Fagopyrum esculentum*)." In J. T. Williams (ed.), *Cereals and pseudocereals*, 85–127. London: Chapman and Hall.

Joyaux, F. 1998. "Jacques-Louis Descemet, premier obtenteur de roses en France." *Hommes et plantes*, 25: 36–44.

Juàrez-Barrera, F., A. Bueno-Hernandez, J. J. Morrone, A. Barahona-Echeverria, and D. Espinoza. 2018. "Recognizing spatial patterns of biodiversity during the nineteenth century: The roots of contemporary Biogeography." *Journal of Biogeography*, 45: 995–1002.

Judelson, H. S. 1997. "The genetics and biology of *Phytophthora infestans*: Modern approaches to a historical challenge." *Fungal Genetics and Biology*, 22: 65–76.

Jussieu, A. L. de. 1789. *Genera plantarum*. Paris: Herissant, Barrois.

Kabbaj, H., A. T. Sall, A. Al-Abdallat, M. Geleta, A. A. Amri, A. Filali-Maltouf, B. Belkadi, R. Ortiz, and F. M. Bassi. 2017. "Genetic diversity within a global panel of durum wheat (*Triticum durum*) landraces and modern germplasm reveals the history of alleles exchange." *Frontiers in Plant Science*, 8: 1277.

Kaiser, H. 2008. "Von der Pflanze zur Chemie—Die Frühgeschichte der 'Rheumamittel'." *Zeitschrift für Rheumatologie*, 67: 252–62.

Kaneda, C. 2010. "History of rice in Japan." In S. D. Sharma (ed.), *Rice: origin, antiquity and history*, 154–82. Boca Raton: CRC Press.

Karch, S. R. 2003. *A history of cocaine: The mystery of Coca Java and the Kew plant*. London: Royal Society of Medicine Press.

Kaufmann, S. H., and U. E. Schnaible. 2005. "100th anniversary of Robert Koch's Nobel Prize for the discovery of the tubercle bacillus." *Trends in Microbiology*, 13: 469–75.

Kelleher, T. 2017. Historian, Old Sturbridge Village, pers. comm. March 9.

Kelly, C. J. 2005. "Effects of theobromine should be considered in future studies." *American Journal of Clinical Nutrition*, 82: 486–8.

Kelly, J. D. 2010. "The story of bean breeding." White paper prepared for BeanCAP and PBG Works on the topic of dry bean production and breeding research in the U. S. Available online: bean.css.msu.edu/_pdf/Story_of_Bean_Breeding_in_the_US.pdf (accessed February 10, 2017).

Kelways Plants Ltd. 2017. Kelways since 1815. Available online: https://www.kelways.co.uk/page/kelway-family/38/ (accessed May 29, 2017).

Kemp, E. 1850. *How to lay out a small garden*. London: Bradbury and Evans.

Kennan, G. F. 1979. *The decline of Bismarck's European order, Franco-Russian relations, 1875–1890*. Princeton, NJ: Princeton University Press.

Keogh, L. 2017. "The Wardian case: How a simple box moved the plant kingdom." *Arnoldia*, 74: 2–13.

Keogh, L. 2019. "The Wardian case: Environmental histories of a box for moving plants." *Environment and History*, 25: 219–44.
Keogh, L. 2020. *The Wardian case: How a simple box moved plants and changed the world.* Chicago: University of Chicago Press.
Khush, G. S. 1987. "Rice breeding: Past, present and future." *Journal of Genetics*, 66: 195–216.
Khush, G. S. 2010. "Taxonomy and origin of rice." In R. K. Singh, U. S. Singh, and G. S. Khush (eds.), *Aromatic rices*, 5–13. New Delhi: Oxford and IBH.
Kilpatrick, J. 2007. *Gifts from the gardens of China: The introduction of traditional Chinese garden plants to Britain 1698–1862*. London: Frances Lincoln.
Kilpatrick, J. 2014. *Fathers of botany: The discovery of Chinese plants by European missionaries*. Kew: Kew Publishing.
Kinahan, G. H. 1884. "Co. Donegal, May Eve." *Folk-lore Journal*, 2: 90–1.
King, J. 1875. *The American dispensatory*. Cincinnati: Wilstach and Baldwin.
King-Hele, D. 1977. *Doctor of revolution: The life and genius of Erasmus Darwin*. London: Faber and Faber.
Kingsbury, N. 2009. *Hybrid. The history and science of plant breeding*. Chicago: The University of Chicago Press.
Kiple, K., and O. Kriemhild (eds.). 2000. *The Cambridge world history of food*, vol. 2. Cambridge: Cambridge University Press.
Kletter, C. 2015. "The civil pharmacopoeias of Austria." Available online: www.histpharm.org/ISHPWG%20Austria.pdf (accessed August 12, 2017).
Knop, W. G. 1865. "Quantitative Untersuchungen über den Ernährungsprozess der Pflanzen." *Landwirtschaftliche Versuchsstation Poland*, 7: 93–107.
Koerner, L. 1999. *Linnaeus: nature and nation*. Cambridge, MA and London: Harvard University Press.
Kraus, G. 1894. *Geschichte der Pflanzeneinführungen in die europäischen botanischen Gärten*. Leipzig: Engelmann.
Krausch, H. -D. 2003. *"Kaiserkron und Päonien rot ...": Entdeckung und Einführung unserer Gartenblumen*. München: Dölling und Galitz.
Kreft, I., G. Wieslander, and B. Vombergar. 2016. "Bioactive flavonoids in buckwheat grain and green parts." In M. Zhou, I. Kreft, S. -H. Woo, N. Chrungoo, and G. Wieslander (eds.), *Molecular breeding and nutritional aspects of buckwheat*, 161–7. Boston: Academic Press.
Krikorian, A. D., and D. L. Berquam. 1969. "Plant cell and tissue culture: The role of Haberlandt." *Botanical Review*, 35: 59–88.
Kumar, S., et al. 2013. "Characterising long-term land use / cover change from 1850 to 2000 using a nonlinear bi-analytical model." *Ambio*, 42: 285–97.
La Barre, W. 1960. "Twenty years of peyote studies." *Current Anthropology*, 1: 45–60.
Labbé, E. 2002. "Japanese regulations." In A. J. Fletcher, L. D. Edwards, A. W. Fox, and P. D. Stonier (eds.), *Principles and practice of pharmaceutical medicine*, 487–507. London: Wiley.
Lacroix, B., and V. Citovsky. 2013. "Crown gall tumors." *Brenner's encyclopedia of genetics*, 2nd edn., 2: 236–9.
Lack, H. W. 1999. "Plant illustrations on wood blocks. A magnificent Japanese xylotheque of the early Meiji period." *Curtis's Botanical Magazine*, ser. 6, 16: 124–34.
Lack, H. W. 2001. "The plant self impressions prepared by Humboldt and Bonpland in tropical America." *Curtis's Botanical Magazine*, ser. 6, 18: 218–29.
Lack, H. W. 2004. *Jardin de la Malmaison. Der Garten von Kaiserin Josephine*. Munich: Prestel.
Lack, H. W. 2006. *Florilegium Imperiale. Botanische Schätze für Kaiser Franz I. von Österreich*. Munich: Prestel.

Lack, H. W. 2010. "Die Kiefern des Meng Hao-Ran." *Kosmos Osterreich*, 34: 15, 19–21.
Lack, H. W. 2016. *A garden Eden / Ein Garten Eden / Un jardin d'Eden*, 3rd edn. Köln: Taschen.
Lack, H. W., with D. J. Mabberley. 1998. *The* Flora Graeca *story. Sibthorp, Bauer and Hawkins in the Levant*. Oxford: Oxford University Press.
Laidò, G., G. Mangini, F. Taranto, A. A. Gadaleta, A. Blanco, L. Cattivelli, D. Marone, A. M. Mastrangelo, R. R. Papa, and P. De Vita. 2013. "Genetic diversity and population structure of tetraploid wheats (*Triticum turgidum* L.) estimated by SSR, DArT and pedigree data." *PLoS ONE*, 8 (6): 67280.
Lamb, M. 2008. https://brocku.niagaragreenbelt.com/listings/76-parks-gardens-a-conservation-areas/341-comfort-maple.html (accessed 15 June 2021).
Lane, A. 2007. "Joseph Conrad's 'The Planter of Malata': Timing, and the forgotten adventures of the silk plant 'Arghan'." *Textile*, 5: 276–99.
Lange, W. 1907. *Gartengestaltung der Neuzeit*. Leipzig: Weber.
Latham, C. 1878. "Some west Sussex superstitions lingering in 1868." *Folk-lore Record*, 1: 1–67.
Laurent, A. 2016. *The botanical wall chart; art from the golden age of scientific discovery*. London: Ilex.
Lausen-Higgins, J., and P. Lusby. 2008. "Pineapple-growing: Its historical development and the cultivation of the Victorian pineapple pit at the Lost Gardens of Heligan, Cornwall." *Sibbaldia*, 6: 29–39.
Law, R. (ed.). 2002. *From slave trade to "legitimate" commerce: The commercial transition in nineteenth-century West Africa*. Cambridge: Cambridge University Press.
Lawrence, F. 2007. "Sugar rush." *Guardian*, February 15.
Leather, E. M. 1912. *The folk-lore of Herefordshire*. Hereford: Jakeman and Carver.
Ledermann, F. n.d. "Switzerland: An account on the story of the Swiss pharmacopoeias." Available online: www.histpharm.org/ISHPWG%20Switzerland.pdf (accessed August 12, 2017).
Lee, M. R. 2005. "Curare: The South American arrow poison." *Journal of the Royal College of Physicians of Edinburgh*, 35: 83–92.
Lee, M. R. 2011. "The history of *Ephedra* (ma-huang)." *Journal of Royal College of Physicians of Edinburgh*, 41: 78–84.
Lee, R. 2006. "Industrial revolution, commerce, and trade." In S. Berger (ed.), *A companion to nineteenth-century Europe*, 44–55. Oxford: Blackwell.
Leeuwenhoek, A. van, and S. Hoole. 1800. *The select works of Antony van Leeuwenhoek, containing his microscopical discoveries in many of the works of nature*. London: Hoole.
Legg, P. 1986. *So merry let us be—the living tradition of Somerset cider*. Bridgwater: Somerset County Library Service.
Lemmon, K. 1968. *The golden age of plant hunters*. London: Phoenix House.
Leonard, M. W. 2015. *Poppyganda: The historical and social impact of a flower*. London: Uniform Press.
Leong-Salobir, C. Y. 2011. *Food culture in colonial Asia: A taste of Empire*. Abingdon, Oxon: Routledge.
Lichtwark, A. 1909. *Park- und Gartenstudien*. Berlin: Cassirer.
Liebig, J. 1840. *Die Organische Chemie in ihrer Anwendung auf Agricultur und Physiologie*. Braunschweig: Vieweg.
Linares, O. F. L. 2002. "African rice (*Oryza glaberrima*): history and future potential." *Proceedings of the National Academy of Sciences of the United States of America*, 99: 16360–5.

Lindley, J. 1840. *Botanical Garden (Kew). Copy of the report made to the committee appointed by the Lords of the Treasury in January 1838 to inquire into the management, andc. of the Royal Gardens at Kew*. London: Parliamentary Papers.

Linnaeus, C. 1753. *Species plantarum*. Stockholm: Salvius.

Lioi, L., and A. R. Piergiovanni. 2013. "European common bean." In M. Singh, H. D. Upadhyaya, and I. S. Bisht (eds.), *Genetic and genomic resources of grain legume improvement*, 11–40. London: Elsevier.

Loadman, J. 2005. *Tears of the tree: The story of rubber—modern marvel*. Oxford: Oxford University Press.

Lockwood, L., J. Wilson, and M. Fagg. 2001. *Botanic gardens of Australia: A guide to 80 gardens*. Sydney: New Holland.

Loftas, T., and J. Ross. (eds.). 1995. *Dimensions of need: An atlas of food and agriculture*. Food and Agriculture Organization of the United Nations. Rome: FAO. Available online: http://www.fao.org/3/u8480e/U8480E01.htm (accessed July 12, 2018).

Londo, J. P., Y. -C. Chiang, K. -H. Hung, T. -Y. Chiang, and B. A. Schaal. 2006. "Phylogeography of Asian wild rice, *Oryza rufipogon*, reveals multiple independent domestications of cultivated rice, *Oryza sativa*." *Proceedings of the National Academy of Sciences*, 103: 9578–83.

Loudon, G. 2015. *Object lessons: The visualisation of nineteenth-century life sciences*. London: Ridinghouse.

Loudon, J. C. 1822. *An encyclopaedia of gardening*. London: Longman.

Loudon, J. C. 1824. *An encyclopaedia of gardening*, 2nd edn. London: Longman.

Loudon, J. C. 1832. "[Review of Gilpin's *Practical hints upon landscape gardening*.]" *Gardener's Magazine*, 8: 700–2.

Loudon, J. C. 1834a. "Growing ferns and other plants in glass cases." *Gardener's Magazine*, 10: 162–3.

Loudon, J. C. 1834b. "Vegetation under glasses, without change of air." *Gardener's Magazine*, 10: 58–9.

Loudon, J. C. 1835. *An encyclopaedia of gardening*, new edn. London: Longman.

Loudon, J. C. 1838. *Arboretum und fruticetum Britannicum*. London: Longman.

Lovell, J. 2011. *The Opium War*. London: Picador.

Lucas, J. A. 1988. *Plant pathology and plant pathogens*, 3rd edn. Oxford: Blackwell.

Lukacs, P. 2005. *American vintage: The rise of American wine*. New York: Norton.

Lukacs, P. 2013. *Inventing wine: A new history of one of the world's oldest pleasures*. New York: Norton.

Lynn, M. 1999. "British policy, trade, and informal empire in the mid-nineteenth century." In A. Porter (ed.), *The Oxford history of the British Empire: The nineteenth century*, 101–21. Oxford: Oxford University Press.

Mabberley, D. J. 1985a. *Jupiter Botanicus: Robert Brown of the British Museum*. Braunschweig: Cramer; London: British Museum (Natural History).

Mabberley, D. J. 1985b. "'Die neuen Pflanzen von Ch. Huber Frères and Co. in Hyères'." *Taxon*, 34: 448–56.

Mabberley, D. J. 1991. Chapter 14 "The problem of 'older' names." In D. L. Hawksworth (ed.), *Improving the stability of names: Needs and options* [*Regnum Vegetabile* 123], 123–34. Königstein: Koeltz.

Mabberley, D. J. 1992. *Tropical rain forest ecology*, 2nd edn. Glasgow: Blackie; New York: Chapman and Hall.

Mabberley, D. J. 2000. *Arthur Harry Church: The anatomy of flowers*. London: Merrell and the Natural History Museum.

Mabberley, D. J. 2017. *Mabberley's Plant-book. A portable dictionary of plants, their classification and uses*, 4th edn. Cambridge: Cambridge University Press.

Mabberley, D. J. 2019. *Botanical revelation: European encounters with Australian plants before Darwin*. Sydney: NewSouth.

MacCaughey, V. 1912. "The guavas of the Hawaiian Islands." *Bulletin of the Torrey Botanical Club*, 44: 513–24.

McClung, C. E. 1929. *Handbook of microscopic technique*. New York: Hober.

McCook, S. 2016. "'Squares of Tropic Summer': The Wardian case, Victorian horticulture, and the logistics of global plant transfers, 1770–1910." In P. Manning and D. Rood (eds.), *Global scientific practice in an age of revolutions, 1750–1850*, 199–215. Pittsburgh, PA: University of Pittsburgh Press.

McCook, S., and J. Vandermeer. 2015. "The Big Rust and the Red Queen: Long-term perspectives on coffee rust research." *Phytopathology*, 105: 1164–73.

McCracken, D. P. 1997. *Gardens of empire: Botanical institutions of the Victorian British Empire*. Leicester: Leicester University Press.

Macdonald, I. A. W., L. Ortiz, J. E. Lawesson, and J. B. Nowak. 1988. "The invasion of highlands in Galapagos by the red quinine-tree *Cinchona succirubra*." *Environmental Conservation*, 15: 215–20.

Macintosh, C. 1853. *The book of the garden*, vol. 1. Edinburgh: Blackwood.

Macintosh, C. 1855. *The book of the garden*, vol. 2. Edinburgh: Blackwood.

McKelvey, S. D. 1928. *The lilac: A monograph*. New York: MacMillan.

MacKenzie, J. M. 2009. *Museums and empire: Natural history, human cultures and colonial identities*. Manchester: Manchester University Press.

Maddock, J. 1792. *The florist's directory*. London: the author.

Mahood, M. M. 2008. *The poet as botanist*. Cambridge: Cambridge University Press.

Main, J. 1835. *The villa and cottage florists' directory*. London: Whittaker and Co.

Mair, V. H., and E. Hoh. 2009. *The true history of tea*. London: Thames and Hudson.

Major, J. 1852. *The theory and practice of landscape gardening*. London: Longman.

Marco, C. P. A. 2003. "Nationalism and science: The 'new' American plants in 19th century North American materia medica." *Curare*, 26: 213–20.

Maries, C. 1881/2. "Rambles of a plant collector." *The Garden*, 20: 84–6, 239–40; 21: 20–1, 101–2, 367–8.

Markel, H. 2011. "Uber Coca: Sigmund Freud, Carl Koller, and cocaine." *Journal of the American Medical Association*, 305: 1360–1.

Marquand, E. D. 1906. "The Guernsey dialect and its plant names." *Transactions of the Guernsey Society of Natural Science and Local Studies*, 5: 31–47.

Marsh, J. 2016. "Health and medicine in the 19th century." Available online: www.vam.ac.uk/content/articles/h/health-and-medicine-in-the-19th-century/ (accessed May 20, 2021).

Martinsson, K. 2001. *Pelargoner: Kulturarv i kruka*. Stockholm: Prisma.

Marwick, E. W. 1975. *The folklore of Orkney and Shetland*. London: Batsford.

Matthews, L. G. 1962. *History of pharmacy in Britain*. London: Livingstone.

Maund, B. 1846. *The botanic garden*, vol. 11. London: Groombridge.

Mayhew, H. 1851. *London labour and the London poor*, vol. 1. London: Woodfall.

Medicus, F. C. 1806. *Beiträge zur Kultur exotischer Gewächse*. Mannheim: Löffler.

Meertens, H. C. C. 2006. "*Oryza sativa* L." In M. Brink and G. E. Belay (eds.), *Cereals and pulses*, 112–20. Plant Resources of Tropical Africa (PROTA) Foundation. Wageningen: Backhuys Publishers/CTA.

Mellilo, E. 2012. "The first Green Revolution: Debt peonage and the making of the nitrogen fertilizer trade 1840–1930." *American History Review*, 117: 1028–60.

Mendel, G. J. 1866. "Versuche über Pflanzen-Hybriden" [Experiments concerning plant hybrids]. *Verhandlungen des naturforschenden Vereines in Brünn*, IV (1865): 3–47. [Reprinted in *Fundamenta Genetica*, ed. Ja. Křiženecký, 15–56. Prague: Czech Academy of Sciences, 1966.]

Mendel, G. 1901. "Experiments in plant hybridization introduced by W. Bateson." *Journal of the Royal Horticultural Society*, 26: 1–32.

Merezhko, A. F. 2001. "Wheat pool of European Russia." In A. P. Bonjean and W. J. Angus (eds.), *The world wheat book: A history of wheat breeding*, 257–88. Paris: Lavoisier.

Meyer, F. S., and F. Ries. 1904. *Die Gartenkunst in Wort und Bild*. Leipzig: Scholtze.

Meyer, F. S., and F. Ries. 1911. *Gartentechnik und Gartenkunst*. Leipzig: Scholtze.

Meyer, G. 1859/60. *Lehrbuch der schönen Gartenkunst*. Berlin: F. Riegel.

Miley, U., and J. Pickstone. 1987. "Medical reform; or, every man his own doctor: Medical botany around 1850." In R. Cooter (ed.), *Studies in the history of alternative medicine*, 139–53. Basingstoke: Macmillan.

Miller, W. 1915. "The prairie spirit in landscape gardening." *Illinois Agricultural Experiment Station Circular*, 184: 5.

Mintz, S. W. 1986. *Sweetness and power: The place of sugar in modern history*, repr. New York: Penguin.

Miracle, M. P. 1965. "The introduction and spread of maize in Africa." *Journal of African History*, 6: 39–55.

Mirbel, C. F. de B. 1815. *Élemens de physiologie végétale et de botanique*. Paris: Magimel.

Modest, W. 2018. "'A period of exhibitions': World's fairs, museums, and the laboring black body in Jamaica." In T. Barringer and W. Modest (eds.), *Victorian Jamaica*, 523–50. Durham, NC: Duke University Press.

Möring, G. M. 1949. "Die Hamburgische Familie Booth und ihre Bedeutung." Diss., Universität Hamburg, Germany.

Molina, J., M. Sikora, N. Garud, J. M. Flowers, S. Rubinstein, A. Reynolds, P. Huang, S. Jackson, B. A. Schaal, C. D. Bustamante, A. R. Boyko, and M. D. Purugganan. 2011. "Molecular evidence for a single evolutionary origin of domesticated rice." *Proceedings of the National Academy of Sciences*, 108: 8351–6.

Molisch, H. 1914. "Über die Herstellung von Photographien in einem Laubblatte." *Sitzungsberichte der Kaiserlichen Akademie der Wissenschaften. Mathematisch-Naturwissenschaftliche Classe*, 123: 923–31.

Molyneux-Berry, D. 1990. *The Sotheby's guide to classic wines and their labels*. New York: Ballantine.

Moody, J. W. T. 1971. "The reading of the Darwin and Wallace papers: An historical 'non-event'." *Journal of the Society for the Bibliography of Natural History*, 5: 474–6.

Moore, J. R. 1979. *The post-Darwinian controversies; a study of the Protestant struggle to come to terms with Darwin in Great Britain and America 1870–1900*. Cambridge: Cambridge University Press.

Morgunov, A. I. 1992. "Wheat and wheat breeding in the former USSR." *Wheat special report*, No 13. CIMMYT Wheat Program.

Morris, D. 1884. *Some objects of productive industry*. Part 4: *Native and other fibre plants*. Kingston, Jamaica: Institute of Jamaica.

Morris, J. P. 1925. "Some old-time superstitions of Devon." *Report and Transactions of the Devonshire Association for the Advancement of Science*, 56: 305–8.

Morris, W. 1882. *Hopes and fears for art.* London: Ellis and White.

Morton, A. G. 1981. *History of botanical science. An account of the development of botany from ancient times to the present day.* London: Academic Press.

Morton, R. S. 1968. "Dr Thomas ('Quicksilver') Dover, 1660–1742." *British Journal of Venereal Diseases,* 44: 342–6.

Morwood, W. 1973. *Traveller in a vanished landscape: The life and times of David Douglas.* London: Gentry.

Moxham, R. 2003. *Tea: Addiction, exploitation and empire.* London: Constable.

Mt. Pleasant, J. 2014. "Indigenous perceptions of biocultural collections." In J. Salick, K. Konchar, and M. Nesbitt (eds.), *Curating biocultural collections: A handbook,* 245–58. Kew: Royal Botanic Gardens, Kew.

Mueggler, E. 2011. *The Paper Road archive and experience in the botanical exploration of west China and Tibet.* Berkeley, CA: University of California Press.

Muijzenberg, E. W. B. van den. 1980. *A history of greenhouses.* Wageningen: the author.

Mukharji, A. 2009. "Pharmacology, indigenous knowledge, nationalism: A few words from the epitaph of subaltern science." In B. Pati and M. Harrison (eds.), *The social history of health and medicine,* 195–212. London: Taylor and Francis.

Murphy, D. 2007. *Plant breeding and biotechnology: Societal context and the future of agriculture.* Cambridge: Cambridge University Press.

Murray, R. G. E., and J. G. Holt. 2015. *The history of Bergey's manual.* New York: Wiley.

Musto, D. F. 1991. "Opium, cocaine and marijuana in American history." *Scientific American,* 265: 40–7.

Nabhan, G. 2009. *Where our food comes from: Retracing Nikolai Vavilov's quest to end famine.* Washington, DC: Island Press.

National Research Council (NRC). 1984. *Amaranth: Modern prospects for an ancient crop.* Washington DC: National Academy Press.

Natter, T. C. 2000. "Gustav Klimt: Frauenbildnisse." In T. G. Natter and G. Frodl (ed.), *Klimt und die Frauen,* 76–147. Köln: DuMont.

Nesbitt, M. 2005. "Grains." In G. Prance and M. Nesbitt (eds.), *The cultural history of plants,* 45–60. London and New York: Routledge.

Nesbitt, M. 2018. "Botany in Victorian Jamaica." In T. Barringer and W. Modest (eds.), *Victorian Jamaica,* 209–39. Durham, NC: Duke University Press.

Nester, E. 2008. "*Agrobacterium*: The natural genetic engineer 100 years later." [Online] APSnet Features. DOI: 10.1094/APSnetFeatures-2008-0608.

Neubert, W. 1857. "Neue englische Gewächshaus-Spritze." *Deutsches Magazin für Garten- und Blumenkunde,* 10: 81–3.

Neubert, W. 1879. "Gartenspritzen." *Deutsches Magazin für Garten- und Blumenkunde,* 32: 7–10.

Nguyen, D. T., and M. Rose. 1987–8. "Demand for tea in the UK 1874–1938: An econometric study." *Journal of Development Studies,* 24: 43–59.

Nicholson, G. (ed.). 1884–8. *The illustrated dictionary of gardening: A practical and scientific encyclopaedia of horticulture for gardeners and botanists.* London: Upcott Gill.

Nickelsen, K. 2015. *Explaining photosynthesis: Models of biochemical mechanisms.* Berlin: Springer.

Nicolson, J. 2006. *The perfect summer: Dancing into shadow in 1911.* London: Murray.

Noble, W. C. 2009. "Chilean trees and shrubs: A history of introduction to the British Isles." *Garden History,* 37: 151–73.

Nobelprize.org. 2014. Emil Fischer—Biographical. *Nobelprize.org. The Nobel Prize in Chemistry 1902 From the Nobel Lectures, Chemistry 1901–1921.* Nobel Media AB 2014.

Available online: http://www.nobelprize.org/nobel_prizes/chemistry/laureates/1902/fischer-bio.html (accessed August 10, 2017).

Noltie, H. J. 2002. *The Dapuri drawings. Alexander Gibson and the Bombay Botanic Gardens*. Sine loco: Antique Collectors' Club.

Noltie, H. J. 2007. *Robert Wight and the botanical drawings of Rungiah and Govindoo*. Edinburgh: Royal Botanic Garden Edinburgh.

Noltie, H. J. 2014. "*Doryanthes excelsa* and *Rafflesia arnoldii*: two 'swagger prints' by Edward Smith Weddell (1796–1858) and the work of the Weddell family of engravers (1814–1852)." *Archives of Natural History*, 41: 189–208.

Noltie, H. J. 2016. *The Cleghorn Collection: South Indian botanical drawings 1845 to 1860*. Edinburgh: Royal Botanic Garden Edinburgh.

Norton, M. 2006. "Tasting Empire: Chocolate and the European internalization of Mesoamerican aesthetics." *American Historical Review*, 111: 660–91.

Nugent, S. L. 2017. *The rise and fall of the Amazon rubber industry: An historical anthropology*. London: Routledge.

O'Brien, L., M. Morell, C. Wrigley, and R. Appels. 2001. "Genetic pools of Australian wheats." In A. P. Bonjean and W. J. Angus (eds.), *The world wheat book: A history of wheat breeding*, 611–48. Paris: Lavoisier.

Offer, A. 1999. "Costs and benefits, prosperity and security, 1870–1914." In A. Porter (ed.), *Oxford history of the British Empire*. Vol. 3: *The nineteenth century*, 690–711. Oxford: Oxford University Press.

Offernans, S., and W. Rosenthal. 2008. "Phenylethylamine." In S. Offermanns and W. Rosenthal (eds.), *Encyclopedia of molecular pharmacology*, 2nd edn., 1219–22. Berlin: Springer.

Olabimtan, K. 2013. "Church Missionary Society projects of agricultural improvement in nineteenth-century Sierra Leone and Yorubaland." In R. Law, S. Schwarz, and S. Strickrodt (eds.), *Commercial agriculture, the slave trade and slavery in Atlantic Africa*, 203–24. Woodbridge: Currey.

Olmstead, A. L., and P. W. Rhode. 2006. "Biological globalization: The other grain invasion." ICER Working Paper No. 9. Available online: https://papers.ssrn.com/sol3/papers.cfm?abstract_id=932056 (cited with permission from the authors; accessed March 11, 2018).

Olver, L. 2009. "The food timeline presents a history of paella." Available online: Foodtimeline.org (accessed February 16, 2018).

Onwueme, I. C. 1978. *The tropical tuber crops: Yams, cassava, sweet potato, and cocoyams*. New York: Wiley.

Orfila, M. 1814. *Traité des poisons*. Paris: Crochard.

O'Rourke, K., and J. G. Williamson. 1999. *Globalization and history: The evolution of a nineteenth century Atlantic economy*. Cambridge, MA: MIT Press.

Osborne, M. A. 1991. "A collaborative dimension of the European empires." In R. W. Home and S. G. Kohlstedt (eds.), *International science and national scientific identity*, 97–119. Dordrecht: Springer Netherlands.

Osborne, M. A. 2000. "Acclimatizing the world: A history of the paradigmatic colonial science." *Osiris*, 15: 135–51.

Osseo-Asare, A. D. 2008. "Bioprospecting and resistance: Transforming poisoned arrows into Strophantin pills in colonial Gold Coast, 1885–1922." *Social History of Medicine*, 21: 269–90.

Osseo-Asare, A. D. 2014. *Bitter roots: The search for healing plants in Africa*. Chicago: University of Chicago Press.

Osterhammel, J. 1999. "Britain and China 1842–1914." In A. Porter (ed.). *The Oxford history of the British Empire: The nineteenth century*, 146–69. Oxford: Oxford University Press.

Paddon, C. J. 2013. "High-level semi-synthetic production of the potent antimalarial artemisinin." *Nature*, 496: 528–32.

Palmer, K. 1976. *The folklore of Somerset*. London: Batsford.

Parker, R. M. 1990. *Burgundy: A comprehensive guide to the producers, appellations and wines*. New York: Simon and Schuster.

Parry, L. 2013. *William Morris textiles*. London: Victoria and Albert Museum.

Parsons, C. M., and K. S. Murphy. 2012. "Ecosystems under sail: Specimen transport in the eighteenth-century French and British Atlantics." *Early American Studies: An Interdisciplinary Journal*, 10: 503–29.

Pascoe, B. 2018. *Dark emu*, 2nd edn. Broome: Magabala.

Pasteur, L. 1909–14. "Scientific papers." Vol. XXXVIII, part 7. The Harvard Classics, New York: Collier. Available online: www.bartleby.com/38/7 (accessed December 27, 2016).

Pavord, A. 1999. *The tulip*. London: Bloomsbury.

Paweletz, N. 2001. "Walther Flemming: Pioneer of mitosis research." *Nature Reviews Molecular Cell Biology*, 2: 72–5.

Pawson, E. 2008. "Plants, mobilities and landscapes: Environmental histories of botanical exchange." *Geography Compass*, 2: 1464–77.

Paxton, J. 1844. "On planting showy shrubs in masses." *Paxton's Magazine of Botany, and Register of Flowering Plants*, 11: 136–9.

Pearman, G. 2005. "Nuts, seeds, and pulses." In G. Prance and M. Nesbitt (eds.), *The cultural history of plants*, 133–52. London/New York: Routledge.

Peate, I. C. 1971. "Corn ornaments." *Folklore*, 82: 177–84.

Peers, D. M. 2004. "Britain and empire." In C. Williams (ed.), *A companion to nineteenth-century Britain*, 53–78. Oxford: Blackwell.

Pellerin, G., and B. Touillon. 1996. *Outils de jardin: objets domestiques*. New York: Abbeville.

Pelletier, J., and J.-B. Caventou. 1818. "Sur la matière verte des feuilles." *Annales de Chimie et de Physique*, 9: 194–6.

Pelletier, P. J., and J.-B. Caventou. 1819. "Mémoire sur un nouvel alcali végétal (la strychnine) trouvé dans la fève de Saint-Ignace, la noix vomique, etc." *Annales de Chimie et de Physique*, 10: 142–76.

Pelletier, P. J., and J. B. Caventou. 1820. "Examen chimique de plusieurs végétaux de la famille des Colchicées et du principle actif qu'ils renferment [cévalille (*Veratrum sabadilla*); hellebore blanc (*Veratrum album*); colchique commum (*Colchicum autumnale*)]." *Annales de Chimie et de Physique*, 14: 69–81.

Pendergrast, M. 2013. *For God, country, and Coca-Cola: The definitive history of the great American soft drink and the company that makes it*. New York: Basic Books.

Perron, R. 2011. "La fabbrica del Duomo e l'invenzione del risotto." *Corriere della Sera* (in Italian). Available online: http://www.corriere.it/cronache/11_gennaio_29/perrone_fabbrica_duomo_c9650902-2b72-11e0-8f5d-00144f02aabc.shtml (accessed February 16, 2018).

Petzold, E. 1847. "Ueber die Anwendung der Farben auf die bildende Gartenkunst." *Allgemeine Gartenzeitung*, 15: 1–6, 9–15.

Petzold, E. 1849. *Beiträge zur Landschafts-Gärtnerei*. Weimar: Hoffmann.

Petzold, E. 1853. *Zur Farbenlehre der Landschaft*. Jena: Frommann.

Petzold, E. 1862. *Die Landschafts-Gärtnerei*. Leipzig: Weber.

Pierenkemper, T., and R. Tilly. 2004. *The German economy in the nineteenth century*. Oxford: Berghahn.

Pim, S. 1984. *The wood and the trees: Augustine Henry, a biography*, 2nd edn. Kilkenny: Boethius.

Piperno, D. R. 2011. "The origins of plant cultivation and domestication in the New World Tropics: patterns, process, and new developments." *Current Anthropology*, 52 (S4): S453–S470.

Pisoni, C. A. 2005. *Amor di pianta: Giardinieri, floricoltori, vivaisti sul Verbano, 1750–1950*, vol. 1. Verbania: Bindoni.

Pita, J. P. n.d. "Portugal: Brief history of Portuguese pharmacopoeias (18th–20th century)." Available online: www.histpharm.org/ISHPWG%20Portugal.pdf (accessed August 12, 2017).

Poole, C. H. 1877. *The customs, superstitions and legends of the County of Somerset*. London: Sampson Low.

Porter, A. 1999. "Trusteeship, anti-slavery, and humanitarianism." In A. Porter (ed.), *The Oxford history of the British Empire: The nineteenth century*, 198–220. Oxford: Oxford University Press.

Porter, R. 1999. *The greatest benefit to mankind. A medical history of the humanity from antiquity to the present*. London: Fontana.

Pradhan, A. 2013. *The nutri-cereals small millets*. Sadar Nagpur: Dattsons.

Prendergast, M. 1999. *Uncommon grounds: The history of coffee and how it transformed our world*, 2nd edn. New York: Basic, London: Texere.

Price, U. 1798. *An essay on the picturesque*, vol. 2. Hereford: Walker.

Priestley, J. 1772. "Observations on different kinds of air." *Philosophical Transactions of the Royal Society of London*, 62: 147–264.

Priyadarsini, K. I. 2014. "The chemistry of curcumin: From extraction to therapeutic agent." *Molecules*, 19: 20091–112.

Pückler-Muskau, H. von. 1834. *Andeutungen über Landschaftsgärtnerei*. Stuttgart: Hallberger.

PwC. 2011. "The Australian grains industry: The basics." PricewaterhouseCoopers International Limited, Australia. Available online: https://www.pwc.com.au/industry/agribusiness/assets/australian-grains-industry-nov11.pdf (accessed February 11, 2011).

Quennell, P. (ed.) 1984. *Mayhew's London*. London: Bracken Books.

Quiles, J. 2006. *Olive oil and health*. Wallingford, UK: Centre for Agriculture and Biosciences International.

Radar Farm Services. 2015. "The combine harvester." Available online: http://www.bushywood.com/farming/Combine_Harvester.htm (accessed February 9, 2018).

Rafinesque, C. S. 1828. *Medical flora: or, manual of the medical botany of the United States of North America*. Philadelphia: Atkinson and Alexander.

Ravindran, P. N., K. Nirmal Babu, and K. Sivaraman (eds.). 2007. *Turmeric: The genus Curcuma*. Boca Raton: CRC Press.

Reed, H. S. 1942. *A short history of the plant sciences*. Boston, MA: Chronica Botanica.

Regel, E. 1855. "Die Blutbuche." *Gartenflora*, 4: 93–94.

Regel, E. 1856. "Reisenotizen, gesammelt auf der Reise von Zürich nach Petersburg." *Gartenflora*, 5: 66–82.

Reider, J. E. von. 1828. *Handbuch der Blumenzucht*. Nürnberg: Zeh.

Reider, J. E. von. 1831. *Das Ganze der Blumenzucht*. Nürnberg: Zeh.

Reider, J. E. von. 1832a. "Ueber Rabatten in großen Gärten." *Annalen der Blumisterei*, 8: 233–8.

Reider, J. E. von. 1832b. *Vollständige Anweisung zum zweckmäßigen Anlegen von Blumen-, Obst-, Gemüse-, Hopfen-, Schul-, Handels-, Haus- und botanischen Gärten*. Berlin: Amelang.

Reinikka, M. A. 1995. *A history of the orchid*, rev. edn. Portland, Oregon: Timber Press.

Rensten, M. 1996. "Royal roses." *The Lady*, June 11–17: 38.
Riedl, H. 1976. "Erklärung der von J. B. Rupprecht erwähnten Pflanzennamen." In Anon., *Johann Knapp Jacquins Denkmal. Ein naturwissenschaftliches Huldigungsbild von 1822*, 14–17. Vienna: Osterreichische Galerie Schloss Belvedere.
Riello, G. 2013. *Cotton: The fabric that made the modern world*. Cambridge: Cambridge University Press.
Rigby, N. 1998. "The politics and pragmatics of seaborne plant transportation, 1769–1805." In M. Lincoln (ed.), *Science and exploration in the Pacific: European voyages to the southern oceans in the eighteenth century*, 81–100. Woodbridge: Boydell.
Riley, C. V. 1893. "Parasitic and predaceous insects in applied entomology." *Insect Life*, 6: 130–41.
Riley, G. (ed.). 2007. *The Oxford companion to Italian food*. Oxford: Oxford University Press.
Rilliet, F., and E. Barthez. 1839. *Dunglison's American medical library. A treatise on the pneumonia of children*. Philadelphia: Waldie.
Ristaino, J. B., and D. H. Pfister. 2016. "'What a painfully interesting subject': Charles Darwin's studies of potato late blight." *Bioscience*, 66: 1035–45.
Roberts, E. 2007. *Real of the Black Mountain*. London: Hurst.
Roberts, H. F. 1929. *Plant hybridization before Mendel*. Princeton, NJ: Princeton University Press.
Roberts, R. 1971. *The classic slum*. Manchester: Manchester University Press.
Robertson, E. G., and J. Robertson. 1977. *Cast iron decoration: A world survey*. Melbourne: Thames and Hudson Australia.
Robinson, J. [1994] 2015. *The Oxford companion to wine*. Oxford: Oxford University Press.
Robinson, W. 1870. *The wild garden*. London: Murray.
Robinson, W. 1883. *The wild garden*, 3rd edn. London: Murray.
Rocco, F. 2003. *The miraculous fever-tree: The cure that changed the world*. London: Harper Collins.
Roll-Hansen, N. 2009. "Sources of Wilhelm Johannsen's genotype theory." *Journal of the History of Biology*, 42: 457–93.
Roll-Hansen, N. 2014. "Commentary: Wilhelm Johannsen and the problem of heredity at the turn of the 19th century." *International Journal of Epidemiology*, 43: 1007–13.
Rosengarten, F. [1984] 2004. *The book of edible nuts*. New York: Dover.
Ross, C. 2017. *Ecology and power in the age of empire: Europe and the transformation of the tropical world*. Oxford: Oxford University Press.
Rowbottom, S. 2008. *Edward Carpenter: A life of liberty and love*. London: Verso.
Rupp, R. 2014. "The history of vanilla." *The Plate* (National Geographic) October 23. Available online: http://theplate.nationalgeographic.com/2014/10/23/plain-vanilla (accessed).
Ruskin, J. 1838. "The poetry of architecture." *Architectural Magazine*, 5: 494.
Sachs, J. 1862. "Uber den Einfluss des Lichtes auf die Bildung des Amylums in den Chlorophyllkornern." *Botanische Zeitung*, 20: 365–73.
Sachs, J. 1864. "Uber die Auflösung und Wiederbildung des Amylums in den Chlorphylkornen bei wechselnder Beleuchtung." *Botanische Zeitung*, 22: 189–294.
Salter, J. 1865. *The chrysanthemum: Its history and culture*. London: Groombridge.
Samuelsson, G., and L. Bohlin. 2009. *Drugs of natural origin. A treatise of pharmacognosy*, 6th edn. Stockholm: Swedish Pharmaceutical Press.
Sanderson, H. 2005. "Roots and tubers." In G. Prance and M. Nesbitt (eds.), *The cultural history of plants*, 61–76. London: Routledge.

Saunders, N. J. 2014. *The poppy: A history of conflict, loss, remembrance and redemption*. London: Oneworld Books.
Sauer, J. D. 1993. *Historical geography of crop plants, a select roster*. Boca Raton: CRC Press.
Savage, H., and E. J. Savage. 1986. *André and Francois André Michaux*. Charlottesville: University Press of Virginia.
Schaaf, L. J. 1985. *Sun gardens: Victorian photograms*. New York: Aperture.
Schaaf, L. J. 2000. *The photographic art of William Henry Fox Talbot*. Princeton: Princeton University Press.
Schiebinger, L. 2004. *Plants and empire: Colonial bioprospecting in the Atlantic world*. Cambridge, MA: Harvard University Press.
Schiebinger, L., and C. Swan (eds.). 2007. *Colonial botany: science, commerce, and politics in the Early Modern world*. Philadelphia: University of Pennsylvania Press.
Schifter, L., and P. Aceves. n.d. "Mexico: The role of Mexican pharmacopeias in the construction of a national identity." Available online: www.histpharm.org/ISHPWG%20 mexico.pdf (accessed August 12, 2017).
Schleiden, M. J. 1838. "Beiträge zur Pathogenesis." *Archiv für Anatomie, Physiologie und Wissenschaftliche Medizin (J. Müller)*, 5: 137–76.
Schleiden, M. J. 1849. *Principles of scientific botany, or botany as an inductive science*. Translated by E. Lankester. London: Longman.
Schmidlin, E. 1843. *Die bürgerliche Gartenkunst*. Stuttgart: Hoffmann.
Schmitz, R. 1985. "Friedrich Wilhelm Sertürner and the discovery of morphine." *Pharmacy in History*, 27 (2): 61–74.
Schmoll, gen., H. Eisenwerth 1994. "Ginkgo biloba im Kunsthandwerk Ostasiens und Europas." In M. Schmid and H. Schmoll, gen. Eisenwerth (eds.), *Ginkgo. Ur-Baum und Arzneipflanze—Mythos, Dichtung und Kunst*, 97–122. Stuttgart: Wissenschaftliche Verlagsgesellschaft.
Schneider, C. 1904. *Deutsche Gartengestaltung und Kunst*. Leipzig: Scholtze.
Scholda, U. 2002. "297. Becher." In G. Frodl (ed.). *Das 19. Jahrhundert*, 576–7. Munich: Prestel.
Schul, J. 2004. "Det Kongelige Danske Haveselskab—175 år." *Fra Kvangård til Humlekule*, 34: 27–40.
Schwann, T. 1839. *Mikroscopische Untersuchungen über die Ubereinstimmung in der Struktur und dem Wachstum des Thiere und Pflanzen*. Leipzig: Engelmann.
Schwann, T. 1847. *Microscopical researches into the accordance in the structure and growth of animals and plants*. Translated by H. Smith. London: Sydenham Society.
Schweikart, L., and M. Allen. 2004. *A patriot's history of the United States*. Sentinel, New York.
Sedding, J. D. 1891. *Garden-craft, old and new*. London: Kegan Paul.
Senebier, J. 1783. *Mémoires physico-chimiques sur l'influence de la lumière solaire pour modifier les étres des trois règnes de la nature et surtout ceux de règne vegetal*. Geneva: Chirol.
Senebier, J. 1788. *Expérience sur l'action la lumière solaire dans la végétation*. Paris: Briande.
Sertürner, F. W. 1817. "Uber das Morphium, eine neue salzfähige Grundlage, und die Mekonsäure, als Hauptbestandteile des Opiums." *Annalen der Physik*, 25: 56–90.
Sharma, S. D. 2010. "Domestication and diaspora of rice." In S. D. Sharma (ed.), *Rice: origin, antiquity and history*, 1–24. Boca Raton: CRC Press.
Sheets-Pyenson, S. 1988. *Cathedrals of science: The development of colonial natural history museums during the late nineteenth century*. Kingston and Montreal: McGill-Queen's University Press.

Shirreff, P. 1873. "Improvement of the cereals and an essay on the wheat fly." Available online: https://archive.org/stream/improvementcere00shirgoog/improvementcere00shirgoog_djvu.txt (accessed June 2, 2018).

Shull, G. H. 1908. "The composition of a field of maize." *Journal of Heredity*, 4: 296–301.

Siebeck, R. 1860. *Die Verwendung der Blumen und Gesträuche zur Ausschmückung der Gärten.* Leipzig: Voigt.

Sikes, W. 1880. *British goblins.* London: Sampson Low.

Silva Tarouca, E. von. 1910. *Unsere Freiland-Stauden: Anzucht, Pflege und Verwendung aller bekannten, in Mitteleuropa im Freien kulturfähigen ausdauernden krautigen Gewächse.* Leipzig: Freytag.

Silva Tarouca, E. von. 1913. *Unsere Freiland-Laubgehölze: Anzucht, Pflege und Verwendung aller bekannten in Mitteleuropa im freien kulturfähigen Laubgehölze.* Leipzig: Freytag.

Silva Tarouca, Ernst von, and C. Schneider. 1922. *Unsere Freiland-Stauden: Anzucht, Pflege und Verwendung aller bekannten, in Mitteleuropa im Freien kulturfähigen ausdauernden krautigen Gewächse*, rev. 2nd edn. Leipzig: Freytag.

Simmonds, N. W. 1959. *Bananas.* London: Longmans.

Simmonds, N. W. 1995. "Potatoes *Solanum tuberosum* (Solanaceae)." In J. Smartt and N. W. Simmonds (eds.), *Evolution of crop plants*, 2nd edn., 466–72. Singapore: Longman.

Simpson, J. 1976. *The folklore of the Welsh border.* London: Batsford.

Simpson, J., and S. Roud. 2000. *A dictionary of English folklore.* Oxford: Oxford University Press.

Sinclair, G. 1816. *Hortus Gramineus Woburnensis.* London.

Smil, V. 2001. *Enriching the earth.* Cambridge, MA: Massachusetts Institute of Technology.

Smith, A. (ed.). 2012. *The Oxford encyclopedia of food and drink*, vol. 2. Oxford: Oxford University Press.

Smith, C. H. J. 1852. *Parks and pleasure grounds.* London: Reeve.

Smith, E. F., and C. O. Townsend. 1907. "A plant tumor of bacterial origin." *Science*, 25: 671–3.

Smith, H. S. 1919. "On some phases of insect control by the biological method." *Journal of Ecoomic Entomology*, 12: 288–92.

Smith, J. 1989. *Fairs, feasts and frolics: Customs and traditions in Yorkshire.* Otley: Smith Settle.

Smith, J. B. 1994. "Comparative notes on traditional German and English names and uses of some plants representing the Palm-Sunday palm." *German Life and Letters*, 47: 242–53.

Smith, W. D. 1992. "Complications of the commonplace: Tea, sugar, and imperialism." *Journal of Interdisciplinary History*, 23: 259–78.

Solman, D. 1995. *Loddiges of Hackney: The largest hothouse in the world.* London: Hackney Society.

Sonnedecker, G. 1986. *Kramer's and Urdang's history of pharmacy*, 4th edn. Wisconsin: American Institute of the History of Pharmacy.

Soret, J. L. 1883. "Analyse spectral: sur le spectre d'absorption du sang dans la partie violette et ultra-violette." *Comptes rendus de l'Académie des sciences*, 97: 1269–73.

Sorge-Genthe, I. 1973. *Hammonias Gärtners: Geschichte des Hamburger Gartenbaues in den letzten drei Jahrhunderten.* Hamburg: Christians.

Soroceanu, V., n.d. "Romanian Pharmacopoiea from the first to the last edition." Available online: www.histpharm.org/ISHPWG%20Romania.pdf (accessed August 12, 2017).

Späth, L. 1930. *Späth-Buch: 1720–1930.* Berlin: Späth.

Spielmaker, D. 2014. "Historical timeline farm machinery and technology 17th to 18th centuries." National Institute of Food and Agriculture (NIFA), United States of Agriculture

(USDA). Available online: https://www.agclassroom.org/gan/timeline/farm_tech.htm#top (accessed February 13, 2018).

Spongberg, S. A. 1990. *A reunion of trees: The discovery of exotic plants and their introduction into North American and European landscapes*. Cambridge, MA: Harvard University Press.

Staehelin, L. A. 2003. "Chloroplast structure: From chlorophyll granules to supra-molecular architecture of thylakoid membranes." *Photosynthesis Research*, 76: 185–96.

Stanfield, M. E. 1998. *Red rubber, bleeding trees: Violence, slavery and empire in northwest Amazonia, 1850–1933*. Albuquerque, NM: University of New Mexico Press.

Stephan, M. 2010. "Der Wintergarten König Ludwigs II. von Bayern." In *Goldorangen, Lorbeer, Palmen—Orangeriekultur vom 16. bis 19. Jahrhundert*. Petersberg: Imhof.

Stephenson, J. 1932. "Robert Brown's discovery of the nucleus in relation to the history of cell theory." *Proceedings of the Linnean Society of London*, 144: 45–54.

Stevens, J. 1847. *Medical reform, or physiology and botanic practice, for the people*. London: Whittaker.

Stevens, P. F. 1994. *The development of biological systematics: Antoine-Laurent de Jussieu, nature, and the natural system*. New York: Columbia University Press.

Stevens Cox, J. 1971. *Guernsey folklore recorded in summer of 1882*. St. Peter Port: Toucan.

Stevenson, T. 2001. *The New Sotheby's wine encyclopedia: A comprehensive reference guide to the wines of the world*, 3rd edn. New York: Dorling Kindersley.

Stewart, A. 2013. *The drunken botanist: The plants that create the world's great drinks*. Chapel Hill, NC, and New York: Algonquin.

Stewart, S. 1987. *Lifting the latch*. Oxford: Oxford University Press.

Strasburger, E. 1876. *Studien über Protoplasma*. Jena: Dufft.

Streatfield, D. 2002. *Cocaine*. London: Virgin Books.

Sumner, J. 2004. *American household botany: A history of useful plants*. Portland: Timber Press.

Sussex, I. M. 2008. "The scientific roots of modern plant biotechnology." *Plant Cell*, 20: 1189–98.

Sutter, P. S. 2014. "The tropics: A brief history of an environmental imaginary." In A. C. Isenberg (ed.), *The Oxford handbook of environmental history*, 178–204. Oxford: Oxford University Press.

Sutton, S. B. 1970. *Charles Sprague Sargent and the Arnold arboretum*. Cambridge, MA: Harvard University Press.

Sutton, W. S. 1902. "On the morphology of the chromosome group in Brachystola magna." *Biological Bulletin of the Marine Biological Laboratory*, 4: 24–39.

Sweeney, M., and S. McCouch. 2007. "The complex history of the domestication of rice." *Annals of Botany*, 100: 951–7.

Taber, G. 2006. *Judgment of Paris*. New York: Scribner.

Tanaka, T. K. 2016. "Buckwheat production, consumption, and genetic resources in Japan." In M. Zhou, I. Kreft, S. -H. Woo, N. Chrungoo, and G. Wieslander (eds.), *Molecular breeding and nutritional aspects of buckwheat*, 61–80. Oxford: Academic Press.

Tansley, A. G. 1935. "The use and abuse of vegetational terms and concepts." *Ecology*, 16: 284–307.

Taylor, D. 2017. "Botanical gardens and their role in the political economy of empire: Jamaica (1846–86)." *Rural History*, 28. 47 68.

Taylor, J. M. 2014. *Visions of loveliness: Great flower breeders of the past*. Athens. Ohio University Press.

Taylor, J. M. 2018. *An abundance of flowers: More great flower breeders of the past*. Athens: Ohio University Press.

Taylor, R. B. 2017. *The amazing language of medicine: Understanding medical terms and their backstories*. London: Springer.
Terzi, E. 2007. *The nineteenth[-]century olive oil industry in Ayvalik and its impact on the settlement pattern*. M. A. thesis, Middle East Technical University, Ankara, Turkey.
Thacker, C. 1979. *The history of gardens*. London: Croom Helm.
Thiselton Dyer, T. F. 1889. *The folk-lore of plants*. New York: Appleton.
Thompson, R. 1906. *The gardener's assistant: A practical and scientific exposition of the art of gardening in all its branches*. 2 vols. London: Gresham.
Thompson, S. P. 1910. *Life of William Thomson, Baron Kelvin of Largs*. 2 vols. London: Macmillan.
Thomson, D. 1868. *Handy book of the flower garden*. Edinburgh: Blackwood.
Thomson, D. 1966. *Europe since Napoleon*. Harmondsworth: Penguin.
Thomson, S. 1822. *New guide to health*. Boston, MA: the author.
Thornton, W. H. 1907. "Concerning some old habits and decaying industries in the west of England, and more particularly in the county of Devon." *Report and Transactions of the Devonshire Association for the Advancement of Science*, 37: 111–21.
Thouin, A. 1805. "Essai sur l'exposition et division méthodique de l'économie rurale." In F. Rozier, *Cours complet d'agriculture*, vol. 11. Paris.
Timiriazeff, C. A. 1877. "Sur la decomposition de l'acide carbonique dans le spectre solaire par le particules verte des végétaux." *Comptes rendus de l'Académie des sciences*, 84: 1236–9.
Timmer, M. 2009. *Bloembollen in Holland 1860–1919: De ontwikkeling van de bloembollensector met een doorkijk naar de 21ste eeuw*. Houten: Hes and De Graaf.
Tirumalai, G., and A. Long. n.d. "United States pharmacopeial convention: respecting the past, moving confidently into the future." Available online: www.histpharm.org/ISHPWG%20 USA.pdf (accessed August 12, 2017).
Topik, Steven. 2004. "The world coffee market in the eighteenth and nineteenth centuries: From colonial to national regimes." London School of Economics, Working Paper 04-04, 2004. Proceedings of GEHN Conference, September 2003, Bankside, London.
Towne, M., and W. Rasmussen. 1960. "Farm gross product and gross investment in the nineteenth century." In the Conference on Research in Income and Wealth (eds.), *Trends in the American economy in the nineteenth century*, 255–316. Princeton, NJ: Princeton University Press. Available online: http://www.nber.org/chapters/c2480 (accessed February 9, 2018).
Trattinnick, L. 1825–43. *Neue Arten von Pelargonien deutschen Ursprungs*. 6 vols. Wien: Tendler.
Trevelyan, M. 1909. *Folk-lore and folk-stories of Wales*. London: Elliot Stock.
Triggs, H. I. 1902. *Formal gardens in England and Scotland*. London: Batsford.
Trinkley, M., and S. Fick. 2003. "Rice cultivation, processing, and marketing in the eighteenth century." Chicora Foundation Inc. Columbia [Liberty Hall (Research Series 62) data recovery project]. Available online: http://www.chicora.org/pdfs/Rice%20Context.pdf (accessed February 19, 2018).
Trocki, C. 1999. *Opium, empire and the global political economy*. London: Routledge.
Truffaut, G. 1896. *Sols, terres et composts utilisés par l'horticulture*. Paris: Doin.
Tswett, M. 1906. "Absorption Analyse und Chromatographische Method. Anwendung auf die Chemie des Chlorophylls." *Berichte der Deutschen Botanischen Gesellschaft*, 24: 384–93.
Tully, J. A. 2011. *The devil's milk: A social history of rubber*. New York: Monthly Review Press.
Turgan, J. F. 1860. *Les grandes usines de France*. Paris: Librarie Nouvelle.
Tuten, J. H. 2010. *Lowcountry time and tide: The fall of the South Carolina Rice Kingdom*. Columbia: University of South Carolina Press.

Udal, J. S. 1922. *Dorsetshire folk-lore*. Hertford: Austin.
Uglow, J. 2004. *A little history of British gardening*. London: Chatto and Windus.
USDA (United States Department of Agriculture). 1863. "Report of the Commissioner of Agriculture the Year 1862." 37th Congress 3d Session, House of Representatives. Ex. Doc. No. 78. Washington: Government Printing Office.
Vadon, C. 2014. *Orchidées du bout du monde*. Paris: La Martinière.
Van Andel, T. 2010. "African rice (*Oryza glaberrima* Steud.): Lost crop of the enslaved Africans discovered in Suriname." *Economic Botany*, 64: 1–10.
Van Damme-Sellier, Joseph.1861. *Histoire de la société royale d'agriculture et de botanique de Gand*. Gand: van Doosselaere.
Van Groeningen, I. 1996. *The development of herbaceous planting in Britain and Germany from the nineteenth to the early twentieth century*. Thesis, University of York, UK.
Vasil, I. K. 2008. "A history of plant biotechnology: From the cell theory of Schleiden and Schwann to biotech crops." *Plant Cell Report*, 27: 1423–40.
Vaughan, J., and C. Geissler (eds.). 1999. *The new Oxford book of food plants*. New York: Oxford University Press.
Veitch, J. H. 1906. *Hortus Veitchii*. London: Veitch.
Vibert, J. -P. 1824. *Essai sur les roses*, vol. 1. Paris: Huzard.
Vickery, R. 1995. *A dictionary of plant-lore*. Oxford: Oxford University Press.
Vickery, R. 2006. "Mrs Byartt and Primrose Day." *FONC* [Friends of Nunhead Cemetery] *News*, n.s. 16: 10.
Vickery, R. 2008. *Naughty man's plaything—folklore and uses of stinging nettles in the British Isles*. London: the author.
Vickery, R. 2010. *Garlands, conkers and mother-die*. London: Continuum UK.
Vickery, R. 2013. "Cowslips—Conservation icon and flower of folklore." *Evolve*, 15: 50–3.
Victoria, Queen. 1868. *Leaves from the journal of a life in the Highlands*. London: Smith, Elder.
Viglione, M. A. n.d. "Argentina: Brief outline of the evolution of pharmacopoeia in Argentina." Available online: www.histpharm.org/ISHPWG%20Argentina.pdf (accessed August 12, 2017).
Vilmorin, H. L. de. 1880. *Les meilleurs blés: Description et culture des principales variétés de froments d'hiver et de printemps*. Paris: Vilmorin Andrieux et cie.
Vilmorin-Andrieux et cie. 1866. *Les fleurs de pleine terre*, 2nd edn. Paris: Vilmorin-Andrieux et cie.
[Vilmorin-Andrieux et cie] Grönland, J., and T. Rümpler. 1873/4. *Vilmorin's illustrirte Blumengärtnerei*, vols. 1–2. Berlin: Wiegandt.
[Vilmorin-Andrieux et cie] André, É. F. 1894. *Les fleurs de pleine terre*, 4th edn. Paris: Vilmorin-Andrieux et cie.
Virchow, R. 1858. *Die Cellullarpathologie im ihrer Begrüngung und physiologische und pathologische Gewebelehre*. Berlin: Hirschwald.
Vöchting, H. 1878. *Uber Organbildung im Pflanzenreich*. Bonn: Cohen.
Vries, H. de 1889. *Intracellulaire pangenesis*. Jena: Fischer.
Wade, O. L. 1986. "Digoxin 1785–1985. 1. Two hundred years of digitalis." *Journal of Clinical Hospital Pharmacy*, 11: 3–9.
Wallace, H. A., and E. N. Bressman. 1956. *Corn and corn growing*. New York: Wiley.
Wallace, H. A., and W. L. Brown. 1956. *Corn and early fathers*. East Lansing, MI.
Walsh, L. 2015. "Yesterday's herbals and today's health fads. Stories from the National Museum of American History." [Blog] February 2. Available online: http://americanhistory.si.edu/blog/yesterdays-herbals-and-todays-health-fads (accessed May 20, 2021).

Walton, J. R. 1999. "Varietal innovation and the competitiveness of the British cereals sector, 1760–1930." *Agricultural History Review*, 47: 29–57.

Warburg, O., and T. Uyesugi. 1924. "Uber die Blackmansche Reaktion." *Biochemische Zeitschrift*, 146: 486–92.

Ward, N. B. 1835/6. "Improved method of transporting living plants." *Companion to the Botanical Magazine*, 1: 317–20.

Warming, E. 1895. *Plantesamfund—Grundtræk af den økologiske Plantegeografi*. Copenhagen: Philipsen.

Warming, E., and M. Vahl. 1909. *Oecology of plants—an introduction to the study of plant communities*. Translated by P. Groom and I. B. Balfour. Oxford: Clarendon Press.

Watson, F. 1872. *Flowers and gardens: Notes on plant beauty*. London: Strahan.

Watt, A. 2017. *Robert Fortune: A plant hunter in the Orient*. Kew, UK: Royal Botanic Gardens, Kew.

Waugh, E. 2009. *Rivers Nursery of Sawbridgeworth: The art of practical pomology*. Ware, Herts: Rockingham.

Welch, R. C. 2000. "From manpower to horsepower: Technological change in the nineteenth century." In Iowa State University Center for Agricultural History and Rural Studies, American Agricultural History Primer. Available online: http://rickwoten.com/ManpowertoHorsepower.html (accessed February 14, 2018).

Went, F. W. 1928. "Wuchstoff und Wachstum." *Receuil des Travaux Botaniques Néerlandais*, 25: 1–116.

White, G. 1822. *The natural history of Selborne*. London: Arch; Longman, Hurst, Rees, etc.

Whitten, D. O. 1982. "American rice cultivation, 1680–1980: A tercentenary critique." *Southern Studies*, 21: 5–26.

Wilde, O., et al. 1999. *Teleny*. London: Prowler.

Wilkinson, J. G. 1858. *On colour and on the necessity for a general diffusion of taste among all classes*. London: Murray.

Williams, M. 2003. *Deforesting the earth: From prehistory to global crisis*. Chicago: University of Chicago Press

Willits, C. O. 1965. *Maple-syrup producers manual*. United States Department of Agriculture Handbook 134.

Willson, E. J. 1982. *West London nursery gardens: The nursery gardens of Chelsea, Fulham, Hammersmith, Kensington and part of Westminster, founded before 1900*. London: Fulham and Hammersmith Historical Society.

Willson, E. J. 1989. *Nurserymen to the world: The nursery gardens of Woking and north-west Surrey and plants introduced by them*. London: the Author.

Willstätter, R. 1915. "Chlorophyll." *Journal of the American Chemistry Society*, 37: 323–45.

[Wilson, E. H.] Sargent, C. S. (ed.). 1913–17. *Plantae Wilsonianae: An enumeration of the woody plants collected in western China for the Arnold Arboretum of Harvard University during the years 1907, 1908, and 1910*. Cambridge, MA: Cambridge University Press.

Wilson, G. F. 1883. "Oakwood, Wisley." *Gardeners' Chronicle*, 19 (476): 178.

Wilson, N. 1855. "On the useful vegetable products, especially the fibres, of Jamaica." *Hooker's Journal of Botany and Kew Garden Miscellany*, 7: 335–40.

Wilson, O. T. 1920. "Crown-gall of alfalfa." *Botanical Gazette*, 70: 51–68 plus 10 plates.

Wimmer, C. A. 1991. "Die Kunst der Teppichgärtnerei." *Die Gartenkunst*, 3: 1–16.

Wimmer, C. A. 1998. "Victoria, the Empress Gardener, or the Anglo-Prussian Garden War, 1858–88." *Garden History*, 26: 192–207.

Wimmer, C. A. 2002. "Die Entwicklung der Pelargonien im 19. Jahrhundert." *Gartenpraxis*, 28: 18–22.

Wimmer, C. A. 2012. *Hippe, Krail und Rasenpatsche: Zur Geschichte der Gartenwerkzeuge.* Weimar: vdg.

Wimmer, C. A. 2016. "Schön zum Verwildern: Die zweifelhafte Karriere der Herkulesstaude." *Zandera*, 31: 65–84.

Wimmer, C. A. 2018. *Lustwald, Beet und Rosenhügel: Geschichte der Pflanzenverwendung in der Gartenkunst*, 2nd edn. Weimar: vdg.

Win, U. K. 1991. *A century of rice improvement in Burma.* Manila: International Rice Research Institute.

Wingler, H. M., and F. Welz. 1975. *O. Kokoschka. Das druckgraphische Werk.* Salzburg: Verlag Galerie Welz.

Wink, M. 1998. "A short history of alkaloids." In M. F. Roberts and M. Wink (eds.), *Alkaloids, biochemistry, ecology, and medicinal applications*, 11–44. New York: Plenum Press.

Winsor, M. P. 2009. "Taxonomy was the foundation of Darwin's evolution." *Taxon*, 58: 43–9.

Wollman, A. J. M., R. Nudd, E. G. Hedlund, and M. C. Leake. 2015. "From *Animaculum* to single molecules: 300 years of the light microscope." *Open Biology*, 5: 150019. DOI: 10.1098/rsob.150019.

Wortmann, C. S. 2006. "*Phaseolus vulgaris* L. (common bean)." In M. Brink and G. E. Belay (eds.), *Cereals and pulses*, 146–51. Plant Resources of Tropical Africa (PROTA) Foundation. Wageningen: Backhuys Publishers/ CTA.

Wright, A. R. 1938. *British calendar customs, England*, vol. 2. London: Glaisher.

Wright, A. R. 1940. *British calendar customs, England*, vol. 3. London: Glaisher.

Wright, H. J. 1914. *Sweet peas*, rev. edn. London: Jack.

WTO [World Trade Organization]. 2013. "The evolution of international trade: Insights from economic history." *World Trade Report*.

Wulf, A. 2015. *The invention of nature: Alexander von Humboldt's new world.* London: Murray; New York: Knopf.

Wyse Jackson, P. 2014. *Ireland's generous nature.* St. Louis: Missouri Botanical Garden Press.

Xi, G. 2014. "Chinese perspectives on medical missionaries in the 19th century: The Chinese Medical Missionary Journal." *Journal of Cultural Interaction in East Asia*, 5: 97–118.

Youngken, H. 1922. "Dehydration and the preservation of foods." *Scientific Monthly*, 14: 332–44.

Zohary, D., M. Hopf, and E. Weiss (eds.). 2012. *Domestication of plants in the Old World: The origin and spread of cultivated plants in West Asia, Europe and Nile Valley*, 4th edn. Oxford: Oxford University Press.

CONTRIBUTORS

Claudia Ciotir, PhD is a Zuckerman Postdoctoral Scholar at the Institute of Evolution, University of Haifa in Israel, and a Research Associate at Missouri Botanical Garden in St. Louis, USA. Her research focuses on crop evolution, plant domestication, and quantitative genetics. She built the Perennial Agriculture Project Database to identify perennial, wild, herbaceous species with potential to be developed into perennial crops for use in sustainable perennial polycultures that mimic natural ecosystems.

Patrick Hunt, PhD (Institute of Archaeology, London UCL) has been teaching at Stanford University, California, USA for twenty-eight years and has authored over twenty published books. National Geographic has sponsored his research and he also works as a National Geographic Expeditions Expert and is an elected Fellow of both the Royal Geographical Society and the Explorers Club. He is a National Lecturer for the Archaeological Institute of America.

H. Walter Lack, FMLS is a plant taxonomist and historian of science at the Freie Universität Berlin, Germany. Until 2014 he was director at the Botanic Garden and Botanical Museum Berlin and is a specialist in the Compositae family and the history of plant taxonomy with an emphasis on the eighteenth and nineteenth centuries. He was a visiting professor at the universities of Coimbra (Portugal), Patras (Greece), Palermo, and Pisa (both Italy), and is a recipient of the Great Cross of Merits of the Republic of Austria and the Cross of Merit 1st Class of the Republic of Germany. He has over four hundred publications, ranging over systematics, history of science, museology and botanical illustration. His most recent books include *The Bauers: Masters of botanical illustration* (2015), *A garden Eden*, ed. 3 (2016), *Alexander von Humboldt and the botanical exploration of the Americas*, ed. 2 (2018) and *Pierre-Joseph Redouté. The book of flowers*, ed. 2 (2020).

David J. Mabberley AM, DSc is Emeritus Fellow, Wadham College, University of Oxford, UK. He was, consecutively, Director of the University of Washington Botanic Gardens, Seattle, USA; Keeper of the Herbarium, Library, Art and Archives at the Royal Botanic Gardens, Kew, UK; Executive Director, Royal Botanic Gardens, Sydney. He held a chair in the University of Leiden, the Netherlands, for twenty-three years and has had visiting posts in the University of Paris; Kuwait University; Universities of Peradeniya and Sri Jawarardenepura, Sri Lanka; University of Sydney; Western Sydney University; and Macquarie University, where he is Adjunct Professor. He has over three hundred publications, ranging over plant ecology and systematics to the history of science and botanical illustration. His most recent books include *Botanical Revelation* (2019); *Mabberley's Plant-book: A Dictionary of Plants, their Classification and Uses* (2017); *Painting by Numbers: The life and art of Ferdinand Bauer* (2017); and *Sir Joseph Banks' Florilegium* (2017).

CONTRIBUTORS

Mark Nesbitt is Senior Research Leader at the Royal Botanic Gardens, Kew, UK, and Visiting Professor at the Department of Geography, Royal Holloway, University of London, UK. His research interests are in the interactions of people and plants through time, especially as represented in museum collections. He studied agricultural botany at the University of Reading and archaeobotany at the Institute of Archaeology, UCL, both UK. After fifteen years of archaeological and ethnobotanical research on prehistoric agriculture in the Near East, he joined Kew in 1999. His current role is as curator of Kew's Economic Botany Collection, developing research into the intersection of botany, trade, and empire in the nineteenth century. His special interests include the histories of botanical materials and medicines, the history of plant exploration, and the curation and use of botanical collections.

Anne Osbourn FRS, OBE is a plant biologist at the John Innes Centre, UK. She also holds an honorary professorship at the University of East Anglia. Her research focus is on plant natural products—how they are made, what they do, and how new biosynthetic pathways are formed. She is also a poet, and her first book of poetry, *Mock Orange*, was recently published. She has also developed and co-ordinates the Science, Art and Writing (SAW) Initiative, a cross-curricular science education program (www.sawtrust.org).

Monique S. J. Simmonds, OBE is Deputy Director of Science, and Director of the Commercial Innovation Unit at the Royal Botanic Gardens, Kew, UK. She researches different aspects of plant and fungal chemistry with an emphasis on their economic value including quality control of medicinal plants and insect–plant interactions. Her commercial focused research aims to increase the diversity of plants entering the trade and supporting human well-being through illustrating the economical important of plants and fungal diversity.

Roy Vickery is a botanist who, from 1965 to 2007, was on the staff of the Department of Botany at the Natural History Museum, London. Although now based at the South London Botanical Institute, he remains a Scientific Associate of the Museum. His major interest since the early 1980s has been the folklore and traditional uses of British and Irish plants (see www.plant-lore.com) and he is author of *Vickery's Folk Flora: An A-Z of the Folklore and Uses of British and Irish Plants*, published in 2019.

Clemens Alexander Wimmer is a well-known garden historian and landscape architect, based in Potsdam, Germany. Trained in garden design in Berlin, he is an honorary librarian in Deutsche Gartenbaubibliothek (German Horticultural Library) there. An established scholar and author, he has around three hundred publications to his name, recent books including *Der Gartenkünstler Peter Joseph Lenné: Eine Karriere am preußischen Hof* (2016), *Lustwald, Beet und Rosenhügel: Geschichte der Pflanzenverwendung in der Gartenkunst* (2014, 2018), and *Ein Gärtner auf Grand Tour: Emil Sellos Tagebuch seiner Europareise 1838–1840* (2020).

INDEX

Abernathy, John 56–7
abstract art 8
acclimatization 164
Aconitum and aconitine 107
aesthetic value of plants 154
Agassiz, Louis 22
agricultural revolution 4
Ainslie, Whitelaw 123
Alaska 25
Albert, Prince Consort 7, 17, 146–7
Albius, Edmond 59
alcoholic drink 46
Alexandra, Princess (later Queen) 141, 144–5
alkaloids 105
allotments 130
Alma-Tadema, Lawrence 8, 175
Amaranthaceae 37–8
Amazon region 74
American Civil War 4, 64, 80, 82
American herbs and herbalists, impact in Britain and Europe of 120–4
Ananas comosus 53
André, Edouard 150
animals, introduction of 83
annuals 170
anthropomorphism 26
Antoine, Franz Jr. 193
apothecaries 121
Appert, Nicolas 52
apple trees 127
applied arts, plants in 184–8
apricots 53–4
Arana, Julio César 74
arboreta 76
Arc de Triomphe 178
"architect's garden" 153
architecture, plants in 186
Argentina 83
Aristotle 21
Arnold Arboretum 70
arsenic 106
Art Nouveau 8, 185
the arts 8

Arts and Crafts movement 150, 185, 188
Ashen Faggot Balls 127–8
Ashridge House 150
aspirin 117, 125
Astrantia minor 194
Atkins, Anna 193
Atropa belladonna and *atropine* 107
Auersperg, Anton Alexander 182
Augusta of Cambridge, Princess 138
Australia 4, 23–5, 162, 181
Austria 122
auxin 90

Backhouse, James 70
bacteria 90–1
"balance of power" in Europe 6
Banks, Sir Joseph 9, 68, 70, 73, 147, 164
Barnes, Charles R. 95
Barrillet-Dechamps, Jean-Pierre 150
Bartram, William 161
Bateson, William 97
Batsaki, Y. 163
Bauer, Ferdinand 189, 191
Baumann, Johann and Franz Josef 161
Bayer, Adolf von 95
beans 38–9
Beauharnais, Eugène 166
bedding plants 153, 170
Bedford, Duke of 155
beekeeping 56
beer 48
Beketova, Ekaterina Andreyena 183
Belgium 162
Bennet, Lady Mary Elizabeth 170
Bentham, George 18, 71
Berg, Alban 183–4
Berkeley, Miles 100
Bernhardt, Sarah 173–4
Bertrand-Molleville, Marquis 157
Bethge, Hans 183
beverages 46–9
Bickham, T. 44
Biffen, Sir Rowland 32, 99

INDEX

Biffen, Harry 32
bilberries 131–2
binomial names 97
"biological control" 101
biology, "new" 26
bioprospecting 69
births 133
biscuits 56–8
Blackman, Frederick Frost 95–6
Blaschika, Rudolf 195
Blaxland, Gregory 46
Blyth, Edward 17
Boitard, Pierre 157
Bolivar, Simon 11
Bollé, Hermann 188
Bonpland, Aimé 192
Bonsack, James 64
Booth, Charles 142–3
Booth, James 160
borders (in gardens) 155
Bosse, J. F. W. 169
botanical gardens 5–6, 10, 67–72
    facilities of 68–70
    imperial role of 164
botany
    economic 68, 71–2
    shifting from observation to investigation of fundamental processes 85
Boulger, G. S. 155
Boussingault, Jean Baptiste 95, 100
bracken 133
Braque, Georges 8
Braun, Armin 91
Brazil nuts 62
breadwheat 31–2
breeding of plants 27, 41, 100, 103, 165–70
Breiter, Christian August 168
British Association for the Advancement of Science 9
British Empire 4–7, 44, 73
British Empire Exhibition (1924–5) 71
British Legion 145
British Medical Association 121
British Museum 70
Britten, J. 132
Brodie, Sir Benjamin Collins 118
Bronte, Emily 142
Brown, H. T. 95, 97
Brown, Robert 8–14, 17–18, 23, 26, 85–6
Browne, Janet 9
Buchanan, James 63
Buchner, Johann Andreas 117

buckwheat 37–8
Buffon, Comte de 23
bulbs 169–70
Bull, William 160
Burr, Enoch 19
Busch, John 163
Bute, Marquess of 74

caffeine 109
caged birds 133
Camellia sinensis 5, 80
Candler, Asa Griggs 114
Candolle, Alphonse 98
Candolle, Augustin-Pyramus de 10, 23, 98
cannabis 64
capital, availability of 73
Carapichea and ipecacuanha 107–8
carbon dioxide 93–6
Carleton, Mark 11
Carpenter, Edward 145
carpet bedding 153–5
Casimir, Anne 98
Catholic Church 64
Cattleya labiata gaskelliana 73
Cavara, Fridiano 91
Caventou, Joseph 94, 108
cell theory 85–90, 103
    recognizing cells as the elementary parts of all living organisms 86, 97
cereals 28–38
Chambers, Robert 15–18
champagne 47
change
    contrast between the forces of change and those of conservatism 3
    in the course of the 19th century 3–4
    drivers of 67
    periods of 171
changelings 133–4
Charlotte, Queen to George III 145
cherry syrup 54
Chimborazo, Mount 23
China 3, 5, 76–7, 80, 82, 115, 124, 165, 183
chlorophyll 94–5
chocolate 49–52
Christison, Robert 106
Christmas decorations 145–6
Christmas festivities 143
chromosomes 97
chrysanthemums 168
church decorations 143–4
Church, Arthur Harry 21, 24

cider 62
cinchona 71, 77, 82, 108–10
Cinchona calisaya 71, 110
Clapham, Arthur 24
classification of plants 21, 97
Clemens, Frederic E. 24
Cliveden 155
coca 112–14
Coca-Cola 62–3, 114
cocaine 114
coconut 63
Cocos nucifera 1
Coffea arabica 5, 48–9, 109, 111
coffee 5, 44, 48–9, 82, 109
coffee rust disease 101
Colchicum autumnale and *colchicine* 110
collecting plants 69–72, 162–4
colonialism 5, 11, 24, 67
    six Bs of (botanist – buccaneer – Bible – bureaucrat – banker –businessman) 5
Colonial and Indian Exhibition (1886) 71
color, use of 156
"Comfort Tree" 55
commercial operations 70, 77
condiments 61
Congo Free State 74
conifers 168
conservation measures 24–5
Constable, John 175
consumerism 7, 9
continental drift 10
"continental system" of trade 2
contraception 133
Convention on Biological Diversity 70
Cook, James 10
Cooper, Frank 53
Cope, Edward 22
Copernicus, Nicolaus 22
corn 27, 130–1
Corn Laws 4
Correns, Carl 20
Corvisart, Jean Nicholas 105
cotton 4–5
Courset, Dumont de 168
courtship 135
Cowles, Henry 24
Cownley, Alfred 108
Crabbe, John 60
Crocus sativus 60
Crosby, Alfred 82
Cross, Robert 79
crown gall disease 90

Crystal Palace 7, 159
Cubism 8
cultivars 27, 31–2
Curcuma longa 60, 111
curcumin 111–12
Curie, Marie 7
Cyclamen persicum 190
Cypripedium reginae 85
Cytisus scoparius 132
cytology and *cytogenetics* 26
Czartoryska, Izabella 149

Dadaism 8
Dadant, Charles 56
Darlington, Cyril 18
Darwin, Charles 9–22, 26–7, 52–3, 90, 97–100
Darwin, Emily Catherine 14–15
Darwin, Erasmus 12, 18, 26, 101
Darwin, Robert Waring (two individuals with the same name) 14
Darwinism 8–9, 18–22
David, Père Armand 77
Dearle, John Henry 185
death and mourning 142–3
death rates 125
deaths in the colonies 74
Decker, Stephan 188
Deegen, Christian 168
deforestation and climate 24
Dehmel, Richard 182
de Maria Bergler, Ettore 186
Denmark 124
Derosne, Louis 115
Descemet, Jacques-Louis 166
de Vries, Hugo 20, 22
diagrams, botanical 193–5
digitalis purpurea 112
Dioscorides 121–2, 192
Disraeli, Benjamin 144
distribution of images 193
Dodel, Arnold and Carolina 191
Douglas, David 76, 164–5
Dover, Thomas 108
Dow Jones index 64
Downing, Andrew Jackson 150
Drake, Friedrich 180
Dreer, Henry Augustus and William F. 162
Drude, Oscar 23
drugs 64, 105–6, 122
Dumas, Alexandre 184
Dundee 80

Durand, Peter 52
"dustbowl" conditions 4
Dutrochet, René Joachim Henri 93

East India Company 5, 73, 77, 80–2
Ebert, Oscar 169
eclectic planting 150, 153–4
"ecological imperialism" 82
ecology 23–4
Ehrenberg, Christian Gottfried 90
Eichler, August Wilhelm 194–5
Einstein, Albert 7
elderberries 132
elderflower syrup 54
Elliott, P. C. 76
emetine 108
Endersby, Jim 70, 75
Engelmann, Theodor W. 88, 95
Engert, Erasmus Ritter von 173
Engler, Adolf 23
Enlightenment thinking 68
Ephedra and *ephedrine* 112
Errington, Robert 155
Escherich, Theodor 119
eserine 117
European civilization 4
exchange of plants 67–70, 84
exhibitions 68, 71
exoticism 127
experimental gardens 148
expressionism 178

Fabaceae (Leguminosae) 38–9
Fanta, Josef 181
Farrer, William 33
Fauvism 8
fermented foods 62
fertilizer 100, 157
fiber plants 79–80
Ficus elastica 78
fieldwork 72
fine arts, plants in 171
Fintelmann, Ferdinand 154
First World War 1, 6–7, 10, 23, 26, 145, 170–1
fish 129
Fisher, Hermann 109
Fitch, Walter Hood 189–90
Fitzroy, Robert 13
"fixed air" 93
flag days 145
Fleming, Jenkin 20

Fleming, John 155
Flemming, Walther 97–8
Flinders, Matthew 23
Flora Danica table service 192
flower beds 150, 153–6
flowers
    language of 135–9
    use in churches 143
folklore 133
formaldehyde 95
Forster, Georg 10
Fortune, Robert 81–2
four-course crop rotation 4
foxglove 133–4
France 1–5, 12, 26, 44, 102, 122, 147, 161, 164
Fraser, Thomas 115–17
Frean, George 57
free trade 4–5, 73–4
Freedman, Paul 60
Freud, Sigmund 114
Friedrich, Caspar David 176–7
Friend, Hilderic 142
fruits 53
funerals 142
Futurismo 8

Galapagos Islands 13, 83
Gallé, Emile 185
garden design 150
garden plants 74–7
gardenesque style 148
gardening
    18th and 19th centuries compared 152, 170
    styles of 147–56
    technical advances' effect on 85, 156–9
    two camps within 170
Garnier-Valletti, Francesco 195
Gaudi i Cornet, Antoni 186
Gauguin, Paul 8, 178–9
Geiger, Philip Lorenz 107
General Medical Council 121
genetics and genetic engineering 93, 97
genres and genre painting 173–6
germ theory of disease 91, 119
Germany 4, 7–10, 24, 26, 43–4, 147, 150, 152, 160–1
Gilbert, Joseph Henry 100
ginger 60
gingerbread 56
glasshouses 159
globalization 1, 27, 44

glycerine and nitroglycerin 1
Goethe, Johann Wolfgang von 150, 156, 173, 183
Goodyear, Charles 78
gooseberries 129–30
Gordon, George James 80
Goss, John 20
Gothic revival 185
government intervention 68–9
grapevines 101
grasses 154–5
Gray, Asa 19–21, 25
Great Exhibition (1851) 7, 71, 159
green revolution 41, 44
greenhouses 170
Gros, Antoine Jean 188
Grove, Richard 83
guava 53
Guérin, Anna 145
Guinard, Hector 186

Haber, Fritz 100
Haberlandt, Gottlieb 88, 90, 96
Haeckel, Ernst 23, 191
Hales, Stephen 93
Hardy, Thomas 135
Hartweg, Carl Theodor 165
harvest festivals 129, 144
Hassam, Childe 173–4
Hawker, Robert Stephen 144
Hedgcock, George C. 91
Heimatschutz movement 152
Heinz, Henry John 58, 61
Henry, Augustine 165
Henslow, George 19
Henslow, John 12–14, 19
herbals 119
herbaria 23, 70–1
Herbert, William 14
heredity 12, 97
d'Hervey, Baron 183
Herzl, Theodor 188
Hibberd, Shirley 169
hickory sauce 61
Hills, Amariah Millar 157
Hirst, Henry 162
Hobson, John 5
Hoffmann, Josef 188
Hofmannsthal, Hugo von 184
Hofmeister, Wilhelm 96
Hogg, J. 96
Holland, R. 132

honey 53, 55–6
Hong Kong 77
Hooke, Robert 86, 90
Hooker, Joseph 14, 18–21, 160, 164
Hooker, Sir William 70–1, 112, 188–9
horticulture and horticultural societies 147–8
hospital care 105
hotels 185–6
hothouses 127, 159
Howard, John Eliot 109
Huber, Carl 161
Humboldt, Alexander von 10–12, 23–6, 117, 150, 192
   Cosmos 10–11
Hungary 122
Huxley, Thomas Henry 19
Hydrastis canadensis 115
hygiene, public 125

illustration, botanical 171, 188–93
imperial preference 67
imperialism 5, 74
impressionism and post-impressionism 8, 177
improvisation over shortages of essential supplies 7
inbreeding 27
indentured labour 74
India 5, 73, 77, 80–3, 123, 165
India Office 73
indigenous knowledge 84
industrial revolution 100, 105
Ingen-Housz, Jan 93–4
inheritance, laws of 27
Introduction of new plants *see* transplantation
invasive plants 83
ipecacuanha 107–8
ironwork 186
Iwaski, Kane'en 191

Jacob, Lambert 162
Jacquin, Nikolaus Joseph von 175
Jäger, H. 154–5
jam 52–3
Japan 5, 37–8, 48, 76–7, 122, 191–2
Jardin du Roi 68
Jarry, Alfred 8
Jefferson, Thomas 11
Jekyll, Gertrude 155–6
Jensen, Jens 154
jewelry, botanical 186
Johannsen, Wilhelm 27, 97
Johnstone, William Grosart 193

Joséphine, Empress 2, 166, 173, 188–92
Jugendstil movement 171–2
Jussieu, Antoine-Laurent de 97
Jussieu, Bernard de 97
jute 80

Karolsfeld, Ludwig Schnorr von 175
Keiga, Kawahara 191
Kelway, James and William 160, 169
Kennan, George F. 171
Kerr, William 164
ketchup 61
Kew Gardens 8, 21, 68–73, 79, 109–14, 164
Klimt, Gustav 171–2, 175, 178, 188
Knapp, Johann 175
Knight, Thomas 12, 20
Knop, W. G. 88
Kny, Leopold 190
Koch, Robert (and Koch's postulates) 91, 119
Kokoschka, Oskar 184–5, 188
Koller, Karl 114
Kraus, Gregor 162
Krelage family (Heinrich, Jacob and Ernst) 155, 161

labor practices 79
Lagoa Santa 24
Lamarck, Chevalier de 12, 17, 20, 26
Lambert, Peer 166
lambs 129–30
landscape gardens 148, 150, 154
landscape painting 175
Lange, Willy 152
Langstroth, Lorenzo 56
La Touche, Rose 138
laudanum 64
Lauder, Thomas Dick 155
Lavoisier, Antoine-Laurent 93
Lawes, John Bennet 100
lawnmowers 157–8
lawns 155
Lawrence, Joseph 119
Lawrence, William 12, 18
Lawson, Nicholas 13
Le Havre 5
Lea, John Wheeley 60
Leclerc, Georges-Louis *see* Buffon
Ledger, Charles and George 109
Lefèvre, Robert 173, 176
legumes (Leguminosae) 38–9
Lemaire, Louis Marie 186
Lémoin, Nicolas and Jean-Nicolas 169

Lemoine, Victor and Emile 161, 166
Lenin, Vladimir 5
Lenné, Peter Joseph 150
Leopold II, King of Belgium 74
Lewis, John Frederick 173
Liebig, Justus von 12, 62, 100
lifecycle, human 133–43
light microscopy 96
lilacs 166
Lilien, Ephraim Mose 188
Lincoln, Abraham 25, 28
Linden, Jean 162
linden trees 182–4
Lindley, John 92, 160, 164
Linnaeus, Carolus 12, 23, 67, 97, 99, 175
Linton Park 153
Lister, Joseph 96, 119
literature 8–9
    plants in 182
Livingstone, David 74
Lobb, Thomas 163–4
Lobb, William 76, 163–4
Locke, John 22
Loddiges family (Conrad, William and George) 160, 163
"long" nineteenth century 171, 195
Lösch, Fyodor 108
Loudon, John Claudius 148, 150, 156–9
Löwig, Karl Jacob 117
Lucien, Henry 181
Ludwig II, King of Bavaria 159
luxury, concept of 44, 65
luxury beverages 46–9
luxury foods 43–65
    condiments and spices 58
    oils 45
    plant-derived 49–52
    range of 44–5
    religious and political associations 45
    sweet items 52–61
Lyell, Charles 12, 17–18

macaroons 57
McChung, Clarence 97
McCormick, Willoughby M. 60
McCracken, Donal 68
McCrae, John 145
McIvor, William 109
Mackintosh, Charles Rennie 178
MacMillan, Conway 95
McVities (company) 57
Magendie, François 106

Mahler, Gustav 183–4
maize 27–30
Malaysia 77, 79
Malthus, Thomas 12–13, 18, 21
Mangifera indica 53
mango 53
maple syrup 54–5
Mariani, Angelo 114
Markham, Clements 78, 108, 113
Marmite 62
marriage 138, 141–2
Martius, Carl von 100, 194
Marx, Karl 21
marzipan 57
Matisse, Henri 8, 178
Mayer, Julius Robert 95
Mayhew, Henry 138
medical advances 119
medicinal plants 105–6
    advances in knowledge of 107–19
Mein, Heinrich F. G. 107
meiosis 86
Mendel, Gregor 12, 20, 26–7, 97
Mendeleev, Dmitri 7
Meng Hao-Ran 183
merchant gardeners 159
Mexico 123
miasma theory 90
Michael, Moina 145
Michaux, François André 161
microscopy 85–8, 94, 96
middle classes 147
migration 4, 73
Millais, John Everett 177
Millardet, Pierre-Marie-Alexis 158–9
Miller, Wilhelm 155
millets 36
Mirbel, Charles Brisseau de 26, 86
mistletoe 145
mitosis 98
"mixed plantations" 153
models, botanical 195
Mohl, Hugo von 94
Mohn, Gottlob Samuel 184–5
molasses 55
molecules 95
Molisch, Hans 94
Monet, Claude 177
monoculture 5, 83
Moore, Thomas 193
Morgan, Hugh 59
morphine 102, 113, 122–3

Morris, Daniel 79
Morris, Francis Orpen 19
Morris, J. H. 95
Morris, William 151–2, 187
Moser, Koloman 180–1, 188
Mucha, Alphonse 173–4
Mueller, Ferdinand von 62
Muir, John 25–6
Müller, Wilhelm 183
murders 106
museums 78, 71, 195
music 8, 182–4
Mutis, José Celestino 192
Myanmar 34

Nagayoshi, Nagai 112
naming of plants 23, 70–1
Napoleon 1–3, 44, 102, 161, 178
Napoleon III 47
national parks 24–5
Nativelle, Claude-Adolphe 112
natural history and the Natural History
    Museum, London 9, 70–1, 191
natural products 102
natural selection 14, 18–22, 99
nature, being true to 171
Nature (journal) 22
nature printing 92–3
nature reserves 25
Naturgarten 152
Naudin, Charles 17–18
Negri, Domenico 53
Nepenthes distillatoria 92
Nesbitt, Mark 5
Netherlands 3, 5, 50, 161, 164, 169
nettles 131
networking 70, 73
Newton, Sir Isaac 22
New Zealand 77, 83
Nicolas I, Tsar of Russia 179
Nicotiana tabacum 64
Niederegger, Johann Georg 58
Niemann, Albert 114
nitrogen, production of 100
Nobel, Alfred 1
Nobel prizes 1, 91, 95, 109
nutrition 125
nuts 62, 134

oak trees 182
Oken, Lorenz 9–10
Olbrich, Joseph Maria 180–1

olive oil 45
Oliver, William 56
opium 64, 77, 80, 102, 115, 182
Opium Wars 77, 115, 165
orchids 74–6, 83, 85, 163
Orfila, Mathieu Joseph Bonaventure 106
ornamental plants 70, 83, 147
Oryza glaberrima 35
Oryza spp. 34–7
Osbeck, Pehr 23
O'Shaughnessy, W. B. 64
Osseo-Asare, Abena Dove 77
Ottoman Empire 122
Owen, Richard 17, 19
Oxford University 9

Pagenstecher, Johann 117
painting styles 176–8
paintings, plants shown in 172–8
Paley, William 19
palisade cells 88
Pall Mall Gazette 22
pandemic, effects of 7
Papaver rhoeas 134
papaver somniferum 115–16
Paris 105
Parkinson Society 25
parsley 129
Parsons, Beatrice 152
Pasteur, Louis 62, 90, 119
patrons, institutional 164
Paul, Benjamin 108
Paxton, Joseph 7, 159
Peek, James 57
pelargoniums 169
Pelletier, Pierre Joseph 94, 108, 112
Pemberton, John 114
Pereira, Jonathan 124
perennials 154
    hardy 169
    tender 168–9
periodicals 148
Perkin, William 109
Perrins, William Henry 60
Petar II, Prince-Bishop of Montenegro 182
Petzold, Eduard 156
peyotl 64
pharmaceutical companies 106
pharmacopeias 122–4
pharmacy 105
photography 193
photosynthesis 91–6, 103

physostigma venenosum 115–17
Piave, Francesco Maria 184
Picasso, Pablo 8
pickling 61
pineapples 53
Piria, Raffaele 117
Plato 21
"pleasure-grounds" 150
Poole, C. H. 128
poppies 64, 145
population figures 1, 3
porcelain 192
portraiture 173, 191–2
Portugal 123
Potato Famine 4–5, 41, 100
potatoes 27, 40
poultry 129
Prairie Style 154–5
Price, Sir Uvedale (and Pricean Revival) 150
Prichard, James 12
Priestley, Joseph 93
Primrose League 144
professional gardeners 74
Prokopovych, Petro 56
Prunus armeniaca 53–4
pseudo-cereals 37–8
Psidium guajava 53
Pückler Muskau, Hermann von 150, 153, 155
Pugin, Augustus 185

Quercus spp. 182
quinine 108–9

Rachmaninoff, Sergei 183
Rafinesque, Constantine 120–4
Rauch, Christian Daniel 179
"recalcitrant" seeds 73, 79
Redouté, Pierre Joseph 175, 189
Regel, Arnold von 151
Regel, Eduard von 164
Řehoř, Abbot 20
"Reid's Yellow Dent" 29
religion 22
Repton, Humphry 76, 150
research papers 103
revivals 185
"rhythmic planting" 155
rice 27, 34–7
    African 35–6
Richter, Ludwig 148
Ridley, Henry 79
Rilke, Rainer Maria 183

ritual plants 63
Rivers, Thomas 52–3
Rizo, Salvador 192
Roach, John 43
Robinson, William 150–1, 154–5
Roezl, Benedict 76
Romania 123
romanticism 154, 176
Rontgen, Wilhelm Konrad von 7
rootcrops 39–41
Rosenegg, Gilm zu 184
roses 165–6
Rothamsted 100
Rousseau, Douanier 8
Rousseau, Henri 178
Roxburgh, William 165
Royal Horticultural Society 147
Royal Society of London 9, 19
rubber 77–9
Rückert, Friedrich 184
rum 48
Runge, Friedlieb 109
Ruskin, John 138, 150, 155
Russia 2–3, 32, 83, 122
Rutherford, Ernest 7

Sachs, Julius von 94
Saccharum officinarum 67
saffron 60
Sageret, Augustin 20
Saint-Gaudens, Augustus 180
St. Helena 83
St. Swithin's Day 130
Salix alba and salicylic acid 117
Salter, John 168
Santos Museum 72
Satie, Erik 9
Sambucus nigra 132
Solanum tuberosum 130
sorghum (*Sorghum bicolor*) 36–7
Schiele, Egon 175, 178, 180
Schimper, Andreas F. W. 24
Schleiden, Matthias 85, 87, 97
Schmeller, Johann Joseph 173
Schmiedeberg, Oswald 112
Schmutzler, Matthias 189
Schneevogt, George Voorhelm 161
Schneider, Camillo 156
Schroff, Karl von 114
Schubert, Franz 183
Schwann, Theodor 85, 87
Schwerin, Fritz 152, 167

science
    of plants 85, 88, 90, 103
    representation of plants in 188–95
scientific advances 7–8, 26, 105
Scotland 40
Scott, Robert Falcon 53
Scott, Walter 131
Scouvian custom 127
sculpture, plants in 178–81
Sedgwick, Adam 12, 19
seed banks 69
seed companies 161
seeds 72–3
"select" plantations 153
Sello, Hermann 154
Semmelweis, Ignaz 119
Senebier, Jean 93
Sertürner, Friedrich 102, 115
sexual selection 12
Shakespeare, William 177
Shirreff, Patrick 32
shortbread 57
Shull, George Harrison 29
Siebold, Philipp Franz von 163–4, 191–2
Sierra Club 26
Signac, Paul 177–8
Simon, Léon 161
single-compound drugs 105–6
Skelton, John 120–1
slavery and the slave trade 4, 11, 45, 64, 74, 79–80
Sloane, Sir Hans 70
Smith, Erwin 91
socialism 3
societies, botanical 70
Soret, Jacques-Louis (and the "Soret band") 95
sorghum 36–7
Spain and the Spanish empire 3, 53, 73, 123
Späth, Franz 160
Spence, William 56
Spencer, Herbert 15–18
spices 58, 60
Spindler, Carl 185
spontaneous generation 88
Sprengel, Christian Konrad 14
Spruce, Richard 71, 108, 114
Squire, Peter 124
Standish, John 160
staple foodstuffs 41
    luxuries transformed into 43–4, 65
steamships 72
Stein zum Altenstein, Carl 147–8

Stein, Gertrude 8
Stevens, J. C. 9
"still life" paintings 175
Strampelli, Nazareno 31
Strasburger, E. 97
Strauss, Richard 183–4
strychnos and strychnine 117–18
Strychnos nus-vomica 118
stylization, degrees of 184
succession 23
Suez Canal 65, 72
sugar 2, 43–5, 52–3, 69, 102–3
"survival of the fittest" 18
"swagger prints" 189
Sweet, Robert 168–9
sweeteners 55–6
Switzerland 48, 123
symbolism 182
Syringa spp. 166
syrups 54

Talbot, Henry Fox 193–4
Tansley, Arthur 24, 99
tariffs 43, 74
Taunton 127
Taxus baccata 144
tea 48–9, 54, 80–2, 99
tea roses 166
Tecumseh, William 180
textiles 188
Theobroma cacao 49–50
Thomas, Keith 82
Thompson, Sylvanus 8
Thomson, David 1, 5
Thomson, Samuel 120
Thoreau, Henry 11
Tiedemann, Friedrich 10
Tiffany, Louis Comfort 185
Timiryazev, Climent Arkad'evitch 95
tobacco 63–4
tools, use of 157
totipotency 86–7
Townsend, Charles 91
toxic plants 106
trade 4–7, 74
Trakl, Georg 182
transplantation 67–8, 72–3, 77–8, 82–4, 162–5
transportation of plants 43, 65, 157, 164, 189
treacle 55
trees 74, 76, 166–8, 175
Triticum spp. 30–3
tropical flora 22–3

Truffaut, Georges 159
Tschermak, Erich von 20
Tsvet (or Tswett), Mikhail 95
tulips 130, 155, 161, 170
"tumor inducing principle" (TIP) 91
Turkish delight 58
turmeric 60, 111–12

Underwood, William 52
United States Department of Agriculture (USDA) 28
universities, agricultural 28
urbanization 127
useful plants 77–82
Uyesugi, T. 96

Vaccinium myrtillus 131
Valais region 54
van Gogh, Vincent 175
van Houte, Louis 162
van Leeuwenhoek, Antonie 23, 90, 96
vanilla 60, 69
Veitch family (James, John Gould and Harry James) 70, 160, 164
vellum 189, 192
Ventenat, Etienne-Pierre 173
Verdi, Giuseppe 184
Verlaine, Paul 182
Vibert, Jean-Pierre 166
Victoria, Queen 5, 138–46
Vignes, Jean-Louis 46
Villoresi, Luigi 166
Vilmorin, André de 161
Vilmorin, H. L. de 32
Virchow, Ludwig Karl 88
Viscum album 145
vitalism 26
Vöchting, Hermann 88
Voght, Caspar von 160
Voltaire 52
Voysey, Charley 188
Vries, Hugo de 97

Wagner, Otto 180–1, 186
Wagner, Richard 18, 184
Waldmüller, Ferdinand 171, 173
wall gardening 155
Wallace, Alfred Russell 18–22
Wallich, Nathaniel 165
wall charts 191
wallpaper 186–8
walnut oil 45

Wang Wei 183
Warburg, Otto 96
Ward, Nathaniel (and Wardian cases) 68, 73, 75, 157
Ware, Thomas S. 160
warfare 1, 3, 6–7, 44
Waring, Edward 123
Warming, Eugen 23–4, 99
wassailing 127
water supply 157
Waterman, Catherine Harbeson 137
Waterson, Charles 118
Watson, Forbes 150–1
Watt, Alistair 82
Wedgwood, John 147
Wegner, Alfred 10
Weitsch, Friedrich Georg 192
Wells, William 12, 18
Went, Frits 90
Westcott, William Wynn 124
wheat 27, 30–3
Whiggish view of history 22
whiskey 47–8
Whitsunday 144
whooping cough 134
Wickham, Henry 79
Wight, Robert 191
Wilberforce, Samuel 19
Wild Flower Preservation Society of America 25
wild foods 127, 131–2
wild gardening 151–3
Willstätter, Richard 95
Wilson, Ernest Henry 70, 164–5
Wilson, George Ferguson 148
Wilson, Nathaniel 79
wine 46–7
Winsor, Mary 22
wintergardens 159
Wisley Gardens 148
Withering, William 112
Woburn Abbey 155
women's roles 129–30
Worcestershire sauce 60
"workshop of the world" 4
Wright, Charles Alder 115
Wuchstoff 90
Wulf, Andrea 10

yarrow 135
Yellowstone 24–5
Yosemite 25–6
Young, James 61
Yuul, feast of 127

Zea mays 28–30
Zingiber officinale 60
Zionism 188
Zonnebloemen 176